A Tale of Three Thirsty Cities

Technology and Change in History

Series Editors

Adam Lucas
Steven A. Walton

VOLUME 14

The titles published in this series are listed at *brill.com/tch*

A Tale of Three Thirsty Cities

The Innovative Water Supply Systems of Toledo, London and Paris in the Second Half of the Sixteenth Century

By

Jaime-Chaim Shulman

BRILL

LEIDEN | BOSTON

Cover Illustrations: The London Bridge Waterworks in the 1730s (front).
Henry Beighton, an English engineer, published in 1731 a detailed description of the London Bridge Waterworks, which included a calculation of the quantity of water raised daily by the waterworks and this accurate technical drawing.
Ramelli's 'inclined plane' water-raising machine, 1588 (back).
Agostino Ramelli, Le diverse et artificiose machine del Capitano Agostino Ramelli Dal Ponte Della Tresia Ingegniero del Christianissimo Re di Francia et di pollonia, Parigi: published by Agostino Ramelli, 1588, plate 95, p. 148.

Library of Congress Cataloging-in-Publication Data

Names: Shulman, Jaime-Chaim (Chaim C.), editor.
Title: A tale of three thirsty cities : the innovative water supply systems of Toledo, London and Paris in the second half of the sixteenth century / edited by Jaime-Chaim Shulman.
Description: Leiden ; Boston : Brill, [2018] | Series: Technology and change in history ISSN 1385-920X ; volume 14 | Includes bibliographical references.
Identifiers: LCCN 2017036389 (print) | LCCN 2017037588 (ebook) | ISBN 9789004312425 (E-book) | ISBN 9789004312401 (hardback : alk. paper)
Subjects: LCSH: Water-supply—Spain—Toledo—History—16th century. | Water-supply—England—London—History—16th century. | Water-supply—France—Paris—History—16th century. | Europe—Social life and customs—16th century.
Classification: LCC TD288.T65 (ebook) | LCC TD288.T65 T355 2017 (print) | DDC 628.1094/091732—dc23
LC record available at https://lccn.loc.gov/2017036389

Typeface for the Latin, Greek, and Cyrillic scripts: "Brill". See and download: brill.com/brill-typeface.

ISSN 1385-920X
ISBN 978-90-04-31240-1 (hardback)
ISBN 978-90-04-31242-5 (e-book)

This book is printed on acid-free paper and produced in a sustainable manner.

Contents

Abbreviations, Date Conventions and Notes

Abbreviations

AN	Archives Nationales, Paris
AGP	Archivo General de Palacio, Madrid
AGS	Archivo General de Simancas, Simancas
AHP	Archivo Histórico de Protocolos, Madrid
AMT	Archivo Municipal de Toledo, Toledo
BNM	Biblioteca Nacional de España, Madrid
BNF	Bibliothèque nationale de France, Paris
BM, DPD	British Museum, London, *Department of Prints and Drawings*
Cal. SPD	Calendar of State Papers, Domestic
C & S. R.	Casas y Sitios Reales
C. R.	Cédulas Reales
CLRO	Corporation of London Records Office
Jour	Journals of the Court of Common Council
LMA	London Metropolitan Archives
PRO	Public Record Office, London

Date Conventions

England did not adopt the Gregorian calendar of 1582 until 1752. Besides, while years begin in the Gregorian calendar on 1 January, in England the year began on the feast of the Annunciation, on March 25. Thus, for example: January, February and part of March 1598 in continental Europe were equivalent to January, February and March 1597 in England. On 25 March, the year 1598 began in England, aligned with Catholic Europe until the next 1 January.[1] Following Deborah Harkness, Elizabethan dates for January, February, and portions of March have been converted to bring them in line with modern usage by altering the year, but the day of the month has not been converted.[2] Thus the Elizabethan February 1560 appears here as February 1561.

1 John A Wagner and Susan Walters Schmid, *Encyclopedia of Tudor England*, Santa Barbara, (CA), ABC-CLIO, 2012, p. 199.

2 Deborah F. Harkness, *The Jewel House, Elizabethan London and the Scientific Revolution*, New Haven and London, 2007, p. xiv.

Notes

- All translations in which the translator's name does not appear, are mine.
- All quotations in the original language are transcribed as they appear in the text.

List of Figures

Introduction

The Subject of Our Research

In the last decades of the sixteenth century and the beginning of the seventeenth, Toledo, London and Paris, the three largest cities in their respective kingdoms, undertook ambitious projects to supply water to their populations and their ruler's palaces by using the flow energy of their adjacent rivers for power-lifting devices which collected and raised the water from the same rivers. By investigating the processes which led these cities to undertake such projects, this book aims to contribute to the body of knowledge pertaining to the construction of urban water supply systems in the Early Modern Era and shed light upon the attitude of city authorities under different regimes to embark on large-scale public works in general.

The quality of a society's water expertise is a commonly accepted barometer of that society's development. Archeology investigations have shown that the earliest developed cultures had sophisticated water technologies that preceded European advances.[1] The areas that pioneered the first successful efforts to control the flow of water were Mesopotamia[2] and Egypt,[3] where vestiges of prehistoric irrigation works still exist. The huge geographical distances that separated the highly developed civilizations in China[4] and

1 For a condensed history of world developments in the field of water technology, see Larry W. Mays, "A Brief History of Water Technology During Antiquity: Before the Romans," in Larry W. Mays, (ed.), *Ancient Water Technologies*, Dordrecht, Heidelberg, New York, London: Springer Science & Business Media, 2010, pp. 1–28 and Giovanni De Feo, P. Laureano, R. Drusiani and A. N. Angelakis, "Water and wastewater management technologies through the centuries," *Water Science & Technology: Water Supply—WSTWS*, 10.3 (2010), pp. 336–349.

2 For a description of water technologies in Mesopotamia, see Aldo Tamburrino, "Water Technology in Ancient Mesopotamia," in Larry W. Mays, (ed.), *Ancient Water Technologies*, Dordrecht, Heidelberg, New York, London: Springer Science & Business Media, 2010, pp. 29–52.

3 For a description of technologies in ancient Egypt, see Larry W. Mays, "Water Technology in Ancient Egypt," in Larry W. Mays, (ed.), *Ancient Water Technologies*, Dordrecht, Heidelberg, New York, London: Springer Science & Business Media, 2010, pp. 57–66.

4 For a comprehensive account of the history of irrigation in China, see P. Du and A. Koenig, "History of water supply in pre-modern China," in Andreas N. Angelakis, Larry W. Mays, Demetris Koutsoyiannis and Nikos Mamassis, *Evolution of water supply throughout the millennia*, London and New York: IWA Publishing, 2012, pp. 169–226. For another extensive account of hydraulic Chinese technology, see Joseph Needham, *Science and Civilization in*

 | DOI 10.1163/9789004312425_002

India[5] from Europe represented almost an unsurpassable hurdle for knowledge circulation. Other societies, in Africa,[6] Southeast Asia[7] and obviously America[8] and Oceania, stayed invisible when looking through European eyes.[9]

This work is focused in Europe, the area to which the civilized communities that developed by the Tigris, the Euphrates and the Nile, which related to the Greco-Roman societies, ultimately bequeathed their technological capabilities. We have chosen to focus on Toledo, London, Paris and their surroundings, in light of their dimensions and their importance in their respective kingdoms, as well as the particular interest represented by their water supply projects. We know more about some of the devices treated in this study than about others and not all those that are treated here have attracted much scholarly attention. The implementation of the new water raising technologies sowed the seeds for many undertakings in the same cities, well into the eighteenth century, contributing to bring about demographic and social changes.

Toledo served until 1561 as the capital city of Castile, where the Royal Court was located. Paris and London had by this time already achieved pre-eminence in their respective kingdoms, and, unlike Toledo, were not to lose their supremacy in the future. It should be noted, however, that in several medium and small-size cities in Germany and Poland, such as Augsburg, Gdańsk and Lübeck, water-lifting devices capable of forcing water to storage tanks installed in towers had already been developed, positively influencing the life of their inhabitants. Nevertheless, it was the long run impact of the systems built in London and Paris, and the systems later derived from them, that became cru-

---

China: Volume 4, Physics and Physical Technology, Part 2, Mechanical Engineering, Cambridge: Cambridge University Press, 1965, pp. 330–368.

5 For the most comprehensive account of the technological prowess of the Indus Valley Civilization, which occupied part of present-day North India and Pakistan, see Rita P. Wright, *The Ancient Indus—Urbanism, Economy, and Society*, New York: New York University, 2010.

6 As an example of these advances, see Johann W. N. Tempelhoff, "Historical Perspectives on Pre-Colonial Irrigation in Southern Africa," *African Historical Review*, Vol 40, No. 1, pp. 121–159.

7 A comprehensive description of the Sinhalese achievements in this field is given by Henry Parker, *Ancient Ceylon*, New Delhi and Madras: Asian Educational Services, 1999, specially Chapter X: "The Earliest Irrigation Works," pp. 347–412 and for the sophisticated irrigations schemes in Cambodia during ancient times, see Charles Higham, *The Civilization of Angkor*, Berkeley and Los Angeles: University of California Press, 2004, pp. 154–161.

8 Tom D. Dillehay, Herbert H. Eling, Jr. and Jack Rossen, "Preceramic irrigation canals in the Peruvian Andes," *Proceedings of the National Academy of Sciences of the United States of America*, (2005), vol. 102, no. 47, pp. 17241–17244.

9 Frances and Joseph Gies, *Cathedral, Forge and Waterwheel, Technology and Invention in the Middle Ages*, New York: Harper, 1995, p. 18.

cial to enable the accelerated growth of both capitals. The water-raising efforts in Toledo, though technically successful, did not provide water in sufficient quantities for over three centuries, thus contributing to the stagnation of its urban development.

The so-called 'Juanelo Turriano's Artifice' (*Artificio de Juanelo Turriano*), which raised water from the Tajo River to the city of Toledo, has known waves of intense scholarly interest and epochs of almost absolute oblivion. In the third decade of the nineteenth century, the first reference to the *Artificio* was made by Eugenio Llaguno y Amirola in his treatise on architects and architecture in Spain since its Restoration.[10] Sixty years of complete neglect followed, from Llaguno's work up to the last two decades of the nineteenth century, when a systematic investigation began, initiated by the Spanish mining engineer Luis de Escosura y Morrogh and followed up by Theodor Beck, a professor at the Technische Universität Darmstadt and a well-known German historian of technology.[11] Both strove to reach a clear understanding of the working principles of the *Artificio*. Another period of oblivion followed until active efforts to study in depth the history of the device and the fundaments of its operation took place in the 1960s, when, mainly for political purposes, attempts were made to glorify the technological capabilities of Spain as an ancient nation. This drive brought about renewed interest in innovative engineering undertakings of its Golden Century (*Siglo de Oro*). Although conflicting details exist about the operation of Juanelo's device, it eclipsed by its brilliant elegance the later and seemingly more practical solution developed by Juan Fernández del Castillo, for which historians have not shown great interest.

The second subject of this book—the London Bridge Waterworks—has been continuously referred to in the many narrations and surveys dedicated to the history of London. Yet the majority of these descriptions pertain to the waterworks after their expansion and renovation, which took place in the early 1700s. The technological information about the earlier, 'first generation,' waterworks, which will be at the center of our interest here, is fragmentary, although the commercial aspects of the enterprise are well documented. The subsequent construction in 1609 of the 'New River' conduit, which contributed to solving the water supply problem of London on a larger scale, has been more

10 Eugenio Llaguno y Amirola and Juan Agustín Ceán Bermúdez, *Noticias de los arquitectos y arquitectura de España desde su restauración*, (4 volumes), Vol. 2, Madrid: Imprenta Real, 1829.

11 Ladislao Reti, "El Artificio de Juanelo en Toledo: su Historia y su Técnica," Separata de *Provincia, Revista de la Diputación Provincial de Toledo*, No. 60, (1967), pp. 1–46, p. 8.

thoroughly researched, both in its technical and legal aspects.[12] The 'New River', as the canal and the company which built it were called, drew water from springs in Hertfordshire and, through a sinuous 60 km course, discharged the water into a reservoir at Islington, in the north of the city. But it was the commercial concept behind those undertakings and their successors that would cause dramatic change in people's attitude towards water as a vital resource.

The historiography of La Samaritaine pumping machine in Paris, which is the third project on which we focus in this work, does not go beyond references in eighteenth-century general histories of Paris and of the Pont Neuf in particular. It was the monumental Marly waterworks, built eighty years later, which cast a giant shadow over all previous water-raising devices in France. Its sheer size and complexity caused scholars to direct their efforts to this extravagantly expensive engineering achievement, as it was considered to be the most extraordinary piece of machinery of the seventeenth century.[13] Consequently, as in the case of the London Bridge Waterworks, the Samaritaine project in its earlier, less spectacular configuration previous to its rebuilding in the second decade of the 1700s, has hardly been studied until now. Nevertheless, it represented the spark that ignited the will to build the Notre Dame huge pumps, and ultimately, Marly.

Almost consensually, historians agree that the almost uncontrolled growth of London and Paris was the driving force behind the initiation of public works in a huge scale, including the crucial search for water supply solutions. The local authorities in Toledo desired desperately to foster the growth of the city and keep the Royal court in it, but failed in their attempt to do so.

When investigating the development of water provision systems through the ages, historians of technology have traditionally turned to available engineering drawings, sketches and descriptions to reconstruct and understand complex devices invented in the past. They have striven to retrieve materials such as books, maps, paintings and archaeological findings to help them in their endeavors.[14] Historians of medicine and sanitation have put special

---

12 The best legal analysis is in Bernard Rudden, *The New River: A Legal History*, Oxford: Oxford University Press, 1985.

13 For a description of its impact, see: Thomas Brandstetter, "The Most Wonderful Piece of Machinery the World Can Boast of: The Water-works at Marly, 1680–1830," *History and Technology*, Vol. 21, No. 2, (2005), pp. 205–220.

14 See for example: Cyril Edward Nowill Bromehead, "The Early History of Water Supply," *The Geographical Journal*, Vol. 99, No. 3, (1942), pp. 142–151 and Tomas A. Ewbank, *A Descriptive and Historical Account of Hydraulic and Other Machines for Raising Water*, New York: D. Appleton and Company, 1842.

emphasis on the influence of the quality of urban water supplies and sewerage infrastructures upon public health, drawing conclusions from public records and literary accounts.[15] Other approaches have been taken by scholars who have directed their interest to contracts between the authorities and the entrepreneurs who put in motion undertakings of water supply, looking for specifications of required performance. Legal historians have analyzed those official documents in their quest for a better understanding of the antecedents of modern law, as far as public works are concerned.[16] It is arguable, however, whether these research avenues have ever gone beyond the technical aspects and health effects of these water supply systems. They generally refrain from addressing wider aspects, such as the social impact of these works, cultural influences related to these projects, and their political and economical aspects. In our study, we shall try to give a rough sketch of the state and municipal bodies that played roles in the decisions leading to, or sometimes interfering with, the water raising projects in each of the subject cities. The interests and motivation of the functionaries manning official positions of influence crucially affected the outcome and success of the three enterprises. Their behavior and willingness, either to improve or impede progress, dictated their attitudes, which became as important as the technical acumen of the entrepreneurs. And most importantly, with the benefit of hindsight, we see that neither of these enterprises, nor their impact, would have come into being without the circulation of technological know-how within Europe and beyond. Arguably, this circulation is the common thread that unites the pieces of the wide mosaic of water supply undertakings in Early Modern Europe.

---

15 As an example: Heikki S. Vuorinen and others, "History of Water and Health from Ancient Civilizations to Modern Times," *Water Science & Technology*, Vol. 7, No. 1, (2007), pp. 49–57.

16 See for example: Paul G. Mahoney, "Contract or Concession? An Essay on the History of Corporate Law," *Georgia Law Review*, Nr. 34, (2000), pp. 873–893; Frederick Clifford, *History of Private Bill Legislation*, (2 volumes), London: Butterworths, 1887; William Hepburn Buckler, *The Origin and History of Contract in Roman Law Down to the End of the Republican Period: Being the Yorke Prize Essay for the Year 1893*, London: C.J. Clay, 1895; David J. Ibbetson, *A Historical Introduction to the Law of Obligations*, Oxford: Oxford University Press, 1999; Paul J. du Plessis, "The Protection of the Contractor in Public Works Contracts in the Roman Republic and Early Empire," *Journal of Legal History*, Vol. 25, Nr. 3 (2004), pp. 287–314 and Thomas Ertman, *Birth of the Leviathan: Building States and Regimes in Medieval and Early Modern Europe*, Cambridge: Cambridge University Press, 1997.

Mechanical Engineers and Know-How Migration

The diffusion of technological awareness and know-how was stimulated by the human mobility of the High Middle Ages, when merchants, university students, pilgrims and crusaders travelled extensively. These wandering individuals would have encountered conduits and fountains in the course of their travels. By the end of the Middle Ages, in some cities water was supplied from public fountains fed by conduits, while other towns overcame topographical limitations that prevented the use of gravity-flow systems by installing animal-powered scoop wheels or other simple lifting devices. The evolution of water supply technology underwent a quantum leap in the Late Middle Ages, achieved in parallel with the beginning of the massive use of water power as a favored replacement for animals and humans as prime movers.

Although during most of the Middle Ages, anonymous building masters had excelled in their work, such as the construction of cathedrals,[17] the social status of the technical worker in Europe was marginal. Even the names of these builders have been lost to us, which is a clear sign that their social role was considered minor. The basic school curriculum covered seven liberal arts, and the mechanical arts were at a lower level than all others.[18] This was both an intellectual and a social distinction. Being trained only in mechanical arts meant that you were someone who worked with your hands, someone who was fit only to be directed by someone else who was better educated. Nonetheless, the sophisticated designs of anonymous medieval builders were passed down, through the apprenticeship system, to succeeding generations, as their enterprises extended for years, decades, or more. Franco Franceschi has identified the main mechanisms of transmission of knowledge and know-how, concentrating, through a number of examples, on three means of knowledge transmission: technical manuals, apprenticeships and wandering foreign specialists.[19] John Raymond Harris elaborates on the strategies employed by

17 For a general discussion on cathedral building, see: Alain Erlande-Brandenburg, *The Cathedral Builders of the Middle-Ages*, London: Thames and Hudson, 1995 and John Fitchen, *The Construction of Cathedrals*, Chicago: University of Chicago Press, 1981.

18 The concept of an educational curriculum composed of the seven liberal arts, organized into the *trivium* (grammar, rhetoric, and logic) and the *quadrivium* (arithmetic, geometry, astronomy, and music), is attributed to Boethius, the last great Roman intellectual (c.480–524), see Noel Harold Kaylor and Philip Edward Phillips, *A Companion to Boethius in the Middle Ages*, Leiden: Brill, 2012, p. 174, fn. 27.

19 Franco Franceschi, "La grande manifattura tessile," in *La trasmissione dei saperi nel Medioevo (secoli XII–XV)*, Atti del XIX Convegno Internazionale di Studi (Pistoia, 16–19 maggio 2003). Pistoia, 2005, pp. 355–390.

governments and other institutions to prevent the exodus of skilled labor and with it the knowledge transfer as its consequence.[20] Penalties were inflicted in Europe on those who smuggled machinery out of a country, and France even imposed the death penalty at the end of the seventeenth century to those who would dare doing so. Drastic punishment was applied to foreigners attempting to bribe or lure skilled workers willing to travel abroad.[21] Stephan R. Epstein elaborated on this point and suggested that: "... the principal endogenous bottleneck to pre-modern technical diffusion and innovation was the cost of person-to-person teaching and demonstration."[22]

The profession of 'engineer' had not previously existed. The roots of the term are in Middle English: denoting a designer and constructor of fortifications and weapons.[23] It only entered into the language in the middle of the fifteenth century, becoming widely used in Leonardo's time.[24] As Ortega y

---

20 John Raymond Harris, "Law, Espionage and Transfer of Technology from Eighteenth Century Britain," in Robert Fox, (ed.), *Technological Change: Methods and Themes in the History of Technology*, Amsterdam: Routledge, 1996, pp. 123–136, pp. 135–136.

21 Stephan R. Epstein, "Property Rights to Technical Knowledge in Premodern Europe, 1300–1800," *The American Economic Review*, Vol. 94, No. 2, Papers and Proceedings of the One Hundred Sixteenth Annual Meeting of the American Economic Association, San Diego (CA), January 3–5, 2004 (May, 2004), pp. 382–387, p. 382.

22 Stephan R. Epstein, "Labor mobility, journeyman organizations and markets in skilled labour Europe, 14th–18th centuries," in Mathieu Arnoux and Piere Monnet, (eds), *Le technicien dans la cité en Europe occidentale, 1250–1650*, Rome: École française de Rome, 2004, pp. 251–269, p. 269.

23 Formerly also as *ingineer*: in early use from Old French *engigneor*, from medieval Latin *ingeniator*, from *ingeniare* 'contrive, devise', from Latin *ingenium* in later use from French *ingénieur* or Italian *ingegnere*, also based on Latin *ingenium*. Extracted from: Angus Stevenson, (ed.), *Oxford Dictionary of English*, Oxford: Oxford University Press, 2010 p. 581. Lynn Townsend White Jr., *Medieval Technology and Social Change*, London and Oxford: Oxford University Press, 1962, p. 160, note 2 to p. 79, refers to: Franz Maria Feldhaus, *Die Technik der Antike und des Mittelalters* Potsdam: Akademische verlagsgesellschaft Athenaion m.b.h, 1931, p. 277, who asserts that the word 'engineer' first appears in Johannes Codagnellus, *Annates placentini*, edited by O.-Holder-Egger, Hannover: Lipsiae, Impansis Bibliopolii Hahniani, 1901, p. 23, which, although written in the early thirteenth century, refers, as of 1196, to an 'Alamannus de Guitelmo,' *enceignerius communis Mediolani*. Feldhaus holds that it is derived from *incingere*, 'to fortify'. However, in 1190–2 Ambroise, *L'Estoire de la guerre sainte*, edited by Gaston Paris, Paris: Imprimerie nationale, 1897, v. 2274, explicitly connects engineers with engines: '*engineors qui savaient d'engins plusors*.

24 Edna Suárez, "El organismo como máquina: Descartes y las explicaciones biológicas," in Carlos Álvarez and Rafael Martínez Enríque, (eds), *Descartes y la ciencia del siglo XVII*, México D. F.: Siglo XXI, 2000, p. 146.

Gasset has written: "The technology of the Middle Ages [was] the technology of the artisan ... [in the Renaissance] the artisan has been split up into his components: the worker and the engineer."[25] By the sixteenth and seventeenth centuries engineers were still, as they had been for centuries, closely linked to traditional techniques, such as those employed in the construction of bridges, dams, roads, fortifications, etc., but the development of machines and technical innovation in general increased the demand for the services of engineers-inventors, bringing about an improvement in their social position, while at the same time, awakening the need to protect the intellectual character of their work. One consequence was the attempt to recognize ownership of inventions through a mechanism of privileges or actual patent systems. In theory, technical knowledge could indeed move around through published texts, through patents and through migrating individuals. But pre-modern patent laws did not demand accurate technical specifications; the descriptions of inventions were generic, drafted obscurely, concealing more than disclosing, thus rendering them ineffective as sources of know-how.[26]

In essence, rulers encouraged the introduction of new machines or processes in their jurisdiction by granting a temporary monopoly over mechanical inventions that involved a high investment and were relatively easy to reproduce. New patents were granted in different geographical areas many decades after the original one for basically the same machine/system had been granted in another location. The process of obtaining a patent demanded upfront fees, while evidence and risk rested squarely on the inventor's shoulders. Some innovations were never patented because they were thought to be better protected as trade secrets or considered part of the collective knowledge of a guild. These guilds traditionally opposed any intent to privatize know-how considered to be already in the craft's domain, and so perceived as restraining trade.[27] In mining, metallurgy and dam-building the inventors committed extensive plagiarism, sometimes refraining from publishing specific valuable details.[28]

25 Jose Ortega y Gasset, "Man the technician," in *History as a system*, New York: Norton, 1961, pp. 151–155. For two analyses of this process, see: Alicia Cámara Muñoz, "La profesión de ingeniero: los ingenieros del rey," in Manuel Silva Suárez, (ed.), *Técnica e Ingeniería en España I: El Renacimiento*, (6 volumes), Vol. 1, Zaragoza: Real Academia de Ingeniería, Institución Fernando El Católico y Prensas Universitarias de Zaragoza, 2004, pp. 125–164, and Nicolás García Tapia, *Ingeniería y arquitectura en el renacimiento español*, Valladolid: Secretariado de Publicaciones, Universidad de Valladolid, 1989, pp. 25–38.

26 Epstein, "Property Rights," p. 384.

27 Ibid.

28 Hermann Kellenbenz, "Technology in the Age of the Scientific Revolution, 1500–1700," in Carlo M. Cipolla, (ed.), *The Fontana Economic History of Europe*, London, 1974, pp. 177–272,

Industrial espionage was rampant and encouraged by the authorities when its target objects were foreign. Already in the 1650s, John Lanyon, who excelled as a conduit of intellectual property, relaying information about technical developments in the Low Countries to England, had his 'ideological' thoughts in favor of industrial espionage recorded by Samuel Hartlib: "It is as good an Invention to observe ... all manner of ... useful things invented and practiced already in other Countries as to invent truly new ones."[29]

Carlo Maria Cipolla has maintained that throughout the ages, the migration of people has been the main means for the diffusion of innovations.[30] Brian Dolan formulated the same idea: "What we have traditionally called 'science' does not travel; people who practice science do."[31] Although, as Cipolla noted, it is impossible to quantify the phenomenon, all evidence suggests that the mobility of skilled labor was widespread and growing in Early Modern Europe. Britain, Switzerland, Sweden and Spain were the recipients of this human capital, which mainly came from the southern Netherlands, France and Germany.[32] Epstein, who has done meticulous research on this issue, summarized it succinctly: "The technical knowledge of pre-modern craftsmen and engineers was largely experience-based; thus, virtually all pre-modern technical knowledge was, and had to be, transferred in the flesh."[33] Epstein adds that this was arguably more easily done in Europe than in other parts of the world, because European technicians did not exclusively belong to specific ethnic or religious

---

p. 186. For an extensive chronological discussion on the development of technological secrecy, see: Pamela O. Long, *Openness, secrecy, authorship: technical arts and the culture of knowledge from antiquity to the Renaissance*, Baltimore: John Hopkins University Press, 2001.

29 Hartlib, *Hartlib Papers 28/2/14B and 15B*, quoted by Mark Jenner, "'Another epocha'?: Hartlib, John Lanyon and the improvement of London in the 1650s," in Mark Greengrass, Michael Leslie and Timothy Raylor, (eds), *Samuel Hartlib and Universal Reformation: Studies in Intellectual Communication*, Cambridge: Cambridge University Press, 1994, pp. 343–356, pp. 347–348.

30 Carlo M. Cipolla, "The Diffusion of Innovations in Early Modern Europe," *Comparative Studies in Society and History*, Vol. 14, No. 1 (Jan. 1972), pp. 46–52, Cambridge University Press, p. 48.

31 Brian Dolan, "Embodied Skills and Travelling Savants: Experimental chemistry in Eighteenth Century Sweden and England," in Ana Simões, Ana Carneiro, María Paula Diogo, (eds), *Travels of learning: a geography of science in Europe*, Dordrecht, The Netherlands: Springer, 2003, pp. 115–142, p. 135.

32 Carlo M. Cipolla, *Máquinas del tiempo y de la guerra*, Barcelona: Crítica, 1999, p. 11.

33 Stephan R. Epstein, "The Generation and Transmission of Technical Knowledge in Premodern Europe, c.1200–c.1800," *Global Economic History Network*, Conference 4, Leiden, The Netherlands, 16–18 September, 2004, pp. 1–39, p. 2.

communities and were thus more mobile, giving pre-modern Europe a comparative advantage in respect to other societies.[34]

The present book intends to go beyond such generalities, aiming at identifying the evolution of these emblematic projects through a combination of a close examination of techniques developed for them and the political, social and economic aspects involved in their implementation. In any case, we shall heed Kenneth J. Knoespel's warning: "we cannot expect to approach Renaissance technology unaffected by our own highly-developed technological culture."[35]

---

34 Ibid.

35 Kenneth J. Knoespel, "Gazing on Technology: Theatrum Mechanorum and the Assimilation of Renaissance Machinery," in Mark L. Greenberg and Lance Schachterle, (eds), *Literature and Technology*, London and Toronto: Lehigh University Press, 1992, p. 99.

CHAPTER 1

Human Water Consumption in England, France and Spain in the Sixteenth and Seventeenth Centuries

It has been estimated that vital human needs for water to drink or ingest through our food is on average about 5 liters per person per day.[1] In order to sustain a basic modern standard of personal and domestic hygiene sufficient to maintain health, 25 liters per day are needed for cooking, bathing and washing. An additional 20–25 liters are required for sewage disposal so a total of about 50 liters per capita daily are essential to cover domestic needs, excluding water for food production or energy.[2] Humans can survive without shelter, even of food for some time, but they cannot exist deprived of water for more than a few days.

The access to water can reveal much about the social and political status of citizens to whom it is—or is not—provided, transforming water, when scarce, into an economic resource.[3] The late sixteenth and seventeenth centuries were characterized by remarkable urbanization in Western Europe in general, and England and France in particular: between 1550 and 1650 London quintupleted its population, from 80,000 to 400,000; Parisian population increased from roughly 250,000 in 1564 to about 500,000 by 1645.[4] Both cities became increasingly urbanized cultures, in which not only populations concentrated, but also activities such as the production and exchange of goods via large-scale

1 Peter H. Gleick, "Basic Water Requirements for Human Activities: Meeting Basic Needs," *Water International*, Vol. 21, Issue 2 (1996), pp. 83–92, p. 84.

2 Ibid., p. 88.

3 James Salzman, "Thirst: A Short History of Drinking Water," *Social Science Network Electronic Paper Collection*, Research Paper Series, paper No. 92, Duke Law School Legal Studies, (December 2005), pp. 1–33, p. 3.

4 Karen Newman, "Towards a Topographic Imaginary—Early Modern Paris," in Carla Mazzio and Douglas Trevor, (eds), *Historicism, Psychoanalysis and Early Modern Culture*, New York and London: Routledge, 2000, pp. 59–81, p. 60. Some estimations are lower—Jeremy Boulton, "London 1540–1700," in Peter Clark, (ed.), *The Cambridge Urban History of Britain*, (3 volumes), Vol. 2, Cambridge, 2000, pp. 315–346, p. 316, gives a figure of 220,000 inhabitants for London in 1600 (exceeded only by Naples) and of 430,000 for 1650, by then the most populated city in Europe. Vanessa Harding, *The Dead and the Living in Paris and London, 1500–1670*, Cambridge: Cambridge University Press, 2002, pp. 15–16, estimates the population of Paris at 300,000 in 1600 and 400,000 in 1640.

 | DOI 10.1163/9789004312425_003

markets and the management of a centralized state.[5] Moreover, those activities fostered urban behaviors that justify, as Jan de Vries puts it, "the apparently nonsensical phrase: the urbanization of the cities."[6] In contrast with London and Paris, Toledo's population decreased in the last third of the sixteenth century, from a maximum of 62,000 in 1571 to about 45,000 in 1597, and continued its decline in the seventeenth century.[7]

Was there water scarcity in the second half of the sixteenth century in Toledo, Paris and London? The significance of this apparent dearth, in fact, is difficult to appraise—were people really suffering from a lack of water or did their culture entail life involving usage of small quantities of water, with very little bathing and washing? Quantitative data are lacking. Significantly, the statistician Gregory King (1648–1712) made no calculations about water consumption though he did for milk, beer and other drinks.[8] Twenty century scholars have tried to quantify water consumption in past times, including the early modern period. For example, Cyril Edward Nowill Bromehead has given a few comparative figures of the number of liters per head per day of public water supply in different cities, estimating that in ancient Rome, which he used as a benchmark, it reached 1200 liters about 100 CE compared to 12 liters in Paris and London in 1823 and 140 liters in London in 1936.[9] He noted that these low amounts were not necessarily far below the demand. However, the validity of comparisons of actual water usage in the Roman era with other periods is questionable, since it is based on an evaluation of the quantities of water that reached Rome via aqueducts, in which it is generally believed that water flow was constant and continuous. If it was, a huge amount of water would have been wasted without being utilized.[10] While religious obligations

5 Newman, "Towards a Topographic Imaginary," p. 61.

6 Jan De Vries, *European Urbanization 1500–1800*, Cambridge (MA): Harvard University Press, 1984, p. 12.

7 Linda Martz, *Poverty and Welfare in Habsburg Spain*, Cambridge: Cambridge University Press, 1983, p. 95.

8 Gregory King, "Natural and Politicall Observations and Conclusions upon the State and Condition of England, 1696," an appendix to George Chalmers, *An estimate of the comparative strength of Britain during the present and four preceding reigns; and of the losses of her trade from every war since the Revolution*, London: John Stockdale, 1804, pp. 29–73. King's original manuscript is kept in the Harleian Manuscripts Collection, Ms. 1898, BM, DPD.

9 Cyril Edward Nowill Bromehead, "The Early History of Water supply (Continued)," *The Geographical Journal*, Vol. 99, No. 4 (April, 1942), pp. 183–193, p. 191.

10 For a discussion on this subject, see: Larry W. Mays, M. Sklivaniotis and Andreas Nikolaos Angelakis, "Water for human consumption through history," in Andreas Nikolaos Angelakis, Larry W. Mays and Demetris Koutsoyiannis, *Evolution of Water Supply through the Millennia*, London: IWA Publishing, 2012, pp. 19–43, p. 33. Different historians give

in Islamic countries required bathing facilities comparable with those of the Romans, not many generations ago Europeans used on average relatively little water for personal hygiene.[11]

Water has to have an adequate quality to maintain health—when collected from unprotected sources it often exposes users to risk. Even when water is of a sufficiently high quality at the point of collection it has to be carried and stored, and sanitary practices have to be kept, which often was and is not the case. The inadequacy of the means to limit pollution in urban concentrations caused the contamination of ground and surface water by bacterial pathogens. The inappropriate disposal of human and animal excreta led to a vicious cycle of contamination and infection, from which European population suffered in the early modern period and well into the nineteenth century. As a result of serious cholera and typhoid outburst urban authorities decided to undertake vast public sanitation projects, in London in the 1850s and in Paris and other major European cities in the 1860s.[12] A common thread to the histories of the three cities subject of this research is the fragile balance between water resources and the effects of pollution released to the same rivers from where water was collected. The contaminated water supply made drinking perilous, causing serious health problems.

Already in the eighth century CE the Arab alchemist Jabir ibn Hayyan had distilled water to free it from 'evil spirits' and in the eleventh century CE the Persian physician Avicenna advised travelers either to filter water with a cloth or preferably to boil it.[13] The possibility of 'purifying' water by boiling it was

---

diverse estimations for the volume of water supply to Rome: Thomas Bell, *History of the Water Supply of the World*, Cincinnati: P. G. Thomson, 1882, p. 5, estimated it to have been in the order of 1500 liters per inhabitant per day in the first century CE; John Frederick Bateman, in his report: "On the Present State of Our Knowledge on the Supply of Water to Towns," *British Association for the Advancement of Science, Report of the Twenty-fifth Meeting*, London, 1856, pp. 62–77, p. 64, estimated the daily quantity at about 1200 liters per capita, though he does not refer to the exact period when this was the amount of water provided by the aqueducts.

11 Bromehead, "The Early History of Water supply (Continued)," p. 191. Obviously, by the use of public bathouses instead of private baths, total water consumption for personal hygiene for the population as a whole was reduced.

12 The results in London were impressive, as shown in end of the century statistics; see Cady Staley and George Spencer Pierson, *The separate system of sewerage: its theory and construction*, New York: D. Van Nostrand, 1899, pp. 29–33. An account on the situation in Paris, see David S. Barnes, *The Great Stink of Paris and the Nineteenth-Century Struggle against Filth and Germs*, Baltimore: John Hopkins University, 2006.

13 Mays, "Water for human consumption," p. 36.

mentioned in 1481 in an Italian treatise dedicated to methods of coping with the pest or avoiding it, authored by Marsilio Ficino (1433–99), who was not only a leading humanist and philosopher, but also a medical doctor. This treatise was later republished under different titles and also translated into several languages, including a Spanish edition that appeared in 1598.[14] Most commonly, however, corrupt air was blamed for epidemics well into the nineteenth century.[15] Nevertheless, as a consequence of the perceived danger, by the 1600s a debate raged among doctors and scientists on the advantages and disadvantages of drinking water, although the popular bias against water as a drink was deeply rooted.[16] Fears that water interfered with digestion by cooling the stomach and its furnace-like operation caused many people to avoid drinking it altogether.[17] Not only popular sentiment was against water: "... physicians warned people who were weak, old, phlegmatic, or melancholic to avoid drinking water ... Water was the medical antithesis of wine. What wine cured, water caused to decay."[18]

Abundant sources of the early modern era indicate that in England the subject of the influence of water on human health was considered highly important, though similar concerns about the subject appear in French and Spanish sources as well. Between the fourteenth and eighteenth centuries the English seem to have drunk very little water. Sir John Fortescue proudly noted about the middle of the fifteenth century that English peasants "... drink no water, unless at certain times, upon a religious score, and by way of doing penance ..."[19]

14 Marsilio Ficino, *Tratado de la peste*, edited by Fabián Alejandro Campagne, original published in Pamplona: Mathias Mares, 1598, BNM, catalogued under R-26692. On the boiling of water, see ibid. fol. 16r. The title of the first Italian edition is: *Consiglio contro la peste*, Florence: Apud Sanctum Jacobum de Ripolis [Convent of San Jacopo di Ripoli], 1481.

15 Roberta J. Magnusson, "Water and wastes in medieval London," in Terje Tvedt, (ed.), *A history of water*, (3 volumes), Vol. 3, London: I. B. Tauris, 2006, pp. 299–313, p. 302.

16 Dan Rabin and Carl Forget, *The dictionary of beer and brewing*, Chicago: National Book Network, 1998, p. xiii.

17 James Salzman, "Is it safe to drink the water?" *Duke Journal of Environmental Law & Policy*, Vol. 19:1, (Fall 2008), pp. 1–42, pp. 11–12.

18 Andrea Snowden Cast, "Women Drinking in Early Modern England," (2001), unpublished Ph.D. dissertation, University of Adelaide; on file with James Salzman, quoted by: James Salzman, Is it safe to drink the water?" p. 12.

19 Sir John Fortescue, *Commendation of the laws of England: the translation into English of "De laudibus legum Angliæ,"* translated by Francis Grigor, London: Sweet & Maxwell, 1917, p. 61.

Henry VII declared the waters of England undrinkable.[20] This view was prevalent in royalty, so much so that before the Infanta Catherine of Aragon was to arrive to England in 1501 to marry Henry VIII, her parents instructed her: "… to accustom herself to drinking wine because the English water was undrinkable, and even if it were not, the climate would make it an impossible liquid for refreshment."[21] According to Andrew Boorde (1490–1549), an English traveler, physician and writer, a house should be located where water was available "to wasshe and wrynge, to bake and to brewe," rather than for refreshment.[22] He stresses that water is not "… holesome, sole by it selfe for an Englysshe man …"[23] He rates rain water as the best, running water upon stones and pebbles follows, next is river water, then standing water refreshed by a spring, and last, standing water.[24] This view explains the motivation for the collection of rain water, even if it was not easy in general and impossible to collect from the roofs of most cottages.[25] However, English ingenuity most probably found efficient ways to collect and store rain water in a country with such an abundant rainfall.

The arrogant English attitude towards continental culinary customs, and especially the fact that foreigners drank water, is exemplified by the poet—William Forrest, who in 1548 wrote: "Our English nature cannot live by roots / By water, herbs, or such beggar baggage / That may well serve for vile, outlandish coats.[26]

The English doctor, William Bullein (c.1525–1576), warned, "to drinke colde water is evyll … and causes melancholy …"[27] William Harrison in the 1570s

---

20 Alan Macfarlane, *The savage wars of peace: England, Japan and the Malthusian trap*, Oxford: Wiley-Blackwell, 1997, p. 128, quoting: John Findlay Drew Shrewsbury, *The plague of the Philistines: and other medical-historical essays*, London: V. Gollancz, 1964, p. 69.

21 Eric N. Simons, *Henry VII, the First Tudor King*, London: Muller, 1968, p. 226.

22 Andrew Boorde, *The fyrst boke of the introduction of knowledge*, originally written c.1541, edited by C. J. Furnivall, London: N. T. Trübner & Co., 1870, p. 233. Shakespeare ironically makes Mistress Quickly articulate this phrase in "The Merry Wives of Windsor," in William Shakespeare, Alexander Pope, *The Works of Shakespeare*, (9 volumes), Vol. 1, Birmingham: R. Martin, 1768, pp. 221–308, p. 237.

23 Boorde, *The fyrst boke*, p. 252.

24 Ibid., p. 253.

25 Norman John Greville Pounds, "*The Culture of the English People—Iron Age to the Industrial Revolution*," Cambridge: Cambridge University Press, 1994, p. 161.

26 Joan Thirsk, *Alternative Agriculture: A History: from the Black Death to the Present Day*, Oxford: Oxford University Press, 1997, p. 48.

27 William Bullein, *The government of health: a treatise written by William Bullein, for the especiall good and healthfull preservation of mans bodie from all noisome diseases, proceeding by the excesse of evill diet, and other infirmities of nature: full of excellent medicines, and*

composes what is really a panegyric to England, lavishly praising its soil, water, building materials and resources in general.[28] However, while praising English waters, he notes nothing about its use as a drink but does rank the different types of water according to their source.[29] Harrison disagreed with the famous physician Girolamo Cardano:

> Cardan writeth that our waters are hurtful to our sheep; ... Certes there is no parcel of the main wherein a man shall generally find more fine and wholesome water than in England; and therefore it is impossible that our sheep should decay by tasting of the same.[30]

It is difficult to reconcile the contradiction between the pride exhibited by early modern writers, such as Harrison, about the quality of their homeland water and the numerous references to the health damages that result from its drinking. In their own words, despite the excellence of their water, abstinence from drinking water is taken as a marker of prosperity, water being left to the poor and other peoples who do not have the means to afford alternative drinks.[31] Nevertheless, Harrison did praise English water as the essential ingredient contributing to the good taste of the ale and beer brewed using it[32]—the beer based on the soft waters of Burton-on-Trent was already famous in the Middle Ages.[33]

---

wise counsels, for conservation of health, in men, women, and children. Both pleasant and profitable to the industrious reader, London: Valentine Sims, 1595, p. 71.

28 William Harrison, *The description of England: the classic contemporary account of Tudor social life* (1577), edited by Georges Edelen, New York: Dover Publications, 1994, pp. xxv–xxvii.

29 Ibid., p. 138–139.

30 Ibid., p. 310, referring to Girolamo Cardano, *Hieronymi Cardani Medici Mediolanensis De Subtilitate: Libri XXI*, Norimbergae: apud Ioh. Petreium, 1550, p. 227.

31 Cedric C. Brown, "Sons of Beer and Sons of Ben: Drink as a Social Marker in Seventeenth-Century England," in Adam Smyth, (ed.), *A Pleasing Sinne: Drink and Conviviality in Seventeenth-Century England*, Cambridge: DS Brewer, 2004, pp. 3–20.

32 Substantial quantities of water are required to produce beer, not only as a component but for different steps in the brewing process itself: 3 to 10 liters of water are consumed for a yield of 1 liter of beer. Extracted from W. Hartmeier and M. Reiss, "Production of beer and wine," in Karl Esser, Joan W. Bennett, Paul A. Lemke, H. D. Osiewacz, (eds), *Industrial Applications*, Berlin, Heidelberg, New York: Springer, 2002, pp. 49–66, p. 51.

33 Ian S Hornsey, *A History of Beer and Brewing*, Cambridge: Royal Society of Chemistry, 2003, p. 295.

John Aylmer (1521–1594), who was consecrated Bishop of London in 1576, strove to strengthen the pride of his fellow countrymen by despising other peoples' drinking habits: "Thei [the Germans] drinck commonly water; and thou [the English] good ale and beare.... Thou livest like a Lorde, and thei like dogges ..."[34] Several other informed witnesses of Early Modern England referred to water quality extensively at the beginning of the seventeenth century, categorizing its different kinds according to their source. These 'experts' included William Vaughan—Doctor of civil law, who published a health education manual: *Natural and artificial directions for health* (1600),[35] Tobias Venner—physician and medical writer, author of *Via recta ad vitam longam* (1620)[36] and Robert Burton—writer, scholar and Anglican clergyman, whose best known work is *Anatomy of Melancholy* (1621).[37] But none of these writings addresses the question whether and how much water was drunk, or how much of it should be drunk.[38] The Welsh writer James Howell (c.1594–1666) wrote to Lord Cliff in 1634: "I do not know or hear of any nation that hath water only for their drink, except the *Japonois* ..."[39] Cesar De Saussure, a Swiss traveller visiting London in the 1720s declared that:

34 F. J. Fisher, "Tawney's Century," in F. J. Fisher, (ed.), *Essays in the Economic and Social History of Tudor and Stuart England*, Cambridge: Cambridge University Press, 1961, pp. 1–14, pp. 12–13.

35 William Vaughan, *Naturall and artificial directions for health: deriued from the best philosophers, as well moderne, as auncient. By William Vaughan, Master of Artes, and student in the ciuill law*, London: Richard Bradocke, 1600.

36 Tobias Venner, *Via recta ad vitam longam, or a plaine philosophical discourse of the nature ... and effects of all such things, as by way of nourishments and dieteticall obseruations make for the preseruation of health ... Wherein also ... the true vse of our famous Bathes of Bathe is ... demonstrated*, London: E. Griffin for R. Moore, 1620.

37 Robert Burton, *The anatomy of melancholy what it is. With all the kindes, causes, symptomes, prognostickes, and seuerall cures of it. In three maine partitions with their seuerall sections, members, and subsections. Philosophically, medicinally, historically, opened and cut vp. By Democritus Iunior. With a satyricall preface, conducing to the following discourse*, Oxford: Iohn Lichfield and Iames Short, for Henry Cripps, 1621.

38 For a general discussion on the role of healthy lifestyles and appropriate food and drink, as considered in the early modern period, see Andrew Wear, *Knowledge and Practice in English Medicine, 1550–1680*, Cambridge (U.K.) and New York: Cambridge University Press, 2000, part I, chapter 4, pp. 154–209.

39 James Howell, *Familiar letters on important subjects wrote from the year 1618 to 1650*, Aberdeen: F. Douglass and W. Murray, 1753, p. 365.

> … though water is to be had in abundance in London and of fairly good quality, absolutely none is drunk. The lower classes, even the paupers, do not know what it is to quench their thirst with water. In this country nothing but beer is drunk.[40]

In our times Francis Chapelle has noted that even among the Pilgrims of the Plymouth Colony in the 1620s, a close identification existed between drinking water and poverty, and only those who had no other choice drank it, notwithstanding the availability of good quality water in New England: "There is one thing all Europeans agreed on: drinking water was bad—very bad—for your health."[41]

The inadequacy and unreliability of supplies was undoubtedly a crucial factor in the limited importance of water as a beverage in England, but other influences played their part. The people had a range of alternative drinks to choose from—not only ale and beer, but spirits, and later tea and coffee, as well as milk. The British historian Peter Clark has proposed another explanation, stating that in medieval England water had religious and magical connotations, expressed in holy water, holy wells and liturgical ceremonies. But when the Reformation attacked such practices as superstitious, the ritual aspect of water declined, whereas beer drinking, traditionally associated with guild fellowship and communal solidarity, became increasingly popular.[42]

It is difficult to assess the extent of water consumption in France in the early modern period because water was a free resource, unlike beer and wine, which were regulated and taxed. Some testimonies indicate that 'common people' in France, who could not have afforded to satisfy their liquid needs by drinking only beer or wine, must have drunk water.[43] Strong supporting testimony for this assumption, though probably biased, was given by the already-mentioned Sir John Fortescue, when comparing the virtues of the laws of England with

40 César de Saussure, *A Foreign View of England in the Reigns of George I and George II. The Letters of Monsieur César de Saussure to his Family*, translated and edited by Madame Van Muyden, London: John Murray, 1902, p. 157.

41 Francis Chapelle, *Wellsprings: a Natural Story of Spring Bottled Waters*, New Brunswick (NJ): Rutgers University Press, 2005, p. 102.

42 Peter Clark, "Water Supply and Drink Consumption Patterns in English Towns in the Early Modern Period," in *Le Acque Interne*, Secc. XII–XVIII, Atti della XV Settimana di Studi, 15–20 aprile 1983, Istituto Internazionale di Storia Economica F. Datini, Prato, May 2000, pp. 1–16.

43 Alan David Francis, *The Wine Trade*, London: Adam & Charles Black, 1972, p. 247 and Georges Duby and Robert Mandrou, *A History of French Civilization*, New York: Random House, 1964, p. 213.

those of France, with reference to French peasants: "… their constant drink is water, neither do they taste, throughout the year, any other liquor; unless upon some extraordinary times, or festival days."[44] Additional evidence that water was widely consumed is the general attitude towards women and children, who should consume alcoholic beverages sparingly. The "Monologue du Bon Vigneron" [The worthy winemaker], a poem written by an anonymous writer from Auxerre in western Burgundy around 1590, illustrates the attitude in which he stresses that water should only be drunk by women, children and the poor.[45] Another contemporaneous witness corroborates that those most likely to have to drink water in France were the poor and the less powerful, in particular women. Fynes Moryson in the second decade of the 1600s reported that "Women for the most part and virgins alwaies (except by stealth they offend against the customs) use to drinke water …"[46]

By the second half of the sixteenth century the first signs of concern about water quality were shown by French authorities.[47] In the 1560s Jacques Besson stated that the quality of water depended on: "… an abundant presence of vines, clover and watercress, and an absence of vermin so to prevent unpleasant smells … [Water] filtered through sand or 'pure' silt could only be of good quality since it would be tasteless and odorless.[48] One of the keenest French observers, the Jesuit Father Jean François (1597–1640), dedicated a whole chapter of his book on fountains to the ways to evaluate the quality of water,

44 Fortescue, *Commendation of the laws*, p. 58.

45 Mack P. Holt, "Wine, Community and Reformation in Sixteenth Century Burgundy," *Past and Present*, 1993; No. 38, pp. 58–93, p. 84. The quote in Holt's text is from: Anonymous, "Le monologue du bon vigneron sortant de sa vigne, et retournant soupper en aison," in Charles Moiset, (ed.), "La poésie auxerroise au XVIe è," *Annuaire historique du département de l'Yonne*, xxi (1857), pp. 73–83, pp. 77–78. The "Monologue du bon vigneron" was written in Auxerre in western Burgundy in the 1590s and published anonymously in 1607.

46 Fynes Moryson, *An itinerary containing his ten yeeres travell through the twelve dominions of Germany, Bohmerland, Sweitzerland, Netherland, Denmarke, Poland, Italy, Turky, France, England, Scotland & Ireland*, (4 volumes), Vol. 4, Glasgow: MacLehose, 1907 (original from 1617), p. 142.

47 For a discussion on the development of the attitude towards water quality in Paris, see: Agathe Euzen and Jean-Paul Haghe, "What kind of water is good enough to drink? The evolution of perceptions about drinking water in Paris from modern to contemporary period," *Water History*, July 2012, pp. 1–14.

48 Jacques Besson, *L'art et science de trouver les eaux et fontaines cachées soubs terre autrement que par les moyens vulgaires des Agriculteurs & Architectes*, Orléans: P. Trepperel, 1569, p. 77.

including its effects in the tone and health of the skin of those who drank it.[49] He judged water according to its taste, smell and color, or the lack of such qualities. He wrote about "… those who taste and distinguish between different waters as others do with wine."[50] The concept of purifying river water from the Loire and the Seine began to be seriously considered only in the second half of the seventeenth century,[51] yet at the beginning of the eighteenth century, Nicolás Delamare in *Traité de la police*, first published in 1707, still relied upon Hippocrates, Pliny and Gallen as authorities for his ideas about water quality, which he maintained should be clear, tasteless and odorless. He ranked water from springs as the purest, followed by rain water collected and purified in cisterns; water from wells was last. Delamare concurred with Andrew Boorde that water from rivers, if separated from the impurities gathered by their flow, was best.[52]

A somewhat different attitude arises from works written in the Iberian Peninsula. The copious Spanish medical literature of the sixteenth and seventeenth centuries[53] considers liquids as both food and medicine,[54] and the

---

49 Jean-François, Père de la Compagnie de Jésus, *L'art des fontaines, c'est à dire pour trouver, esprouver, assembler, mesurer, distribuer & conduire les sources dans les lieux publics & particuliers; d'en rendre la conduite perpétuelle; et de donner par art des eaux courantes aux lieux, où elles manquent par nature*, Rennes: Chez Hallaudars, 1665, chapter 4, section 3, pp. 44–52.

50 Ibid., p. 49.

51 Jean-Pierre Goubert, *The Conquest of Water*, translated by Andrew I. Wilson, Cambridge: Polity Press, 1989, p. 171.

52 Nicolás de La Mare, *Traité de la police, où l'on trouvera l'histoire de son établissement toutes les loix et tous les règlemens qui la concernent*, (5 volumes), Vol. 1, Paris: Jean et Pierre Cot, 1705, p. 486.

53 For a comprehensive list of the Spanish medical literature from Roman times up to 1801 see: Joaquín de Villalba y Guitarte, *Epidemiología española o Historia cronológica de las pestes, contagios, epidemias y epizootias que han acaecido en España desde la venida de las cartagineses hasta el año 1801: Con noticia de algunas otras enfermedades de esta especie que han sufrido los españoles en otros reynos, y de los autores nacionales que han escrito sobre esta materia*, Madrid: Imprenta de Fermín Villalpando, 1803. For a corpus of Spanish Medical Texts, which includes 55 medical manuscripts and early printed books on the whole field of medicine see the website of the Hispanic Seminary of Medieval Studies (HSMS): http://www.hispanicseminary.org/t&c/med/index-en.htm.

54 For a detailed article on this subject see: Andrea María Bau y Gabriela Fernanda Canavese, "'Agua que cura, agua que alimenta'. La dietética para sanos y el uso del agua en la sociedad española bajomedieval y moderna," *Cuadernos de historia de España*, Vol. 80, Buenos Aires, enero/diciembre (2006), pp. 127–146.

consumption of water among the factors that influence health. A pediatric text from 1604 declares: "Even if water does not nourish, it is useful as medicine."[55] Probably due to the much warmer Spanish climate, it seems to have been the only region in which the medical debate centered in the convenience or the dangers of drinking liquids cooled by ice or snow, which were stored since winter in deep wells or in well insulated huts located high in the mountains.[56] This obsessive interest in the benefits and dangers of cooling drinks, such as water, beer or wine, is shown by the disproportionate attention dedicated to it in Spanish medical texts from the beginning of the sixteenth to the eighteenth century.[57] Opinions were widely opposed: very cold water was either considered a wonderful remedy for all kinds of maladies or a threat to the health of elderly, unfit people. As a rule, medieval tradition was not in favor of the consumption of beverages cooled by snow; if drunk, it was advised to do it in small

---

55 ["Y aunque el agua no aproueche por alimento, a lo menos les sera vtil por medicamento,"] in Cristóbal Pérez de Herrera, *Texto y Concordancias de la Defensa de las criaturas de tierna edad*, Valladolid: Luis Sánchez, 1604, original in BNM, edited by Andrea María Bau, New York: Hispanic Seminary of Medieval Studies, 2004, fols. 44v–45r, extracted from the Website of the Hispanic Seminary of Medieval Studies (HSMS). http://www.hispanicseminary.org/t&c/med/dct/framconc.htm, accessed 13 August 2012. For an opposite view on the benefits of the utilization of water as a remedy and for almost any kind of treatment involving water, see: Isidro de la Pastora y Nieto, *Una verdad histórica relativa al uso del agua por los médicos españoles en el tratamiento de las enfermedades: Discurso que en el acto de recibir la solemne investidura de doctor en Medicina y Cirujía pronunció ante el claustro de la Universidad Central Central de Madrid, Facultad de Medicina*, Madrid: Imprenta de Díaz y Compañia, 1834.

56 For an extensive description of the use and construction of the storing facilities, commonly called pozos de nieve (snow wells), see: Ángel María Calvo Barco, "Los neveros, una actividad desaparecida en nuestras montañas," in *Zainak, Cuaderno de Antropología y Etnografía*, No. 14, (1997), pp. 203–213; and Eduardo Martínez de Pisón, Carlos Caballero Casado, Sonia Fernández Esteban, Nuria de Andrés de Pablos, David Palacios Estremera, Carlos Caballero Casado, Sonia Fernández Esteban, Amparo Martín Espinosa y Ana Mas Hernández, *Encerrar el frío—El Pozo de Nieve de La Granja y el aprovechamiento tradicional de la nieve en el Guadarrama*, Segovia: Diputación de Segovia, Ayuntamiento del Real Sitio de San Ildefonso, 2014.

57 For an account of the development of Spanish thinking on this subject during the sixteenth and seventeenth centuries, see: Fernando Beltrán Cortés, *Apuntes para una Historia Del Frío en España*, Madrid: CSIC-Dpto. de Publicaciones, 1983. Natalia Juan García, "Prácticas alimentarias en los siglos XVII y XVIII en el clero regular de Aragón—los manjares de la comunidad de monjes de San Juan de la Peña," *Studium: Revista de Humanidades*, No. 15, (2009), pp. 165–198, refers to this curious and heated debate, giving a detailed bibliography, in footnote No. 49, pp. 191–192.

gulps, a recommendation still popular at present. Too cold water was blamed for impairing digestion and cooling the blood.

Alonso Díez Daza,[58] a physician from Seville, wrote in 1576 a small volume, entitled *The book of the benefits and damages that are caused by drinking exclusively water, how to choose the best and correct the one which is not, and how it should be drunk in warm times without causing damage.*[59] He dedicated the second chapter (fols. 13–21) to explain how to recognize 'good water'. His advice is just plain common sense: "… [water] has to have good taste, good color, good odor, as Galen stated …"[60] In a curious discussion he grades the quality of river water according to the direction of the flow: "… best from rivers flowing from west to east, second best when flowing from east to west, worst when flowing north to south or south to north."[61] Not surprisingly, and although cautioning that "not because it is his 'homeland and own nation' … Seville's water is excellent, as its river flows to the east."[62] The eighth chapter of the first book (fols. 57–64) is dedicated to recommended modes of water consumption. His advice, which he bases on Hippocrates, is once again logical: "… drink moderately—as the body regulates hunger, so it does with thirst …"[63]

Along the entire volume, the same author extensively notes the danger of drinking cold water, as he had seen people die instantly while doing so. Certain conditions had to be met in order to make cold drinking acceptable, such as being young, doing it in summer and having healthy intestines. The most complete work on the subject was written in 1576 by Francesc Micó (1528–1592), a Catalan doctor, originary from Vic—*Relief of thirst, where it is discussed the need we have to cool and refresh our beverages with snow, the conditions required doing so, and which bodies can tolerate it.*[64] Micó described thirst as a physiological need, and cited testimonies from ancient philosophers and

58 Sometimes spelled Daça.

59 Alonso Díez Daza, *Libro de los provechos y daños que provienen con la sola bebida del agua; cómo se ha de escojer la mejor y rectificar la que no es tal, y cómo se ha de beber fria en tiempo de calor sin que haga daño,* (2 volumes), Vol. 1, Sevilla: Alonso de la Barrera, 1576.

60 Ibid., fol. 13 v.

61 Ibid., fol. 15 v.–fol. 18 v.

62 Ibid., fol. 19 r.

63 Ibid., fol. 57 v.–fol. 58 r., 65 v.

64 Francesc Micó, *Alivio de los sedientos: en el qual se trata la necesidad que tenemos de beber frio y refrescado con nieve, y las condiciones que para esto son menester, y quales cuerpos lo pueden libremente soportar*, Barcelona: Matheo Barceló, 1792 (originally published in Barcelona by Diego Galán, 1576.).

physicians, who considered beneficial to have beverages cooled or at least refreshed. His main conclusion was that if cooled drinks are recommended for feverish sick people, they surely should be good for healthy people.[65]

However, at least as far as those who could afford it was concerned, there is a clear answer to historian James Salzman's rhetorical question: "if not water, then what did people drink? The answer in ancient times was alcohol. The drink of choice in Egypt was beer, and in ancient Greece was wine."[66] And many could indeed afford beer—the resale price of beer at an alehouse in England at in the seventeenth century ranged between 0.25 d. to 0.5 d. per pint [1/8th of a gallon—about half a liter], according to the quality of the beer,[67] while the average daily craftsman's wage was about 18 d.[68] Obviously, beer prices at wholesale level were much lower, and even lower when home brewed. Thus, even if drinking water, when drawn from conduits, was free, the bias against water was such that only the poorest people drank it. Other historians had reached similar conclusions—according to Fernand Braudel alcohol consumption in the early modern era was very high. As an indication, he estimated that Polish peasants drank up to three liters of beer per day.[69] In Coventry, the average amount of beer and ale consumed reached about 9 liters per person per week, compared to about two liters today,[70] while by the early fifteenth century harvest workers received between 3 and 4.5 liters, depending on the ale strength.[71] On the average, throughout England consumption was about half a liter per day per capita, while English sailors received a ration of a gallon (four and a half liters) of beer per day and soldiers received two-thirds of a gallon (three liters).[72] In general, the alcoholic content of beer was much

65 Micó, *Alivio de los sedientos*, pp. 35–36.

66 Salzman, "Is it safe to drink the water?" p. 12. For a concise history of alcohol consumption in the Western world, see: David J. Hanson, *Preventing Alcohol Abuse: Alcohol, Culture, and Control*, Westport (CT): Greenwood Publishing Group, 1995, chapter 1, pp. 1–19.

67 Stuart Peachey, *The Tipler's Guide to Drink and Drinking in the Early 17th Century*, Bristol (UK): Stuart Press, 1992, p. 9.

68 Edward Anthony Wrigley and Roger S. Schofield, *The Population History of England: 1541–1871: a Reconstruction*, Cambridge: Cambridge University Press, 1989, p. 638.

69 Fernand Braudel, *Capitalism and Material Life, 1400–1800*, translated by Miriam Kochan, New York: Harper and Row, 1974, p. 168.

70 Herbert Anthony Monckton, *A History of English Ale and Beer*, London: Bodley Head, 1966, p. 95.

71 Christopher Dyer, *Everyday Life in Medieval England*, London: A&C Black, 2001, pp. 84–85.

72 Hanson, *Preventing Alcohol*, p. 9.

lower than nowadays.[73] Throughout the Continent many types of beer were produced, with different strengths, including 'small beer', which was obtained by diluting beer with good quality water, thus lowering the alcohol content but anyhow contributing to the elimination of harmful pathogens in the water.[74]

In the sixteenth century the annual consumption of wine per capita in Spain varied according to the region: in Valladolid it reached 100 liters,[75] though in relative terms, is seems that at least in Catalonia, drunkenness was not a problem, as remarks from 1595 state: "in the dives of the port quarter [of Barcelona] one encounters a hundred prostitutes, but not a single drunkard."[76] In the Venetian Arsenal, arguably the biggest industrial establishment in the world in the early modern era, wine was considered a fringe benefit, to which all workers were entitled. The quantities consumed were huge, reaching about five liters per worker per day in 1635.[77] The wine drank by French peasants—'pin-pin'—was the product of the weakest straining resulting from second or third pressings.[78] We can probably assume that most the wine consumed in Spain had a lower content of alcohol than about 10%. Should it have had more, it would have caused net dehydration, aggravating the drunkenness effects, notwithstanding the fact that the water content ingested would have been 90% of the liquid.[79] Overall, the water quantities drank through beer and wine, complemented with the water contained in soups or broths,[80] are consistent with the total human drinking needs detailed at the beginning of this chapter.

In contrast with wine and beer, spirit was drunk in the sixteenth century mainly for medicinal purposes and its use expanded at a much lower pace. This pace would drastically increase in the next two centuries: "... the sixteenth century created it [distilled alcohol]; the seventeenth century consolidated it; the

---

73 Meg Lota Brown and Kari Boyd McBride, *Women's Roles in the Renaissance*, Westport (CT): Greenwood Publishing Group, 2005, p. 113.

74 Salzman, Is it safe to drink the water?" p. 12. For an in depth history of beer in Europe, see: Richard W. Unger, *Beer in the Middle Ages and the Renaissance*, Philadelphia: University of Pennsylvania Press, 2007.

75 Braudel, *Capitalism and Material Life*, p. 165.

76 Emmanuel Le Roy Ladurie, *The Peasants of Languedoc*, translated by John Day, Chicago: University of Illinois Press, 1976, p. 64.

77 Robert Charles Davis, "Venetian shipbuilders and the fountain of wine," *Past and Present*, 156/1 (1997), pp. 55–86.

78 Francis, *The Wine Trade*, p. 247.

79 Charles R. Thomas, *Practical Wine Talk: A Physician-Winemaker Examines Wine*, Bloomington (IN): AuthorHouse, 2013.

80 Victoria R. Rumble, *Soup through the Ages: A Culinary History with Period Recipes*, Jefferson (NC): McFarland, 2009, chapters 7–15, pp. 30–86.

eighteenth popularized it."[81] By the mid-sixteenth century a rare sort of tacit agreement existed between the Catholic Church and Luther, Calvin, the leaders of the Anglican Church and even the Puritans, by which restrained alcohol drinking was tolerated. Churches of all denominations had the position that: "… alcohol was a gift of God and created to be used in moderation for pleasure, enjoyment and health; drunkenness was viewed as a sin."[82] Intoxication was thought to be a sign of hedonism, threatening the spiritual health of society as a whole: "… inconsistent with the emerging emphasis on rational mastery of self and world and on work and efficiency."[83] However, two centuries later, radical Reformist leaders in England, blaming alcoholic beverages for almost all society maladies, would press to abolish their consumption, determined to forbid alcohol completely.[84]

It should be here stressed that in the Middle Ages and the early modern period most water in villages and towns was not used for drinking and washing of human beings, but rather for a great variety of domestic, industrial and other purposes. Water served various functions in walled towns: drinking and cooking water for the population, filling the protecting moats, irrigating the fields close to the urban area, drinking water for domestic animals and as a power source for grain milling and other industries. Public hygienic and medical bathing, indoors and outdoors, was quite common in major Western European towns and cities until about the thirteenth century, when the Church began to exert its influence to close unisex bathhouses, on morality grounds.[85] On the other hand, bathing at home was the privilege of the rich, who could afford the expensive fuel used to heat the water.

Animal breeding required comparatively big quantities of water, a constant need that included towns and cities, where horses and other pack animals were needed for transportation of people and goods. An average horse will intake 20 to 40 liters of fresh water per day, depending if it feeds on daily grazing on fresh pasture grasses, which contain large amounts of water, in summer, or forage in winter. In cold seasons a horse depends upon the forage of dried

81 Braudel, *Capitalism and Material Life*, p. 170.

82 Hanson, *Preventing Alcohol*, 1995, p. 9.

83 Austin, *Alcohol in Western Society*, pp. 129–130.

84 Hanson, *Preventing Alcohol*, 1995, p. 11.

85 Madeleine Pelner Cosman and Linda Gale Jones, *Handbook to Life in the Medieval World*, (3 volumes), Vol. 1, Infobase Publishing, 2009, pp. 123–126. For a comprehensive discussion on this subject, see: Alexandra Cuffel, "Polemicizing Women's Bathing Among Medieval and Early Modern Muslims and Christians," in *The Nature and Function of Water, Baths, Bathing, and Hygiene from Antiquity Through the Renaissance*, Volume 11 of Technology and Change in History, (ongoing series), Leiden and Boston: Brill, 2009, pp. 171–190.

grasses or hay, which contains a lower amount of water. Therefore, a horse may need an increase of offered water in the winter months, more so than in the grazing periods. Cows require about 40 liters per day, and twice as much for milking cows.[86] Like humans, animals drink more in hot weather and while doing heavy work. Water quality for livestock is also to be considered as it can have an impact on the volume of water consumed. Foul odors or tastes, for example, may discourage animals from drinking. Typically, poultry is more sensitive to the taste and mineral content of the water than other livestock types.[87]

Another important use was the care of orchards and gardens, which, in many cases, were still part of the urban landscape, serving for self consumption as well as for sale. The production of bread necessitated big amounts of good quality water, as did keeping fish alive in fishmonger's shops. The manufacturing of bricks and paper also required huge quantities of water, as did filling fire extinguishing reserve cisterns. During events such as market-days and fairs water demand increased drastically, and even more so water pollution.[88]

A broader perspective about the way that municipal and state authorities of the three kingdoms related to large-scale public works in general can be obtained by examining other types of hydraulic undertakings, such as land reclamation by drainage and sewerage projects. Land drainage undertakings were made feasible by engineers and their financial backers, just as they created new opportunities in urban water supply.[89] A crucial motivation for the

86 Data extracted from "How much water do animals need each day?" University of California, Website of the Division of Agriculture and Natural Resources, http://ucanr.edu/sites/placernevadasmallfarms/files/90846.pdf, accessed 12.12.2015.

87 Daniel Ward and K. McKague, "Water Requirements for Livestock Factsheet," Ontario (Canada): Ministry of Agriculture and Food and Rural Affairs, May 2007.

88 Maria Bogucka, "The Largest Polish Towns Water Supply in the XVIth–first Half of the XVIIIth Centuries," in *Le Acque Interne*, Secc. XII–XVIII, Atti della XV Settimana di Studi, 15–20 aprile 1983, Istituto Internazionale di Storia Economica F. Datini, Prato, May 2000, pp. 1–15, pp. 3–7.

89 For an in-depth examination on the subject of drainage and irrigation, see: Salvatore Ciriacono, "Investimenti capitalistici e cultura irrigue: La congiuntura agricola nella Terraferma veneta (secoli XVI–XVII)," in *Venezia e la Terraferma attraverso le relazioni dei rettori*, Milan: Giuffrè, 1981, pp. 123–158; Salvatore Ciriacono, *Land drainage and irrigation*, Aldershot (UK): Ashgate, 1998; Clive Holmes, "Drainage Projects in Elizabethan England: The European Dimension," in *Eau et développement dans l'Europe moderne*, edited by Salvatore Ciriacono, Paris: Ed. de la Maison des sciences de l'homme, 2004, pp. 87–102 and D. R. Curtis and M. Campopiano, "Medieval land reclamation and the creation of new societies: comparing Holland and the Po Valley, c.800–c.1500," *Journal of Historical Geography*, Volume 44, (2014), pp. 93–108. The bibliographies included in Holmes' and Curtis' articles detail extensive lists of primary and secondary sources on this subject.

drive for land reclamation during the fifteenth and up to the seventeenth centuries was the recovery of ground that had been abandoned by the decline in population after 1300 and the subsequent agricultural crisis. In sharp contrast with the efforts invested in water supply and land drainage enterprises, European cities of the High Middle Ages seldom and inadequately attempted to organize solid and liquid wastes removal, in spite of the rapidly growing population.[90] Nevertheless, even if it is widely accepted that the concepts of hygiene in general—toilets, baths, sewage systems, and basic sanitation vanished with the fall of the Roman Empire, modern studies have proven that the notions of cleanliness and sanitation were present in Medieval cities as well.[91] In the early modern period these basic notions of personal hygiene evolved—people were expected to wash their face and hands, comb their hair, brush their

90 For a condensed history of world developments in the field of sanitation, see Andrew I. Wilson, "Drainage and Sanitation," in Örjan Wikander, (ed.), *Handbook of Ancient Water Technology*, Technology and Change in History, (ongoing series), Vol. 2, Leiden: Brill (2000), pp. 151–179; Giovanni De Feo, P. Laureano, R. Drusiani and A. N. Angelakis, "Water and wastewater management technologies through the centuries," *Water Science & Technology: Water Supply—WSTWS*, 10.3 (2010), pp. 336–349 and Martin V. Melosi, *The Sanitary City: Environmental Services in Urban America from Colonial Times to the Present*, Pittsburgh: University of Pittsburgh Press, 2008, pp. 4–9. For an overview of the status of sewerage in Europe in medieval times, see: Roberta Magnusson, "Water Supplies and Sewerage," in Thomas F. Glick, Steven J. Livesey and Faith Wallis, (eds), *Medieval Science, Technology, and Medicine*, pp. 505–506; Magnusson, *Water Technology in the Middle Ages* and Giovanni De Feo, George Antoniou, Hilal Franz Fardin, Fatma El-Gohary, Xiao Yun Zheng, Ieva Reklaityte, David Butler, Stavros Yannopoulos and Andreas N. Angelakis, "The Historical Development of Sewers Worldwide," *Sustainability, 6*, (2014), pp. 3936–3974. For the most colorful and vivid account, see David Nicholas, *Urban Europe 1100–1700*, Basingstoke (UK) and New York: Palgrave Macmillan, 2003, pp. 155–167.

91 See for example: Steven J. Burian, Findlay G. Edwards, "Historical Perspectives of Urban Drainage," in E. W. Strecker, W. C. Huber, (eds), *Global Solutions for Urban Drainage, Proceedings of Ninth International Conference on Urban Drainage*, Portland (OR), USA, September 2002, pp. 1–16; C. Taylor, "The disposal of Human Waste: A comparison between Ancient Rome and Medieval London," *Past Imperfect*, 11, (2005), pp. 53–72; E. L. Sabine, "Latrines and cesspools of Mediaeval London," *Speculum*, 9, (1934), pp. 303–321; D. Jorgensen, "Cooperative sanitation: Managing streets and gutters in the late medieval England and Scandinavia," *Technology and Culture*, Vol. 49, No. 3, (2008), pp. 547–567; D. Jorgensen, "Local government response to urban river pollution in late medieval England," *Water History*, 2, (2010), pp. 35–52 and N. J. Ciecieznski, "The Stench of Disease: Public Health and the Environment in the Late-Medieval English towns and cities," *Health Culture Society*, 4, (2013), pp. 91–104.

teeth, and wash their hands before and after meals. Frequent full baths were generally looked upon with uneasiness, as if they would be a kind of bodily extravagance or some kind of medical treatment. In general, attitudes toward public defecation were tolerant and humorous scatological language was common.[92]

92 Allison P. Coudert, "Sewers, Cesspools, and Privies: Waste as Reality and Metaphor in Pre modern European Cities," in Albrecht Classen, (ed.), *East Meets West in the Middle Ages and Early Modern Times: Transcultural Experiences in the Premodern World*, Berlin and Boston: Walter de Gruyter, 2013, pp. 713–733, p. 718.

CHAPTER 2

Water-Lifting Technology in the Graeco-Roman World and Its Development through the Renaissance

The Invention of Water-Lifting Devices

The invention of the waterwheel and its applications, such as the water mill, is generally ascribed to the Greek world, between the third and first century BCE. In fact, all water-lifting devices based on subdivided wheels and cylinders seem to have originated in that period.[1] The reference to water-raising devices in the extant writings of Ctesibius (285–222 BCE),[2] Philo of Byzantium (c.280–220 BCE), Marcus Vitruvius Pollio (c.75–15 BCE) and Hero of Alexandria (c.10–70 CE) bear testimony to the fact that lifting water constituted an important challenge for the outstanding mechanical technologists

1 For discussions on this subject, see: Adam Robert Lucas, *Wind, water, work: ancient and medieval milling technology*, Leiden: Brill, 2006, Technology and Change in History, (ongoing series), Vol. 8; John Peter Oleson, *Greek and Roman Mechanical Water-Lifting Devices: The History of a Technology*, Toronto: University of Toronto Press, 1984, pp. 325–331; John Peter Oleson, "Water-Lifting," in Örjan Wikander, (ed.), *Handbook of Ancient Water Technology*, Technology and Change in History, (ongoing series), Vol. 2, Leiden: Brill (2000), pp. 217–302, specially pp. 229–241; Örjan Wikander, "The Water-Mill," in Örjan Wikander, (ed.), *Handbook of Ancient Water Technology*, Technology and Change in History, (ongoing series), Vol. 2, Leiden: Brill (2000), pp. 371–400; K. Donners, M. Waelkens and J. Deckers, "Water Mills in the Area of Sagalassos: A Disappearing Ancient Technology," *Anatolian Studies*, Vol. 52, (2002), pp. 1–17; K. D. White, "'The Base Mechanic Arts'? Some Thoughts on the Contribution of Science (Pure and Applied) to the Culture of the Hellenistic Age," in Peter Green, (ed.), *Hellenistic History and Culture*, Los Angeles and London: University of California Press, 1993, pp. 211–237; Larry W. Mays, (ed.), *Ancient Water Technologies*, Dordrecht, Heidelberg, London and New York: Springer, 2010; and Stavros I. Yannopoulos, Gerasimos Lyberatos, Nicolaos Theodossiou, Wang Li, Mohammad Valipour, Aldo Tamburrino and Andreas N. Angelakis, "Evolution of Water Lifting Devices (Pumps) over the Centuries Worldwide," *Water*, 7 (2015), pp. 5031–5060.

2 Although no Ctesibius' writings have survived, Vitruvius specifically refers to Ctesibius' works as available in his time. See: Marcus Vitruvius Pollio, *The Ten Books on Architecture*, translated by Morris Hicky Morgan, Cambridge (MA): Harvard University Press, 1914, book X, chapter 7, pp. 297–298.

 | DOI 10.1163/9789004312425_004

of Antiquity.[3] Yet most of the writings of the period have been lost, and much of our information about the scholars of the Hellenistic period comes from remarks made by later authors. When texts have been saved, the selection process was fortuitous, arguably resulting in the preservation of some of the worst and the destruction of some of the best.

The Italian historian Lucio Russo attributes the birth of this inventiveness to a Hellenistic scientific and technological revolution that achieved depths that had not been reached in the classical times, when Greek culture had reached great heights in art, literature and philosophy.[4] This so-called 'revolution', although lasting about two centuries, actually covered a period that was shorter than what is generally considered to be the Hellenistic age—the period between the death of Alexander the Great in 323 BCE to the conquest of Egypt by Rome in 30 BCE.[5] For Russo, the end of the Hellenistic period was linked to the end of the quantum leap of scientific knowledge in the second century BCE, when scientific studies declined and eventually disappeared. Russo maintains that the science developments achieved during the era labeled by him as the scientific revolution were lost with the Roman conquest and during the Middle Ages, because the scholars of those periods did not have the capability to understand them. The Hellenistic period was almost immediately forgotten, whereas interest in classical philosophers like Aristotle and Plato revived. In Russo's view, the legacy of Hellenistic science was one of the bases of the scientific revolution of the sixteenth century, when ancient texts in these fields started once again to be available in Europe.[6]

---

3 See Thorkild Schioler, *Roman and Islamic Water-Lifting Wheels*, Odense: Odense University Press, 1973, pp. 110–174. Schioler presents a detailed chronological account of the sources relevant to study the evolution of water-raising devices. Schioler's book remains a frequently quoted study in this area by scholars, but his systematic approach to standardize the terminology characterizing the different devices has not been adopted by other historians, and confusion among them is rampant in this respect.

4 For a controversial discussion on this subject, see: Lucio Russo, *The Forgotten Revolution: How Science Was Born in 300 BCE and Why It Had to Be Reborn*, translated by Silvio Levy, Berlin and Heidelberg: Springer, 2004. The ensuing paragraphs summarize succinctly some ideas from his thesis.

5 This periodization is according to German historian Johann Gustav Droysen in *Geschichte des Hellenismus*, (2 volumes), Vol. 1, Cambridge: Cambridge University Press, 2011 (originally written between 1836 and 1843).

6 Russo, *The Forgotten Revolution*, chapter 11, pp. 329–397, specially pp. 385–307.

The Use of Water-Lifting Devices in the Roman Era

The most complex and effective water supply systems of Antiquity were built by the Romans, who achieved a degree of sophistication that may have been unequaled until modern times.[7] Aqueducts entering into the city were piped to deliver water according to a specific set of priorities: first to public basins that were used by citizens for fetching free water for domestic use, then to private use and finally to bath-houses.[8] It is estimated that about forty percent of the water flowing into Rome went to the private buildings of high status citizens or officials, who paid a special tax—*vectigal*—to have their residences connected with pipes from the main system. The amount levied varied according to the size of the supply-pipe dispensing nozzle. The tax funds were used to cover the cost of maintenance of the system. Legislation assured free access to water, guaranteed by the emperor. It was one of the duties of the *cura aquarum* (water commissioner) to see to it that the public fountains and water basins were continuously supplied.[9] Thus, water for the affluent was considered a priced good whereas for the average citizen, water was free and available by right.[10]

Although water was used widely for drinking, cooking and hygiene, the Roman drive to build aqueducts was powered by social goals. These impressive public works were intended, first and foremost, as political statements, to remind the common people that they received their water by imperial

7 For an account of the sophistication of Roman hydraulic technology, see Deborah Chatr Aryamontri, "Running Water: Advances In Urban Water Supply During The Roman Empire," in *The Nature and Function of Water, Baths, Bathing, and Hygiene from Antiquity Through the Renaissance*, Volume 11 of Technology and Change in History, (ongoing series), Leiden and Boston: Brill, 2009, pp. 319–336 and Larry W. Mays, "A Brief History of Roman Water Technology," in Larry W. Mays, (ed.), *Ancient Water Technologies*, Dordrecht, Heidelberg, New York, London: Springer Science & Business Media, 2010, pp. 115–134.

8 Harry B. Evans, *Water distribution in ancient Rome: the evidence of Frontinus*, Ann Arbor (MI): University of Michigan Press, 1997, p. 9. According to Norman John Greville Pounds, *The Culture of the English People—Iron Age to the Industrial Revolution*, Cambridge: Cambridge University Press 1994, p. 53, in Roman cities in Britain, public baths were the first priority.

9 Christer Bruun, "Roman emperors and legislation on public water use in the Roman Empire: clarifications and problems," *Water History* 4 (2012), pp. 11–33 and Christer Bruun, "Water Supply, Drainage and Watermills," in P. Erdkamp, (ed.), *The Cambridge Companion to the City of Rome*, Cambridge: Cambridge University Press, 2013, pp. 297–313.

10 Salzman, "Thirst: A Short History," pp. 11–12.

beneficence, while providing flamboyant displays of technological prowess to visitors to the city.[11] Although no pictorial or textual descriptions of water-lifting devices have survived before the Renaissance period, archaeological findings of batteries of draining waterwheels, stacked in order to reach deep galleries in antique Roman mining sites,[12] such as the Rio Tinto copper mines in Spain,[13] the Dolaucothi gold mine in Wales[14] or the Alburnus Maior site in the Western Carpathian Mountains of Romania,[15] confirm their extensive use by the Romans at widely distanced locations, a testimony to their expertise in this area and the practicability of their know-how.[16] The involvement of the Roman Army in the mines and the similarity of the technical solutions in different sites, indicate a high probability that military experts travelled and advised, or even imposed, accepted designs throughout the Empire. This same kind of 'standardization' of water supply solutions may have been imposed in army camps.[17] In Strabo's *Geography* the author refers to the water supply of a military camp in Egypt, which combined different water-lifting methods:

> At present it is an encampment for one of the three legions which garrison Egypt. There is a mountainous ridge, which extends from the encampment as far as the Nile. At this ridge are wheels and screws, by which water is raised from the river, and one hundred and fifty prisoners are [thus] employed.[18]

---

11 Steven Mithen, *Thirst: For Water and Power in the Ancient World*, Cambridge (MA): Harvard University Press, 2012, p. 288.

12 Alfred Michael Hirt, *Imperial Mines and Quarries in the Roman World: Organizational Aspects 27 BCE-CE 235*, Oxford: Oxford University Press, 2010. For a detailed description of the role played by water in mining the gold ore of the Rio Tinto and Dolaucothi mines, see Kevin Greene, *The Archaeology of the Roman Economy*, Berkeley and Los Angeles: University of California Press, 1990, pp. 145–146.

13 An account of these wheels is given in John Gray Landels, *Engineering in the Ancient World*, Berkeley and Los Angeles: University of California Press, 1978, pp. 63–70.

14 George C. Boon and Colin Williams, "The Dolaucothi Drainage Wheel," *Journal of Roman Studies*, Volume 56, Issue 1–2 (November 1966), pp 122–127.

15 Hirt, *Imperial Mines*, p. 33.

16 The devices used for the purpose of draining flooded mines included human driven Archimedean screws, mechanical bucket chains and double-acting pumps. See Greene, *The Archaeology of the Roman Economy*, p. 146.

17 Oleson, "Water-Lifting," pp. 293–294.

18 Strabo (c.63 BCE–c.24 CE), *Geography*, written c.20 CE, literally translated (the first six books) by H. C. Hamilton, the remainder by W. Falconer, (3 volumes), Vol 3, Book XVII, chapter 30, London: Henry H. Bohn, 1857, p. 287.

In sharp contrast with aqueducts and dams, extensively studied and admired by early historians, Greek and Roman devices to lift water were only recently studied and recognized for their value by scholars. Practical considerations partly explain this early neglect—pumping devices were less massive and, as wood was the main building material, much more likely to be destroyed. The few surviving relics can only be found in difficult to reach abandoned mines galleries or shipwrecks.[19] However, it is clear that authorities in the Greek and Roman societies had strong motivation to rend support for water-lifting: the harvests of grain from the Nile valley, which fed Rome, would have been poorer, flooded mines in Spain, Britain and Romania could not have yielded enough metals and the hazard of fire in Roman cities would have increased.[20]

The pioneer historians of technology—Marc Bloch in 1935[21] and Lewis Mumford in 1934[22]—, developed what is commonly referred to as 'the Roman technological stagnation thesis, in itself based largely on a supposed dearth of Roman water mills, and inferring that this scarcity characterized water-lifting devices as well. They argued that the classical era was characterized by a static picture of motionless economic and technological activity. Adam Lucas[23] and George Brooks[24] strongly challenge Bloch's and Mumford's ideas, presenting archeological evidence that vertical and horizontal water mills were relatively common in the ancient world, even if literary references are unusual. Lucas sustains that Roman advances in water powered milling technology not only directly influenced the medieval Europe but probably migrated eastwards toward China, thus contributing to undermine the notion of technological passivity in the ancient world. He specifically includes water supply as one of the areas in which "technological advances took place in the Graeco-Roman world between the third century BCE and the second century CE."[25] Brooks attributes the invention of the horizontal watermill to the medieval period. In his view, even if there were more mills during the early medieval period than during the Roman's, the ownership of a substantial part of them was transferred

19 Oleson, "Water-Lifting," p. 286.

20 Ibid., p. 287.

21 Marc Bloch, "Avènement et conquêtes du moulin à eau," *Annales d'histoire économique et sociale*, T. 7, No. 36, (1935), pp. 538–563.

22 Lewis Mumford, *Technics and Civilization*, London: Routledge & Kegan Paul Ltd, 1955, pp. 112–118.

23 Lucas, *Wind, water, work*, pp. 42–47.

24 George Brooks, "The 'Vitruvian' Mill in Roman and Medieval Europe," in Steven A. Walton, (ed.), *Wind & Water in the Middle Ages: Fluid Technologies from Antiquity to the Renaissance*, Tempe (AZ): Arizona Center for Medieval and Renaissance Studies, 2006, pp. 1–38.

25 Lucas, *Wind, water, work*, p. 49.

by the Romans, to be maintained and developed by medieval carpenters for generations.[26]

In 1965, in an influential article by M. I. Finley,[27] he maintained that although the Greeks and Romans built high, powerful and intellectual civilizations, they bequeathed to their successors few new inventions. Thirty five years later, Kevin Greene disputed upfront Finley's case by reconsidering texts and analyzing new archeology evidence, reaching the conclusion that during the five centuries of steady rise of Greek states and gradual expansion of the Roman Empire a corresponding growth in infrastructure, equipment and innovation took place.[28] There is evidence supporting Greene's position. A water driven stone cutting sawmill is mentioned by Ausonius in his poem *Mosella* (CE 370–71), although Lynn White, Jr. rejected the passage both on literary and technical grounds. It is now considered authentic.[29] An archeological relic, found in 2005, in Hierapolis, located in present day Pamukkale (Turkey), represents a sophisticated water powered saw-mill. The carving, on a stone sarcophagus of the second half of the third century CE, clearly describes the machine operation, thus providing unquestionable evidence of the technology being available at the time.[30]

When looking at water powered mills, so closely related to water lifting technologies, Bloch attributed a vital role to medieval monasteries in the use of watermills and their technical development. He sustained that religious institutions, as members of the landowners elite, promoted the use of efficient watermills in order to strengthen their domination over the lower classes of the population. Mumford saw in the monks the keepers of a tradition of con-

26 Brooks, "The 'Vitruvian' Mill," p. 34.

27 M. I. Finley, "Technical Innovation and Economic Progress in the Ancient World," *The Economic History Review*, New Series, Vol. 18, No. 1, Essays in Economic History Presented to Professor M. M. Postan (1965), pp. 29–45.

28 Kevin Greene, "Technological Innovation and Economic Progress in the Ancient World: M. I. Finley Re-Considered," *The Economic History Review*, New Series, Vol. 53, No. 1 (2000), pp. 29–59.

29 D. L. Simms, "Water-driven saws, Ausonius, and the Authenticity of the *Mosella*," *Technology and Culture, Vol. 24*, (1983), pp. 635–634 and D. L. Simms, "Water-driven Saws in Late Antiquity," *Technology and Culture*, Vol. 26, (1985), pp. 275–276.

30 Klaus Grewe, "La máquina romana de serrar piedras. La representación en bajorrelieve de una sierra de piedras de la antigüedad, en Hierápolis de Frigia y su relevancia para la historia técnica," translated by Miguel Ordóñez, in Isaac Moreno Gallo, (ed.), *Las técnicas y las construcciones de la Ingeniería Romana*, Madrid: v Congreso de las Obras Públicas Romanas, 2010, pp. 381–401.

stant technology improvements and inventiveness. These powerful, almost consensual, thesis has influenced several more recent historians,[31] but has been challenged the last two decades by revisionist scholars. In "The Role of the Monasteries in the Development of Medieval Milling,"[32] Adam Lucas painstakingly researched archaeological and textual evidence. In his conclusions he does not totally refute the 'monastic innovation thesis' but provides convincing evidence that the role of Christian monasteries in the development of mills, with the exception of the Benedictines in the ninth through eleventh centuries, has been overstated. Lucas questions as well the validity of the 'seigneurial monopoly model' for medieval water mills, as he found many cases in which mills were owned and operated by those of the lower classes, particularly from the thirteenth century.[33]

The character of social attitudes in the ancient world toward technological invention and innovation remains the subject of scholarly dispute as well. Until recently, the standard analyses of ancient technology suggested that the widespread presence of slavery and the denigrating attitude of the elite toward manual labor discouraged and suppressed innovation in prime movers utilization.[34] More recently this position is being revised by scholars such as Michael Jonathan Taunton Lewis, whose research discovered, among other applications, several cases of cam-operated tilt-hammers in the Roman Empire,[35] and Örjan Wikander, who gives the example of a military innovator in the fourth century CE that recommended the construction of a ship to be powered by oxen turning a paddle wheel through a right angle gearing—an idea that in effect shows deep understanding of water power.[36] Their common conclusion

31 For more recent authors adopting the Bloch thesis see; Lynn White Jr., *Medieval Technology and Social Change*, Oxford: Oxford University Press, 1964, p. 82; Gimpel, *The Medieval Machine*, pp. 6–10; Gies, *Cathedral, Forge and Waterwheel*, p. 35; and Norman John Greville Pounds, *An Economic History of Medieval Europe*, New York: Longman Publishing, 1994, p. 198.

32 Adam Robert Lucas, "The Role of the Monasteries in the Development of Medieval Milling," in Steven A. Walton, (ed.), *Wind & Water in the Middle Ages*, pp. 89–128.

33 Adam Robert Lucas, *Ecclesiastical lordship, seigneurial power and the commercialization of milling in medieval England*, Farnham (UK): Ashgate, 2014.

34 Marc Bloch, "Medieval Inventions," in *Land and Work in Medieval Europe*, translated by J. E. Anderson, 1967, (1935), pp. 169–185, p. 182.

35 Michael Jonathan Taunton Lewis, *Millstone and Hammer: The Origins of Water Power*, Hull (UK): Hull University Press, 1997, pp. 84–115.

36 Örjan Wikander, "Industrial Applications," in Orjan Wikander, (ed.), *Handbook of Ancient Water Technology*, Technology and Change in History, (ongoing series), Vol. 2, Leiden: Brill (2000), pp. 401–410.

is that the use of water-power in the Roman world was more pervasive than had previously been thought and that social attitudes and pressures did not overcome the will of individuals to innovate.[37]

Islamic Technological Influence

In spite of copious, well preserved documentation, which testifies to the importance given to water power in the Islamic world between the seventh and the fourteenth centuries, some earlier Western historians have disregarded the role of Arab engineers in influencing European mechanical technology. Until recent years, an analysis of the research findings of leading historians and archaeologists in what concerns the transfer of knowledge from Islam to Western Europe in the area of water-lifting technology shows a consensus about the complete lack of influence exerted by Arab technology over Europe. Thomas Ewbank, the author of the important book, *A Descriptive and Historical Account of Hydraulic and Other Machines for Raising Water*, writing in 1842, does not mention any impact of Arabic water supply know-how over Europe—in fact he does not mention Arab technology at all. In his *Engineering in History*, Richard Shelton Kirby does not refer to any Arab contribution in a relatively extense chapter on the subject of water power utilization technology,[38] and neither Charles Singer in the chapter dedicated to water lifting in *A History of Technology*, refers at all to Arab contribution to technological progress in this area during the medieval period.[39] Lyon Sprague de Camp stresses the Arabic engineers' skill in building water clocks and the quest for perpetual motion that arose from this expertise, but he does not refer to any input in the field of water-raising technology.[40] On what irrigation is concerned, according to Andrew M. Watson, "The Islamic contribution was less in the invention of new devices than in the application on a much wider scale of devices

37 Oleson, "Water-Lifting," p. 287. Kevin Greene, in his analysis of the views of Ö. Wikander, J. P. Oleson and K. D. White as they transpire from their work, unequivocally joins the ranks of scholars who believe the attitude of the Roman elite was favourable to technological innovation. See Kevin Greene, "Perspectives in Roman Archeology," *Oxford Journal of Archaeology*, Volume 9, Issue 2 (1990), pp. 209–219.

38 Richard Shelton Kirby and others, *Engineering in History*, New York, Toronto and London: McGraw-Hill Book Company, 1956.

39 Charles Singer, E. J. Holmyard, A. R. Hall and Trevor I. Williams, *A History of Technology* (3 volumes), Vol. 2, Oxford: Clarenton Press, 1956, pp. 675–694.

40 Lyon Sprague DeCamp, *The Ancient Engineers*, New York: Ballantine Books, 1963, pp. 306–307.

which in pre-Islamic times had been used only over limited areas and to a limited extent."[41] Michael Decker, who strongly differs with Watson on his main 'Green Revolution' thesis, agrees with him that Islamic know-how in irrigation consisted mainly of gradually perfecting existing techniques, by showing that new developments overlapped earlier Roman networks,[42] thus contesting Thomas Glick's assertion that radical transformation in irrigation took place.[43] Robert James Forbes in his *Studies in Ancient Technology* gives short credit to the Arab role in improving irrigation technology and its impact in conquered Spain's agriculture, but none to other applications.[44] Bernard Gille also notes the intense effort put by the Arabs in developing advanced water clepsydras and automata, but makes no reference to water-lifting devices.[45]

In this controversial subject on the one hand, Arabs, when referred to, are often described as having been no more than librarians and archivists for preserving Greek science till Europe was in a position to take its heritage back; on the other, books such as *The Genius of Arab Civilization: Source of Renaissance*,[46] as its title implies, present the richness, sophistication and intellectual achievement of the Arab-Islamic civilization, drawing attention to its original inventiveness, thus expanding its role to much more than only a link between the Hellenistic past and the Renaissance future. Historians of Arabic technology tenaciously emphasize Islam's contribution to science and technology. Ahmad Y. Al-Hassan gives examples in the field of water-raising, astronomy, clock-making, glass manufacturing, architecture and the adoption of Arab words for technical subjects in different European languages.[47] The numerous

---

41 Andrew M. Watson, *Agricultural Innovation in the Early Islamic World*, Cambridge: Cambridge University Press, 1983, p. 108.

42 Michael Decker, "Plants and Progress: Rethinking the Islamic Agricultural Revolution," *Journal of World History*, Vol. 20, No. 2 (2009), pp. 187–206, p. 190.

43 Thomas F. Glick, *Irrigation and Society in Medieval Valencia*, Cambridge (MA): Harvard University Press, 1970.

44 Robert James Forbes, *Studies in Ancient Technology*, (9 volumes), Volume 2, 3rd. edition, Leiden: Brill, 1993, pp. 49–50.

45 Bertrand Gille, *The Renaissance engineers*, London: Lund Humphries, 1966, pp. 22–23.

46 John Richard Hayes, *The Genius of Arab Civilization: Source of Renaissance*, Cambridge (MA): Taylor & Francis, 1983.

47 Ahmad Y. Al-Hassan and Donald R. Hill, *Islamic Technology, an Illustrated History*, Cambridge: Cambridge University Press, 1986, pp. 31–35. See also Mohammed Abattouy, "The Arabic-Latin Intercultural Transmission of Scientific Knowledge in Pre-Modern Europe: Historical Context and Case Studies," in Nayef R. F. Al-Rodhan, *The Role of the Arab-Islamic World in the Rise of the West: Implications for Contemporary Trans-Cultural Relations*, London: Palgrave Macmillan, 2012, pp. 167–219, pp. 200–202.

primary sources used by Schioler in chapters III and V of his book on water wheels present a picture of Islamic inventiveness and deep understanding of the workings of the waterwheels and auxiliary mechanisms.[48] D. Fairchild Ruggles has researched the use of waterwheels and water technology, both for irrigation and aesthetic purposes in Islamic gardens. Though almost no archeological work has been done in the Islamic world on the subject, literary evidence is abundant, such as contained in the *Bayad wa Riyad* manuscript (thirteenth–fourteenth centuries) and *The One Thousand and One Nights* (oral traditions beginning in the tenth century, written down in the second half of the thirteenth). In Spain plentiful archeological testimonies confirm the widespread use of water technology in Medieval Islam and the importance of gardens to that culture.[49] In conjunction with these capabilities, whole manuscripts are preserved dedicated to 'wheels that turn by themselves'—the language used to describe perpetual motion.[50]

Recent historiography, supported in this endeavor by a renewed look at Islamic accomplishments in this field, refutes the almost exclusive role that Russo attributed to Hellenistic inanimate power technology as antecedents to Western Science. Thomas Glick, who has dealt with this subject extensively,[51] when referring to the movement of ideas and techniques and specifically to technological transfer, stresses that know-how diffusion in the High Middle Ages mostly followed a trail from China and India westwards, through the mediation of Persia, in itself a center of technological innovation. Though this was a very slow process, the Arab conquests in Iberia significantly accelerated the migration of Eastern techniques to Europe.[52] In an in-depth analysis of Islamic hydraulic technologies, Andrew Wilson argues that Islamic culture transmitted the Egyptian, Hellenistic and Roman methods to medieval Europe, Spain being the link for their further diffusion northwards. His interpretation of the archaeological and literary evidence he collected for irrigation and water lifting is that the use of classical water techniques was continuous in both shores

48 Schioler, *Roman and Roman and Islamic Water-Lifting Wheels*, chapter 3: "Islamic Waterwheels and Perpetual Motion," pp. 56–83 and chapter 5: "Waterwheels in Ruined Islamic Castles," pp. 90–96.

49 D. Fairchild Ruggles, "Waterwheels and Garden Gizmos: Technology and Illusion in Islamic Gardens," Steven A. Walton, (ed.), *Wind & Water in the Middle Ages*, pp. 69–88.

50 Ibid., pp. 180–185.

51 Thomas F. Glick, *Islamic and Christian Spain in the early Middle Ages*, Princeton: Princeton University Press, 1979, Part Two, Movement of Ideas and Techniques, chapter 7, Technology, p. 217–247.

52 Ibid., p. 217.

of the western Mediterranean, with possible gradual improvements caused by populations migration that were aware of the same technology previously.[53]

George Saliba's research has contributed to clarify how Islamic knowledge disseminated to Europe during the second half of the Middle Ages and the Renaissance.[54] He describes what he calls the classical alternative, i.e. dating the beginnings of Islamic science in times of the early Abbasid caliphate (from about 750 to 850 CE), subsequently developing thanks to Greek material translations, through Byzantine or Persian intermediates. According to this thesis, Islamic science flourishes between the ninth and eleventh centuries, until its decline in the eleventh century as a result of a strong anti-scientific religious movement and in the thirteenth century as a result of the Mongol invasion. The main thrust of history of science then migrates to Europe. Saliba proposes a revisionist alternative, in which he dates the beginning of Islamic science, especially astronomy, much earlier and extends its prime into the 16th century. He minimizes the possibility of either the Byzantines or the Persians having awakened Arabic interest in Greek materials, while giving a comprehensive description of the wider social, political and economic factors created a critical mass of elite Arabic-speaking bureaucrats, who had the will and the capabilities to absorb and develop Islamic knowledge. Saliba's ideas on the origins of the Islamic sciences and their influence on the development of the European sciences have been by and large accepted by present day scholars.

Although Arab books in the fields of algebra, chemistry, geology, trigonometry and medicine were translated into Latin, apparently no mechanical technological writings found their way into western languages and though it can always be claimed that knowledge was transmitted through other channels, no concrete evidence has been found on which to base such contention.[55] Nevertheless, accepting the fact that the Muslim world technological prowess in the Middle Ages is undeniable, it leads to the crucial question: to what

53 Andrew I. Wilson, "Classical water technology in the early Islamic world," in C. Bruun and A. Saastamoinen, (eds), *Technology, ideology, water: from Frontinus to the Renaissance and beyond*, Roma: Instituti Romani Finlandiae, 2004, pp. 115–141.

54 George Saliba, *Islamic science and the making of the European Renaissance*, Cambridge (MA): MIT Press, 2007.

55 F. Jamil Rapeg, "The Translation of Greek Natural Philosophy into Arabic: Background and Motivations," in David C. Lindberg and Michael H. Shank, (eds), *The Cambridge History of Science*, (8 volumes), vol. 2, New York: Cambridge University Press, 2013, pp. 27–61, pp. 34–38; David C. Lindberg, "The Transmission of Greek and Arabic Learning to the West," in David C. Lindberg, (ed.), *Science in the Middle Ages*, Chicago: Chicago University Press, pp. 52–90, pp. 62ff; A. I. Sabra, "Situating Arabic Science: Locality versus Essence," *Isis*, 87 (1996), pp. 654–70 and Al-Hassan, *Islamic Technology*.

extent this know-how influenced the western world? Even if it had not been aware of the Hellenistic science, would early modern European water-raising technology still have come up the way it did, catalyzed by the development of inanimate power know-how in the Middle Ages and by Islamic advanced knowledge?

The Diffusion of the Waterwheel in Europe

The fall of the Western Roman Empire did not cause an immediate disappearance of Roman hydraulic technology. The know-how of gravity-flow systems of channels, subterranean pipes, hydraulic cement, and even inverted siphons was preserved, mainly by religious institutions, for uses such as baptistery fonts and atrium fountains. Yet for several centuries, towns and cities lagged behind monasteries where water supply technology is concerned and, arguably (the subject is vividly debated among historians of different generations), by the rapid expansion of new religious orders, such as the Premonstratensians, Dominicans, Franciscans and Cistercians. The plans and the lay-out of these institutions were practically identical, and so were the waterpowered systems they operated.[56] Authorities did not even attempt to find solutions for the urban water supply for general population. There were a few instances in which Roman aqueducts were restored: in Rome by the popes of the eighth and ninth centuries, but it was only by the second half of the sixteenth century that extensive repairs took place in the city water supply system.[57] Sophisticated waterworks were built in the Carolingian palaces at Ingelheim and Aachen. These were exceptions, as during Late Antiquity and the Early Middle Ages the majority of the complex Roman water systems gradually decayed as a result of poor maintenance and were finally abandoned. At the end of the first millennium, communities were forced to obtain their water from rivers, wells and cisterns or by conveying water from rivers with the help of pack animals. However, in the eleventh and twelfth centuries numerous new water conduits were constructed throughout Western Europe, most of them to supply religious houses and royal palaces. The technological revival soon spread from monasteries and palaces to the burgeoning towns, as well as to a few hospitals, castles and gardens.

56 Jean Gimpel, *The Medieval Machine*, New York: Penguin Books, 1976, p. 5.

57 Pamela O. Long, "Hydraulic Engineering and the Study of Antiquity: Rome, 1557–70," *Renaissance Quarterly*, vol. 61, no. 4, (winter 2008), pp. 1098–1138.

For several decades, mainly in the mid-twentieth century, historiography has been dominated by an extremely influential thesis according to which the second half of the European Middle Ages witnessed an industrial revolution based on waterpower. The most evident signs of this revolution, according to its supporters—scholars such as Marc Bloch, Eleanora Carus-Wilson,[58] Robert Forbes and Lynn White, Jr.—would be a dramatic increase in the number of mills powered by water and wind, and a drastic growth in their industrial applications. This technological paradigm has been challenged by a reassessment of mill historiography, on the basis of mill scholarship and empirical evidence, which greatly exceeds the material that was available to the proponents of the 'medieval industrial revolution' thesis.[59] This new approach has revised, and in some cases, refuted, several long-held historiographical assumptions. A 2005 paper by Adam Lucas on this subject is the most systematic survey compiling all of the relevant evidence to date.[60] Lucas shows that the widespread application of water and wind power is recorded only in a few regions ('pockets of innovation')[61] of what today are France and Italy, characterized by substantial regional differentiation in the diffusion and social contexts of mill technologies. Asia and the Islamic world seem to have been the sources of many innovations, later conveyed to Western Europe, thus diminishing the importance of what was considered the 'unique' inventiveness of medieval Europeans.[62] The most important textual testimony about the widespread use of watermills in *Hispania* is the *Líber ludiciorum* (*Lex Visigothorum*), or *Fuero Juzgo*, a recopilation, written in Latin, of the laws of the Visigoths Kingdom in Spain.[63]

58 E. M. Carus-Wilson analyzes with great detail the economic and social impact that water powered fullling mills caused in England in the thirteenth century. See E. M. Carus-Wilson, "An Industrial Revolution of the Thirteenth Century," *The Economic History Review*, Vol. 11, No. 1, (1941), pp. 39–60.

59 For a comprehensive description and analysis of the different approaches to this subject, see Adam Robert Lucas, "Narratives of Technological Revolution in the Middle Ages," in A. Classen, (ed.), *Handbook of Medieval Studies: Terms—Methods—Trends*, Berlin: De Gruyter, 2010, pp. 967–990.

60 Adam Robert Lucas, "Industrial Milling in the Ancient and Medieval Worlds: A Survey of the Evidence for an Industrial Revolution in Medieval Europe," *Technology and Culture*, Vol. 46, No. 1, (2005), pp. 1–30.

61 Lucas, *Wind, Water, Work*, p. 231.

62 Lucas, "Industrial Milling." See also Adam Robert Lucas, *Wind, water, work*, pp. 61–68.

63 *Fuero Juzgo o Libro de los Jueces, Cotejado con los más antiguos y preciosos códices por la Real Academia Española*, Valladolid: Lex Nova, 1980.

Enacted in 654 by the king Recesvinto, and revised in 681 and 693, the *Fuero Juzgo* harshly punishes theft of watermills or parts of them.[64]

What is indisputable is that the waterwheel caused labor savings in several industries and massive increases in production, enabling the existence of enterprises previously served by manual labor or draught animals, which could not have been possible without the concentrated water energy provided.[65] Some primary graphic testimonies of industrial applications survived, such as a self-operated waterwheel driven saw-mill, included in the famous sketchbook of *Villard de Honnecourt*,[66] dated about 1235 (fig. 2.1). Yet no graphic record of water-lifting devices possessing such level of sophistication has survived from medieval Europe.

Historians writing in *Working with Water in Medieval Europe*,[67] a compilation of in-depth articles by leading scholars about water utilization in Ireland, England, France, Germany, the Netherlands, Italy and Spain during the Middle Ages, imply that with the exception of Spain, medieval Europeans do not appear to have used Archimedean screws or other ancient water-lifting devices at all. In Islamic Spain, on the other hand, the evidence of their use is overwhelming.[68] Thus, we can appreciate two distinctive and unconnected paths in the evolution of water-powered applications in the Muslim world and

---

64 See in ibid. for example: Libro VII, Título II, XII (Concerning the Theft of Mill Machinery), quoted in an extended article by Ignacio Gonzalez Tascon, "La Difusión Medieval del Molino Hidráulico," in Pedro Navascués Palacio e Ignacio González Tascón, (eds.), *Ars mechanicae: ingeniería medieval en España*, Madrid: Fundación Juanelo Turriano, Ministerio de Foment, 2008, pp. 99–130.

65 Terry S. Reynolds, *Stronger than a Hundred Men*, Baltimore and London: John Hopkins University Press, 1983, p. 4 and Richard Holt, "Medieval England's Water-Related Technologies," in Paolo Squatriti, (ed.), *Working with Water in Medieval Europe: Technology and Resource-use*, Leiden: Brill, 2000, pp. 51–100, pp. 69–79. For another example in metallurgy in England, see Jane Geddes, "Iron," in *English Medieval Industries: Craftsmen, Techniques, Products*, in John Blair, W. John Blair and Nigel Ramsay, (eds), London and Rio Grande: Hambledon Press, 1991, pp. 167–188, pp. 172–173.

66 J. B. A. Lassus and Alfred Darcel, (eds), *Album de Villard de Honnecourt*, facsimile of a XIII century manuscript by Villard de Honnecourt, Paris: Imprimerie Impériale, 1858. For a commented facsimile of the manuscript, see: Carl F. Barnes, *The Portfolio of Villard de Honnecourt*, Farharm (UK): Ashgate Publishing, Ltd., 2009.

67 Paolo Squatriti, (ed.), *Working with Water in Medieval Europe, Technology and resource-use*, Leiden, Boston, Koln: Brill, 2000.

68 Thomas F. Glick and Helena Kirchner, "Hydraulic Systems and Technologies in Islamic Spain: History and Archaeology," in Squatriti, (ed.), *Working with Water*, pp. 267–330.

FIGURE 2.1 *Self-operating saw-mill* (1235).[69]

The text beneath the drawing is translated as: 'How to make a saw operate itself?' ... it is evident that the saw reciprocates and is held at its base by pinned levers, driven by a waterwheel, and the top end is fastened to a spring (a bent branch). Cams (four levers) fastened to the axletree of the waterwheel cause the saw to reciprocate and a wheel fitted with spikes provides for the forward motion of the sawn log.

in Europe in the medieval period—scant agricultural use of water-lifting technologies in Europe and few industrial skilled uses in the Islamic world.

Several reasons can be raised to explain the existence of the advanced Arab water-lifting technology when compared with Western European technology during the Middle Ages. Water was a much scarcer and precious commodity in the Moslem world—a strong motivating factor in the development of water-raising devices. Supporting this view is the evidence of the wider use of dams for irrigation and water-lifting wheels, both uses fulfilling needs in areas of extremely dry climates. Lack of water stimulated innovation and special interest in hydraulic devices. Machines built on the basis of these techniques were adapted to the characteristics of the terrain and the availability of suitable materials nearby, so that the final result differed from place to place. Culture related factors, such as the Moslem industrial tradition, which emphasized

69 Lassus, *Album de Villard de Honnecourt*, planche XIII. Explanation based on Bryan Lawton, *Various and Ingenious Machines, Power Generation and Transport*, (2 volumes), Vol. 1, Leiden and Boston: Brill, 2004, p. 247.

carefully hand-manufactured goods, probably caused disregard for products resulting from water power. Likewise, Moslem societies had norms against polluting water, for example, as was common to all Middle Eastern cultures. The technological capabilities in Islam could have enabled it to build water-power industrial machines, but the use of water-power seems to have been focused on two tasks—grinding grain and raising water.[70]

The Improvement of Water Lifting Technology in Europe, Evolution or Quantum Leaps?

According to Jonathan Sawday, by the beginning of the 1400s Western Europe faced an energy shortage, which could be attributed to several factors: the availability of human labor was drastically curtailed by the plague, incentivizing the invention of mechanisms and devices to substitute manpower; the depletion of forests, which further deteriorated during the sixteenth century, brought about the extension of the use of coal instead of firewood.[71] The exhaustion of the easier reachable strata of lignite[72] necessitated the excavation of deeper mines. But mines were prone to flooding, making effective work impossible, thus increasing the need to create powerful and reliable pumping systems to drain them.[73] Reynolds concurs, noting that it was only by the fifteenth century that dramatic technological improvements were applied to and caused the rapid development of water-raising machines.[74] Lucien Lefbre states that with the invention of the movable type printing the demand for

70 Reynolds, *Stronger than a Hundred Men*, pp. 119–120.

71 Jonathan Sawday, *Engines of the Imagination: Renaissance Culture and the Rise of the Machine*, New York: Taylor & Francis, 2007, pp. 34–35. For an account of the long term effects of the plague in the diminishing size of the working population in Europe, see: Norman F. Cantor, *In the Wake of the Plague*, New York: Free Press, 2001, pp. 24–25. William H. McNeil, *Plagues and Peoples*, London: Penguin Books, 1985, p. 184, maintains that the population growth after the plague was swifter in the fourteenth century than in any other pandemic in the past. It can be inferred that this recovery weakened the motivation to invent labor-saving devices. This view is contested by other scholars: W. T. S. Gould, *Population and Development*, New York: Routledge, 2013, p. 78, estimated that it took about 100 years to reach the pre-plague population. For an account of the grave effect of deforestation upon French industry, see: Henry Heller, *Labour, Science and Technology in France, 1500–1620*, Cambridge: Cambridge University Press, 2002, pp. 85–87.

72 A soft form of coal, found close to the earth surface.

73 Sawday, *Engines of the Imagination*, p. 35.

74 Reynolds, *Stronger than a Hundred Men*, chapter 3, pp. 47–121, especially table 2.2.

paper increased drastically, which often brought about conversions of former water powered corn grain mills into paper mills.[75] The location preferred was in the higher reaches of a big river or its tributaries, where water was cleaner and it was relatively easy to cut a diverted canal in which the waterwheel was installed.[76]

Nevertheless, as we have seen, this step function in the timing of new developments has been challenged, with opponents detecting instead a linear, gradual advance. When analyzing other types of water driven devices, archeological evidence supports much greater continuity: Colin Rynne has found that the horizontal mills of western Ireland and Scotland and some Scottish islands represent the sustained use of a technology from the early medieval period to modern times.[77] According to Michael Kucher, beginning in the eleventh century, towns adjacent to rivers that could be channeled to drive waterwheels, such as Bologna, Milano and Firenze, became rich by implementing innovations in water power for industrial utilization. On the other hand, Sienna, a typical hill town, suffered from the disadvantage of the lack or sufficient water resources. Nevertheless, as Reynolds before him, Kucher maintains that full exploitation of water as a prime mover would be realized only by the fourteenth century.[78] Carlo Cipolla gives an account of technology developments during the Middle Ages, and shows to what an extent industrial water power was a crucial impulsing factor in this progress, with Italy at the economic and technology forefront in the period of the twelfth to the fifteenth centuries, later diffusing its expertise to France, England and Germany.[79] Although concrete evidence is scarce, it seems likely that a similar continuous line of utilitarian use development and innovation connects water raising mechanisms during the same centuries throughout Europe. This assumption necessarily embodies the generational transfer of the specific engineering skills, not only to build, but to maintain those structures.

---

75 Lucien Febvre and Henri-Jean Martin, *The Coming of the Book: The Impact of Printing 1450–1800*, translated by David Gerard, London and New York: Verso, 1976, p. 33.

76 Ibid., p. 34.

77 Colin Rynne, "Technological Continuity, Technological 'Survival': the Use of Horizontal Mills in Western Ireland, c.1632–1940, *Industrial Archaeology Review*, 33: 2, (2011), pp. 96–105, p. 104.

78 Michael P. Kucher, *The Water Supply System of Siena, Italy: The Medieval Roots of the Modern Networked City*, New York: Routledge, 2005, p. 36.

79 Carlo Cipolla, *Before the Industrial Revolution: European Society and Economy 1000–1700*, New York and London: W. W. Norton & Company, 1994, pp. 137–154.

Basic Water Raising Technologies

We shall now refer to the basic operation principles of the earlier methods to raise water, as first described by ancient authors. The following descriptions relate to the first machines operating in a continuous rotational motion, which could be driven either by an animal or a natural energy resource, such as water flow or, in exceptional cases, wind. We shall limit ourselves here only to the different water-raising methods relevant to the ones implemented in the cities that are the subject of our study.

Direct Water-Lifting with Compartmented Waterwheels

These wheels have two basic configurations: the *tympanum*—a wheel with a compartmented body, and the *noria*—a wheel with separate, attached containers or with a compartmented rim.[80] Though these wheels could be turned either by men treading on their outside or by animals, the wheels that would become relevant to urban water-lifting in the early modern period were mostly driven by energy derived from the flow of a river.

The *tympanum* consists of a series of gutters united at their open ends to a horizontal shaft, which is made hollow at one end and placed a little higher than the point to which the water should be elevated; the gutters are arranged radially, and are of sufficient length to extend from the shaft to a short distance below the surface of the water (fig. 2.2).[81]

A *noria* consists of a narrow undershot vertical waterwheel, to whose rim a series of containers is attached, thus lifting water from a river to a small aqueduct at the top of the wheel. It can raise water to somewhat less than its full diameter. The more sophisticated *norias* are arranged with swing buckets or pots, pivoted just above their centers, and with the catch trough so fixed as to tip the buckets at the highest point, thus giving this wheel the greatest possible advantage as to height of discharge for a given diameter (fig. 2.3). The word *noria* originates from the Arabic word *naurah*—used in Syria for bucket-type waterwheels and it means literally 'the wailer,' due to the wailing sound made by its wooden bearings during operation.[82] Although the discharge capacity of

80 Lucas, *Wind, water, work*, pp. 24–29.

81 Ewbank, *A Descriptive and Historical Account*, p. 110.

82 Mahmod Samman, "Rediscovering the Waterwheel—'Noria Al-Muhammadiyya,' an ASME International Historic Landmark," a presentation to the American Society of Mechanical Engineers, Northwest Houston Subsection meeting, August 2005. He describes the sound

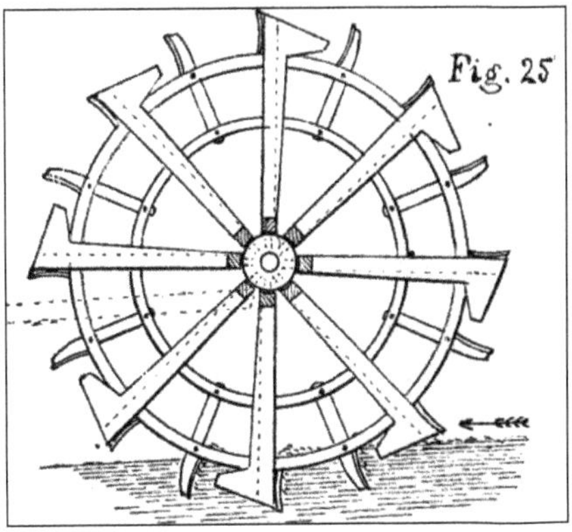

FIGURE 2.2 *Sketch of a* Tympanum.[83]

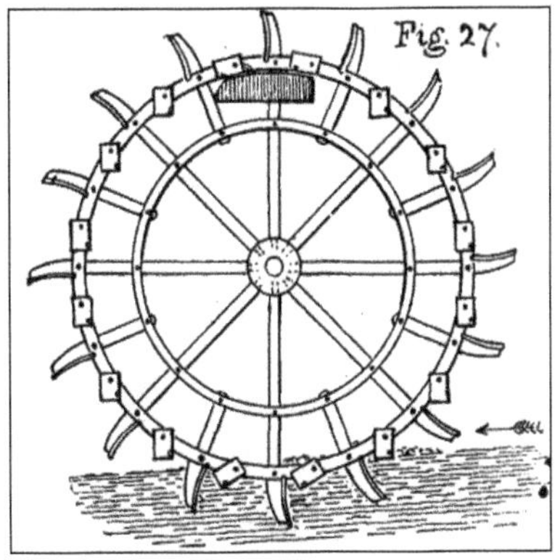

FIGURE 2.3 *Sketch of a* noria.[84]

the *tympanum* is large, it can lift water only to less than the height of its own radius, while requiring a high torque to turn it. The *noria* is generally of a lighter design and dispenses water at almost twice the height of the tympanum.

The earliest published archaeological evidence—in our opinion very flimsy—of an animal-driven waterwheel in Greece comes from the ruins of the Perachora in Kórinthos (c.300 BCE), excavated during the early 1970s.[85] The attempted reconstruction shown in fig. 2.4 is highly speculative.

Though several Greek papyri of the third to second century BCE mention the use of compartmented wheels, the invention itself could not be traced with certainty until recent years. According to John Peter Oleson and Andrew Wilson, the simultaneous appearance of written and archeological evidence in the third-century BCE Egypt reinforces the assumption that these water-lifting wheels may have been invented in the fourth century BCE in Alexandria or its whereabouts.[86] Oleson sustains that the wheel with a compartmented

---

as a mixture of noise and musical notes that is often compared to organ music. Its deepest notes are in the range of 120–170 Hz.

83 G. D. Hiscox, "The Power of Water, or Hydraulics Simplified," *Scientific American Supplement*, No. 799, April 25, 1891, fig. 25.

84 Ibid, fig. 27.

85 R. A. Tomlinson, "The Perachora Waterworks: Addenda," *The Annual of the British School at Athens*, Vol. 71, (1976), pp. 147–148.

86 Oleson, "Water-Lifting," p. 235; Oleson, *Greek and Roman devices*, pp. 325–329; Andrew I. Wilson, "Machines in Greek and Roman Technology," in John Peter Oleson, (ed.), *The Oxford Handbook of Engineering and Technology in the Classical World*, Oxford: Oxford University Press, 2008, pp. 337–368, p. 352. For a concise account of the probable developments that resulted in these inventions, see: George Brooks, "The 'Vitruvian Mill'," pp. 2–9.

FIGURE 2.4
Reconstruction of the Peracora waterwheel (*c.300* BCE).[87]

rim became widely used about 300 CE, when the wooden compartments were replaced with inexpensive ceramic pots that were tied to the outside of an open-framed wheel.[88] The earliest descriptive written reference to a water-moved, compartmented wheel appears in chapter 61 of the Arabic translation of *Pneumatics*, attributed to the Greek engineer Philo of Byzantium.[89] Although some modern scholars attribute some of the *Pneumatics* chapters to later Arabic additions, the philological studies of the British historian of technology, Michael Lewis, have demonstrated that the chapters relevant to water-raising are direct translations of the Greek third-century BCE original into Arabic. Lewis found that a Greek papyrus of the second century BCE known as *Laterculi Alexandrini* mentions a man by the name of Abdaraxos and states that "… he completed the machines of Alexandria"[90] In another Arabic

87 Kostas Kotsanas, "Reconstruction of the Perachora waterwheel," implemented by Augusta Stylianou for the *Ancient Greek Technology Exhibiton*, that took place January 23–March 10, 2009 in Limassol, Cyprus. Courtesy of Prof. Kostas Kotsanas, Director of the Museum of Ancient Greek Technology.

88 Oleson, "Water-Lifting," p. 235.

89 Philon de Byzance (c.280–220 BCE), *Le livre des appareils pneumatiques et des machines hydrauliques*, edité d'après les versions arabes d'Oxford et de Constantinople et traduit en français par le Baron Carra de Vaux, Paris: Imprimerie nationale, Librairie C. Klincksieck, 1902, pp. 177–180. See also Oleson, "Water-Lifting," p. 235.

90 Extracted from Rafael Frankel, "Water Mills in Israel," in Jean-Pierre Brun and Jean-Luc Fiches, (eds), *Énergie Hydraulique et Machines Élévatrices D'eau Dans L'antiquité*, Napoli: Collection du Centre Jean Bérard, 2007, pp. 215–224, p. 217.

text Lewis found a section that reads: "the Byzantine mill or the newly invented vertical mill of Abdaraxos."[91] According to Lewis, Philo seems to know of the bucket chain, the overshot wheel, and perhaps the *noria* and the *saqiya*.[92] Lewis contradicts John Peter Oleson, who claims that: "... Philo shows no evidence of having known the *saqiya* gear."[93]

The first known representation of a fully developed *saqiya gear* is in a first century BCE Hellenistic wall painting in Ptolemaic Egypt, showing two oxen harnessed to a *saqiya*, driving a compartmented wheel through an unseen gear mechanism (fig. 2.5). Norman Alfred Fisher Smith has argued that the vertical watermill evolved from water-lifting devices operating in a vertical plane, such as the animal-driven saqiya, and that it predated the invention of the horizontal mill.[94]

Oleson describes several applications mentioned in Hellenistic sources for the two types of compartmented wheels, driven by water power, trod or by a *saqiya gear*. During Ptolemy IV's reign (221–205 BCE) compartmented wheels appear to have been used to drain dry docks in Alexandria.[95] In Philo's *Parasceuastica*, chapter 91, (late third century BCE) they are recommended as tools to flood siege works,[96] and in other sources they are indicated for irrigation, draining flooded mines and other inundated areas in general. Wheels were also used to raise water for drinking and filling baths—the huge thermal baths in Ostia included whole batteries of compartmented-wheels installations.[97]

---

91 Ibid.

92 Lewis, *Millstone and Hammer*, p. 32, quoted and commented by Andrew I. Wilson, "Machines, Power and the Ancient Economy," *The Journal of Roman Studies*, Vol. 92 (2002), pp. 1–32, p. 7. Although the term *saqiya* is widely used by scholars, it has ambiguous meanings for different authors. Following Thorkild Schioler, *Roman and Islamic Water-Lifting Wheels*, p. 11, we shall denominate *saqiya* as machines in which the traction power of a draught animal is transmitted to a potgarland or a waterwheel by means of a right angle gear.

93 Oleson, *Greek and Roman Devices*, p. 349. In Oleson's terminology '*saqiya-gear*' has an identical meaning with the term *saqiya* in Schioler's terminology.

94 John Langdon, *Mills in the Medieval Economy: England 1300–1540*, Oxford: Oxford University Press, 2004, extracted from *eBook Collection* (*EBSCOhost*), accessed 7 June 2016, citing Norman Alfred Fisher Smith, "The Origins of Water Power: A Problem of Evidence and Expectations," *Transactions of the Newcomen Society*, No. 55 (1), (1983), pp. 67–84, p. 75.

95 Oleson, *Greek and Roman Devices*, p. 326.

96 Ibid., p. 349.

97 Ibid., pp. 349–350. Sketches of the Ostia complex are shown in the same book, figs. 93–98 (pages not numbered). Fot an in-depth study of Roman and Greek baths, see Hubertus Manderscheid, "The Water Management of Greek and Roman Baths, " in Örjan Wikander,

FIGURE 2.5 *The earliest known representation of a* saqiya gear *(First century BCE).*[98]

In the first century BCE Vitruvius described both types of waterwheels in his *De Architectura.*[99] Depictions based on these descriptions appeared for the first time in the sixteenth century in an edition published in 1511 by Fra Giovanni

(ed.), *Handbook of Ancient Water Technology*, Technology and Change in History, (ongoing series), Vol. 2, Leiden: Brill (2000), pp. 467–538.

98 Painting from the Wardian tomb, depicting a *saqiya* driven by two oxen, original kept in the Graeco-Roman Museum, Alexandria (Egypt), location 83.. This subject has been studied by several scholars. See: Schioler, *Roman and Islamic Water-Lifting Wheels*, pp. 152–153; Marjorie Susan Venit, "The Painted Tomb from Wardian and the Decoration of Alexandrian Tombs," *Journal of the American Research Center in Egypt*, Vol. 25 (1988), pp. 71–9, the painting is reproduced in p. 74; Marjorie Susan Venit, "The Painted Tomb from Wardian and the Antiquity of the Sāqiya in Egypt," *Journal of the American Research Center in Egypt*, Vol. 26 (1989), pp. 219–222; Myrto Malouta and Andrew Wilson, "Mechanical Irrigation: Water-Lifting Devices in the Archaeological Evidence and in the Egyptian Papyri," in Alan Bowman and Andrew Wilson, (eds), *The Roman Agricultural Economy: Organisation, Investment, and Production*, Oxford: OUP, 2013, pp. 273–306, painting reproduced in p. 283; Maureen Carroll, *Earthly Paradises: Ancient Gardens in History and Archaeology*, Los Angeles: Getty Publications, 2003, painting reproduced in p. 74 and Anette Schomberg, "Ancient Water Technology: between Hellenistic Innovation and Arabic Tradition," *Syria*, 85, (2008), pp. 119–128, painting reproduced in p. 126.

99 Marcus Vitruvius Pollio, *De Architecture*, translated by Joséph Gwilt, London: Priestley and Weale, 1826, Book X, chapters IX and X, pp. 312–314.

Giocondo, who illustrated the book as well (fig. 2.6).[100] Clear interpretations of the *noria* are found in Daniele Barbaro's Italian and Latin translations of Vitruvius' book, both illustrated by Andrea Palladio[101] (fig. 2.7).

Archimedean Screws

The invention of the water-lifting screw has traditionally been credited to Archimedes (c.287–212 BCE). Diodorus Siculus writes in the first century BCE:

FIGURE 2.6
Tympanum, *Fra Giocondo's of Vitruvius'* De Architectura (*1511*).[102]

100 Fra Giovanni Giocondo, *M. Vitruvius per Iocundum solito castigatior factus cum figuris et tabula ut iam legi et intelligi possit*, Venetias: Giovanni da Tridentino, 1511. For accounts of the circumstances surrounding this edition, see: Ingrid Rowland, "The Fra Giocondo Vitruvius at 500 (1511–2011)," *Journal of the Society of Architectural Historians*, Vol. 70, No. 3 (September 2011), pp. 285–289 and Lucia A. Ciapponi, "Fra Giocondo da Verona and His Edition of Vitruvius," *Journal of the Warburg and Courtauld Institutes*, Vol. 47, (1984), pp. 72–90.

101 Stefano Brusaporci and Romolo Continenza, "Historical watermills for cereals grinding: cultural heritages in Aterno valley (Italy)," in Teun Koetsier and Marco Ceccarell, (eds), *Explorations in the History of Machines and Mechanisms: Proceedings of HMM2012*, Dordrecht (Netherlands): Springer Science & Business Media, 2012, pp. 261–275, p. 264. Identical depictions appear in the Italian translation: Marcus Vitruvius Pollio, *I dieci libri dell'architettura*, tradutti e commentati da Monsignor [Daniele] Barbaro, Venegia: Francesco Marcolini, 1556, p. 265; and in the Latin translation: M. Vitruvii Pollionis, *De Architectura, Libri Decem*, cum commentariis, Danielis Barbari [translation and commentary], Venetiis: Apud Franciscum Franciscum Senensem, & Ioan, Crugher Germanum, 1567.

102 Giocondo, *M. Vitruvius*, p. 100.

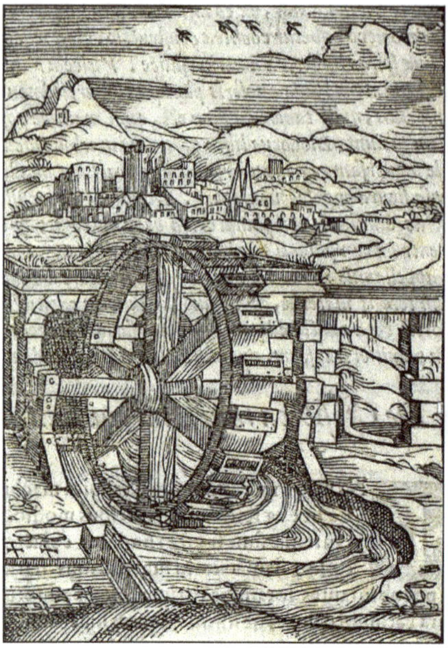

FIGURE 2.7
Noria, *Daniele Barbaro's Latin and Italian editions of* De Architectura (1567).[103]

> ... men easily irrigated the entire region [an island in the delta of the Nile] by means of a [water raising] machine which Archimedes of Syracuse invented, which from its shape is named snail shell—*cochlea*.[104]

The Greek historian Athenaeus of Naucratis (c.200 CE) writes: "And the bilge water [of the ship Syracusia], although of a most enormous depth, was pumped out by one man, by means of the screw, an engine which was the contrivance of Archimedes."[105] No reference to the device is found in Archimedes' extant works, but no mention of the water-raising screw exists before his time.[106] Archeological findings appear to support this dating, as can be seen in figs. 2.8 and 2.9.

---

103 Vitruvii Pollionis, *De Architectura, Libri Decem*, p. 348. It can be appreciated that the water is raised to twice the height that a tympanum of the same diameter would achieve.

104 Diodorus Siculus, *The antiquities of Egypt: a translation with notes of book I of the Library of History by Diodorus Siculus*, translated by Edwin Murphy, New Brunswick (NJ): Transaction Publishers, 1990, p. 43.

105 Athenaeus of Naucratis, *Deipnosophistae*, (15 Books), Book 5, translated by C. D. Yonge (1854), Loeb Classical Library, Cambridge, (MA): Harvard University Press, 1928, p. 208.

106 A discussion of the case for and against Archimedes as the original inventor of the device attributed to him can be found in Stephanie Dalley and John Peter Oleson, "Sennacherib, Archimedes, and the Water Screw. The Context of Invention in the Ancient World," *Technology and Culture*, Vol. 44, No. 1, (2003), pp. 1–26.

FIGURE 2.8
Fresco recovered from the House of the Ephebus in Pompei (*79 CE*).[107]

FIGURE 2.9
Fieldworker turning a cochlea (*c. first century BCE/CE*).[108]

107 A man driving an Archimedes screw as a treadmill, fresco recovered in the mid-1920s from the House of the Ephebus (Casa di P. Cornelius Tages) in Pompeii, National Museum, Naples.

108 Hollow terracotta Figure, Egypt, height: 18.5cm, width: 9.6cm, depth: 4.1cm, late Ptolemaic or Roman, BM, Registration number: OC.680.

The Archimedes screw can be constructed in several ways: sometimes by winding one or more flexible tubes around a cylindrical core (fig. 2.10) and in other instances by winding an inclined plane around a core, and introducing it inside a hollow casing, thus practically creating a tube (fig. 2.11). When the device is tilted, small 'compartments' are formed, which trap water between the blades or inside the coiled tubes. These 'compartments' appear to move upward when the screw is rotated, carrying the water within them, thus collecting water from the lower reservoir into the upper reservoir, where the 'compartments' are unformed.[109] In its simplest form it is rotated by a crank or by a man walking around the circumference of the outer cylinder, in a treadmill manner (as in figs. 2.12 and 2.13).

Vitruvius gave a detailed and informative description of the construction of an Archimedes screw in his *De Architectura*.[110] His description contributed greatly to keeping the device known throughout the ages. However, it is not clear in the drawings of the water-driven water-lifting screw in different versions of the book—in the 1511 edition by Fra Giovanni Giocondo, as well in

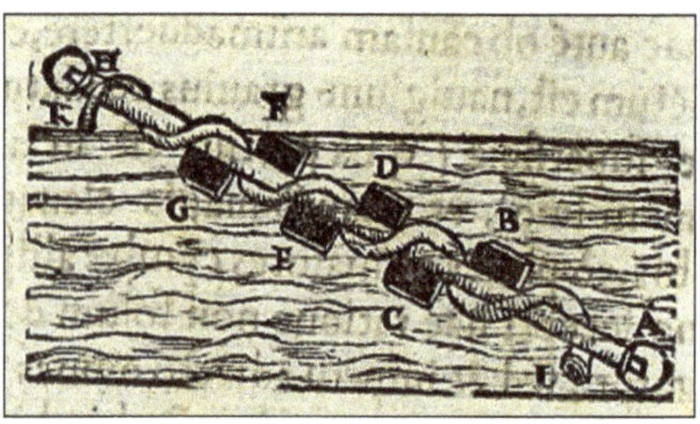

FIGURE 2.10 *Wound tube Archimedean screw*.[111]

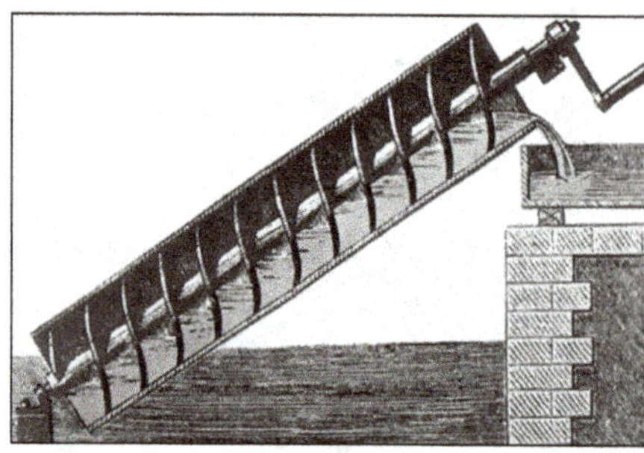

FIGURE 2.11 *Bladed Archimedean screw*.[112]

109 The operation description adapted from Chris Rorres, "The Turn of the Screw: Optimal Design of an Archimedes Screw," *Journal of Hydraulic Engineering*, Vol. 126, No. 1, January, 2000, pp. 72–80, p. 72.

110 Marcus Vitruvius Pollio, "The water screw" in *De Architecture*, Book X, chapter XI, pp. 315–316.

111 Cardano Hieronymi Cardani Mediolanensis, *De Subtilitate*, Libri XXI, Lugduni: apud Stephanum Michaelem, 1580, p. 31.

112 William Dwight Whitney, *The Century dictionary and cyclopedia; a work of universal reference in all departments of knowledge, with a new atlas of the world*, 10 Volumes, Vol. 1, New York: The Century Co., 1904, p. 297.

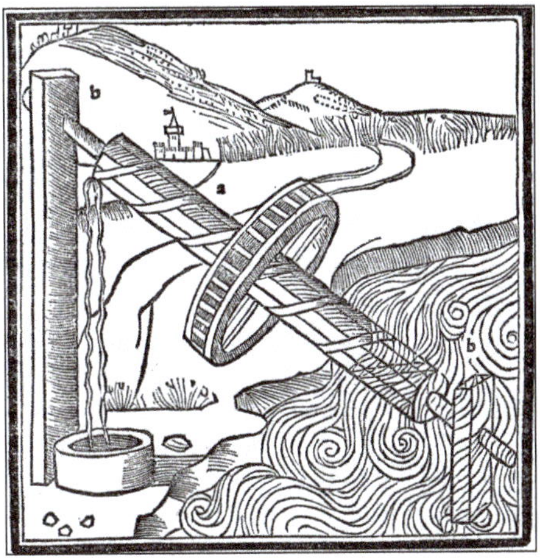

FIGURE 2.12 *Archimedean screw in Fra Giocondo's version of* De Architectura.[113]

FIGURE 2.13 *Archimedean screw in Barbaro's translations of* De Architectura (*1567*).[114]

Daniele Barbaro's Italian[115] and Latin translations,[116] if the large diameter driving wheel is meant to be turned by human power or by the water stream itself. In the last case, since the driving waterwheel appears out of the water flow, no motion can be imparted to the screw and the depiction would be incorrect.

Ctesibius' Double-Acting Force Pumps

The device consists of two ordinary single-acting force pumps, connected to an air vessel and one discharging pipe (fig. 2.14).

It was first described in Vitruvius' *Ten Books on Architecture*, who was the only ancient author to attribute the invention to Ctesibius (fig. 2.15).[117] It has

113 Giocondo, *M. Vitruvius*, p. 102. Giocondo, *M. Vitruvius*, p. 100.

114 Vitruvii Pollionis, *De Architectura, Libri Decem*, p. 348.

115 Vitruvius Pollio, *I dieci libri dell'architettura*, p. 265.

116 Vitruvii Pollionis, *De Architectura, Libri Decem*, p. 348.

117 Vitruvius Pollio, *The Ten Books*, (1914), book x, chapter 7, pp. 297–298.

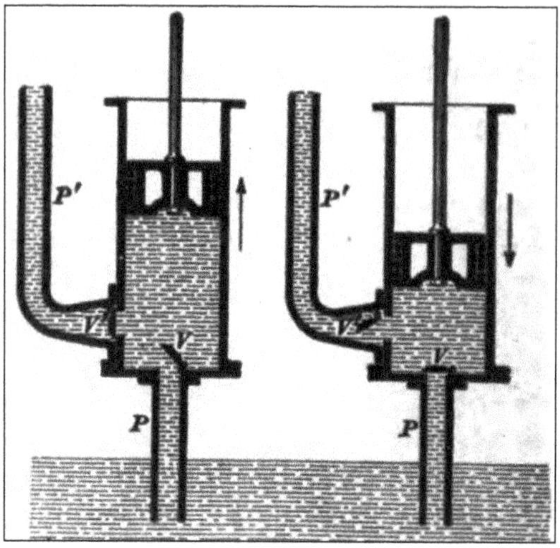

FIGURE 2.14
Single-acting force pump.[118]
As the piston ascends, as shown in the left-hand figure, the pressure of the atmosphere forces the water up the suction pipe P; the water opens the suction valve V and flows into the pump cylinder. When the piston moves down, as shown in the right-hand figure, the suction valve is closed and the delivery valve V' opened. The water in the pump cylinder is now forced up the delivery pipe P'.

FIGURE 2.15
Force pump driven by the river flow in Barbaro's Italian and Latin translations of De Architectura.[119]

118 "Mechanics of Fluids," in *International Correspondence Schools Reference Library*, Scranton (PA): International Textbook Company, 1904, Section 6, pp. 50–51.

119 Vitruvii Pollionis, *De Architectura*, p. 349.

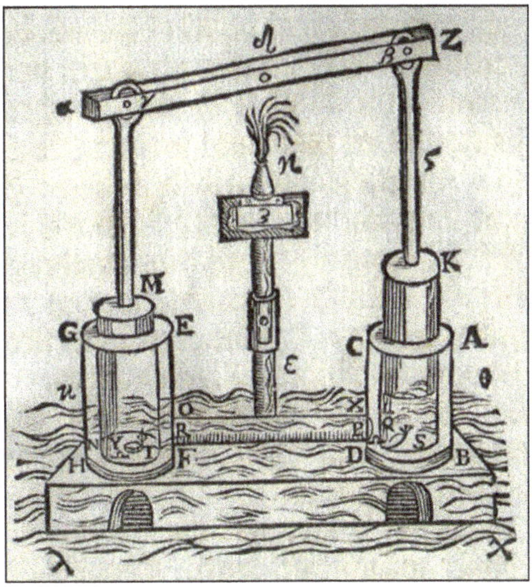

FIGURE 2.16
Force pumps in a fire fighting apparatus, Hero of Alexandria, (c.50 CE), as depicted in Alexandrini's book (1688).[120]

also been described in detail by Hero of Alexandria—arguably, Ctesibius' pupil (fig. 2.16).[121]

Examples have been found at various Roman sites, such as at Silchester in Britain,[122] which has been dated to the second or third centuries CE by some authors,[123] and as late as the fifth century CE by others.[124] Another remarkable relic was found in Bolsena in Italy (fig. 2.17).

---

120 Heronis Alexandrini, *Buch von Lufft- und Wasser-Kuensten*, Franckfurt: Ammon, 1688, p. 67.

121 For a discussion on the available data that helps situate Hero in time and location, and contributes to clarify the relation between Ctesibius and Hero, see the preface of Hero of Alexandria, *The Pneumatics of Hero of Alexandria*, translated from the original Greek and edited by Bennet Woodcroft, London: Carus & Maberly, 1851, pp. ix–xv.

122 For an extensive description of the state of the art of knowledge in this respect, see: R. J. B. Stein, "The Silchester pump," in Michael Gordon Fulford, (ed.), *Silchester and the study of Romano-British urbanism*, Journal of Roman archaeology, Supplementary series, No. 90, pp. 151–164. For a picture reconstructing the pump, based on the relic, see: K. D. White, *Greek and Roman Technology*, London: Thames and Hudson, 1984, p. 16.

123 Maria Vittoria Antico Gallina, "Lo sfruttamento delle acque ipogee e la loro gestione: dall'Oriente all'Occidente attraverso il Medieraneo tre esempificazioni di trasferimenti tecnologici nell lunga durata," in Michèle Merger, (ed.), *Transferts de technologies en Méditerranée*, Paris: Université Paris Sorbonne, 2006, pp. 87–102, p. 93.

124 Oleson, *Greek and Roman Mechanical Water-Lifting Devices*, p. 303.

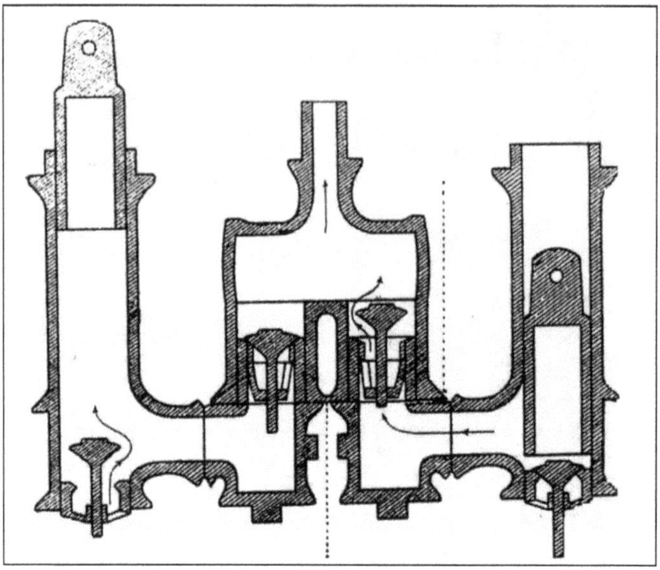

FIGURE 2.17 *Sketch of the force pump found in Bolsena, Italy (3rd century* CE*).*[125]

In all cases (figs. 2.15, 2.16 and 2.17) the device consists of two cylinder pumps operated by a single rocking arm, thus increasing efficiency. Water enters each cylinder through a flap valve at the bottom of the cylinder, then the descending piston forces the water through a second flap valve, located laterally, into the receiver and then up the rising pipe to the outlet. As the piston rises, water enters the chamber, forced in chiefly by the pressure of the water surrounding the base of the pump, and in small part by a slight suction effect.

A careful look at the drawings of water-raising machines conceived by Al-Jazari (1136–1206) in the late twelfth century reinforces the thesis of the Islamic influence. Although the model machine shown in fig. 2.18 seems to have been built as a decorative attraction—a miniature erected besides an ornamental pool; it was nonetheless a practical design.[126] The device is powered either by

125 A. H. Smith, *A guide to the exhibition illustrating Greek and Roman life*, 2nd edition, London: British Museum, 1920, p. 121.. The description for figs. 2.15, 2.16 and 2.17 is based on Sheldon Shapiro, "The Origin of the Suction Pump," *Technology and Culture*, Vol. 5, No. 4 (1964), pp. 566–578, p. 566.

126 Thomas Glick, "Technology in the Middle Ages," in Albrecht Classen, (ed.), *Handbook of Medieval Studies: Terms—Methods—Trends*, 3 volumes, vol. 2, Berlin and New York: De Gruyter, 2010, pp. 1305–1310, p. 1310.

FIGURE 2.18
Al-Jazari's third water-raising machine (c.1200).[127]

a real cow or by the scoop-wheel in the lower level; the upper horizontal axle drove the yellow belt with its attached water-pots to lift water to the upper level. Several of its operational components were scaled up in a utilitarian machine dating from the thirteenth century, located in the banks of the Yazid River, a tributary to the Barada River that crosses Damascus. It was reconstructed in the 1980s by academics and students of the Aleppo University.[128] The riverside

127 Richard Covington, "The Art and Science of Water," *Saudi Aramco World*, vol. 57, number 3, (May/June 2006), p. 19. Copy of one of some of Al-Jazari's extant manuscripts can be seen in Schioler, *Roman and Islamic Wheels*, pp. 184–185. Originals are kept in the Topkapi Palace Museum, Istambul, shelf No. A 3472.

128 Donald Hill, *A History of Engineering in Classical and Medieval Times*, London: Routledge, 1984, p. 148, quoting Ahmed Y. Al-Hassan, *Taqi al-Din and Arabic Mechanical Engineering* (in Arabic), Aleppo: University of Aleppo Press, 1978, pp. 57–70. For further details on this machine and other medieval water-raising devices found in the same area in Damascus, see: Adriana de Miranda, *Water Architecture in the Lands of Syria: The Water-wheels*, Rome: L'Erma di Bretschneider, 2007, pp. 248–260.

path in which it was erected is known as the 'Lane of the Norias', due to the number of waterwheels installed along its length.[129]

The concept of this design is replicated in a much later drawing by Francesco di Giorgio Martini (1439–1502), depicting a chain of buckets, executed in a much more advanced drafting technique (fig. 2.19).

Another two earlier drawings from about 1430, by Mariano Iacopo (Taccola), are based exactly on the same principle. In one of these drawings, rather exceptionally, the mover is wind (fig. 2.20).

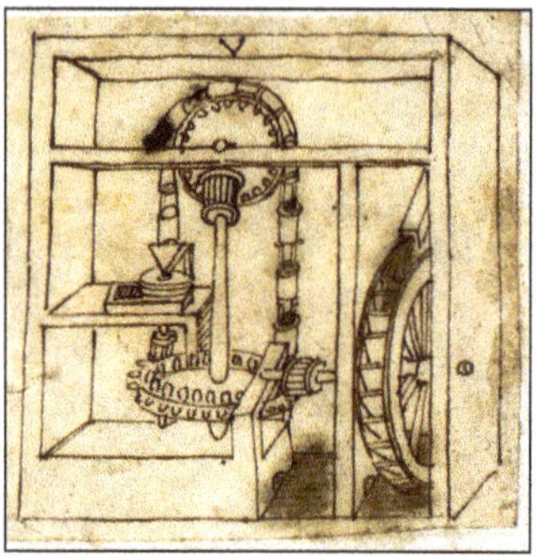

FIGURE 2.19
Francesco di Giorgio Martini, flour-mill and chain of buckets (c.1480).[130]

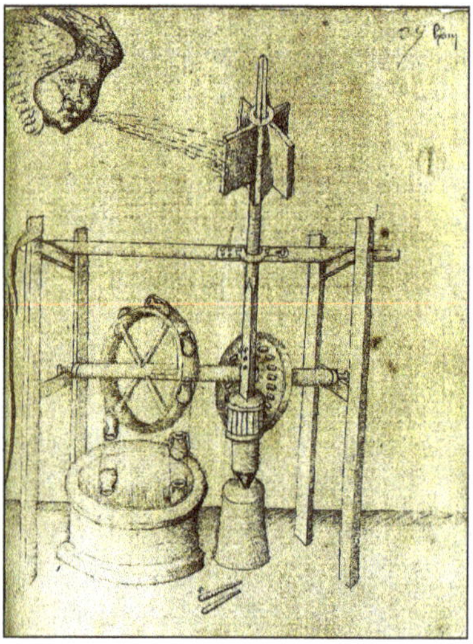

FIGURE 2.20
Wind moved chain of buckets.[131]

129 Al-Hassan, *Islamic Technology*, p. 45.

130 Francesco di Giorgio Martini, *Trattati di architettura, ingegneria e arte militare* (c.1480), Codice Saluzziano 148, fol. 35 v., tav. 66, original manuscript in Biblioteca Reale, Torino (Italy).

131 Mariano di Iacopo (detto il Taccola), *Liber tertius de ingeneis ac edifitiis non usitatis*, Ms. Palatino 766, fol. 37r., Biblioteca Nazionale Centrale, Firenze, 1430.

The practical know-how accumulated during the Late Middle Ages was collected in fifteenth-century technical treatises, including Conradus Kyeser's *Bellifortis*,[132] which despite being dedicated to war and military equipment, has several designs dealing with water supply. This work is considered to be the first of its kind.[133] Likewise, many early designs and applications for water-lifting from the fifteenth century have survived, as exemplified by figs. 2.21 to 2.23.

The painter Antonio Pisanello (c.1395–1455) correctly depicted, shortly before his death, a pair of piston pumps operated by rocking levers raised and lowered by connecting rods from two cranks fitted 180º apart on the two sides of an overshot water-wheel (fig. 2.22).[134] Francesco di Giorgio Martini illustrated around 1480 several practical water-raising machines, such as a workable force pump, shown in fig. 2.23.

FIGURE 2.21
Conradus Kyeser, noria, *c.1405*).[135]

132 Conradus Kyeser, *Bellifortis*, Ms. germ. qu. 15, (originally written c.1405), Elsaß, copy from 1460, kept in the Universitätsbibliothek Johann Christian Senckenberg, Frankfurt am Main.

133 Lynn White, Jr., "Kyeser's Bellifortis: The First Technological Treatise of the Fifteenth Century," *Technology and Culture*, Vol. 10, No. 3 (1969), pp. 436–441.

134 Joseph Needham, *Clerks and Craftsmen in China and the West*, Cambridge: Cambridge University Press, p. 195.

135 Kyeser, *Bellifortis*, fol. 121 r.

FIGURE 2.22
Pisanello, pump driven by a waterwheel (c.1450).[136]

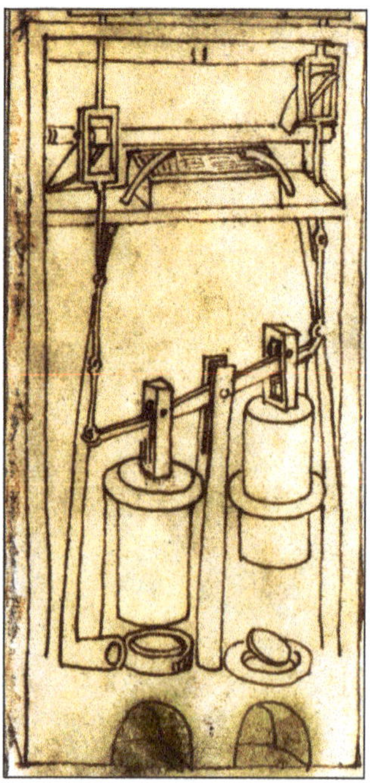

FIGURE 2.23
Giorgio Martini, force pump (c.1480).[137]

136 Bernhard Degenhart, *Antonio Pisanello*, Wien: Schroll, 1942, fig. 147, original in the Louvre Codex Vallardi, drawing No. 2286 (15.9×23.5 cm).

137 Di Giorgio Martini, *Trattati di architettura*, fol. 41v., Tav. 82.

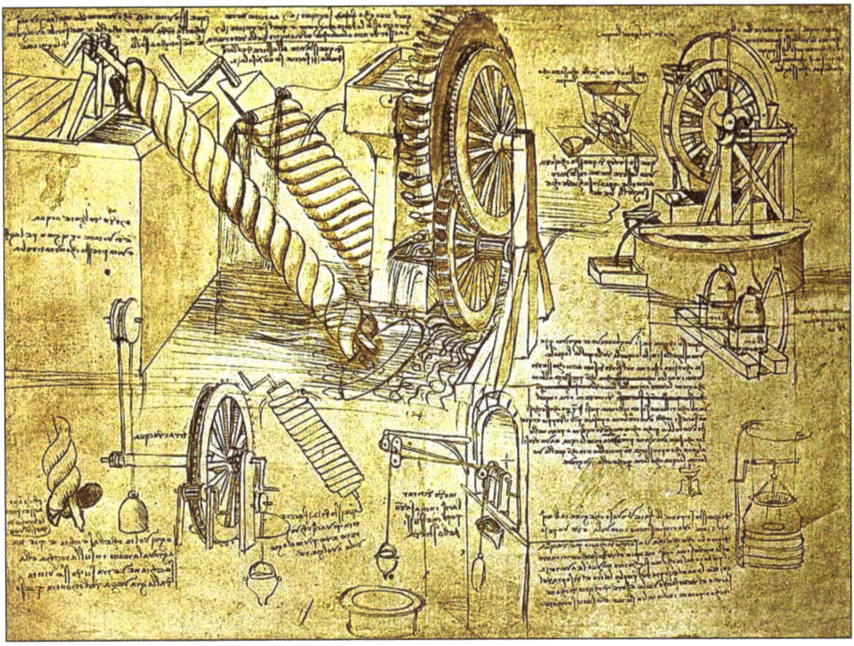

FIGURE 2.24 *Leonardo da Vinci, Codex Atlanticus, drawings of water-lifting devices.*[138]

A vast variety of devices for raising water can be found in Leonardo da Vinci's mechanical sketches—designs of pumps outnumber his more famous designs such as flying-machines, or mechanical instruments in general. Some of these hydraulic sketches, dating from his first Florentine period (1469–82), are shown in fig. 2.24.[139]

These exquisite technical drawings exemplify the interdependence of art and technology, which was a primary characteristic of the Renaissance. Although we commonly think of Leonardo da Vinci's drawings of his inventions as the first modern illustrations of mechanical devices, they were not well known during the Renaissance, and no specific implementation of his ideas is known.[140] Jonathan Sawday also claims that the same lack of realization of designs characterized Leonardo's contemporaries and successors.[141] Specifically on what concerns pumps Bertrand Gille disregards Da Vinci's contribution,

138 Leonardo da Vinci, "Drawings of Water-raising Devices," *Codex Atlanticus*, (12 volumes), Vol. 1, fol. 26, v., Pen and ink on paper, (61×44 cm), Biblioteca Ambrosiana, Milan.

139 Sawday, *Engines of the Imagination*, p. 33.

140 Knoespel, "Gazing on Technology," p. 101.

141 Sawday, *Engines of the Imagination*, p. 31.

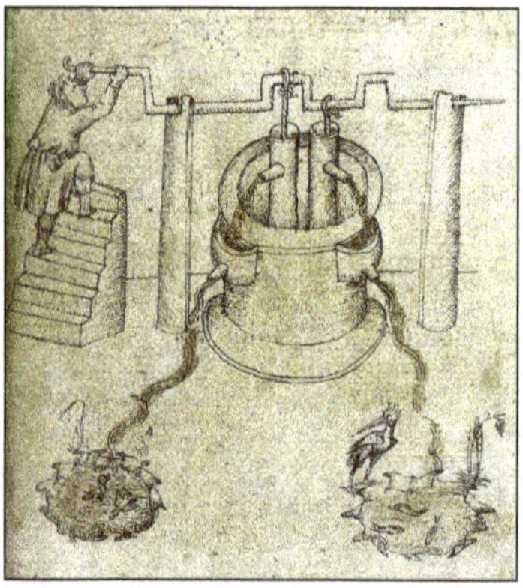

FIGURE 2.25
Taccola, piston-pumps worked by an unaligned crank-shaft (c.1450).[142]

stating that they had been known and demonstrated since Mariano di Jacopo (better known as Taccola).[143] Notwithstanding this view, two of Taccola's drawings, dated between 1441 and 1458, show that he either misunderstood, or intentionally misrepresented, the concept of rod and crank—as can be seen in fig. 2.25: the axis of the depicted crank is not correctly aligned, neutralizing the movement of the second piston.[144]

In general, the diffusion of know-how among these masters seems to have been rather faulty, as it is difficult to discern a pattern of convergent expertise towards better technological solutions.[145] We even perceive in certain famous authors a regression in the degree of sophistication of the water-raising devices. Such is the case of Philipp Mönch's representation of a scoop-wheel, in which water is raised to the full height of the wheel, but freely falls to the shaft level, thus wasting energy (fig. 2.26).

In many instances it is not clear whether the 'creator' of flawed technology was intentionally misleading the receptors of his theories or honestly believed them himself. As an example, even 80 years after Agricola, a depiction by Giovanni Branca has a gross geometrical mistake in one of his drawings (fig. 2.27). It can be seen that the plane of the acting rods cannot be geometrically aligned with the cylinder in the rear, thus impeding its smooth reciprocating movement.

142 Mariano di Iacopo (detto il Taccola), "Liber tertius de ingeneis ac edifitiis non usitatis," in *Codex Latinus Monacensis 197 II*, tlt 03., Bayerische Staatsbibliothek München.

143 Bertrand Gille, *The Renaissance engineers*, London: Lund Humphries, 1966, p. 155.

144 White, *Medieval Technology*, p. 113.

145 Knoespel, "Gazing on Technology," p. 101.

FIGURE 2.26
Philipp Mönch, inefficient scoop-wheel (1496).[146]

FIGURE 2.27
Giovanni Branca, unworkable design of a battery of force pumps (1629).[147]

146 Philipp Mönch, *büch der stryt und buochßen* (manuscript 1496), Universitätsbibliothek, Heidelberg. Almost 90 years later, Jean Errard de Bar-le-Duc made exactly the same mistake: Jean Errard de Bar-le-Duc, *Le premier livre des instruments mathematiques mechaniques*, Nancy: J. Janson, 1584, Reproduction en fac-similé de l'édition de 1584 avec une introduction d'Albert France-Lanord, Paris: Berger-Levrault, 1979, planche 22.

147 Giovanni Branca, *Le machine: volume nuovo et di molto artificio da fare effeta maravigliosi tanto spiritali quanto di Animale operatione arichito di bellissime figure con le dichiarationi a ciascuna di esse in lingua volgare* et latina, Roma: Iacomo Mascardi, 1629, plate A11, p. 11.

Theatres of Machines

A real leap forward in the technological culture of Europe, and more specifically in water-raising technology, was a consequence of the flood of books entirely dedicated to machinery, popularly called 'theaters of machines,' during the second half of the sixteenth and first half of the seventeenth centuries.[148] The fifteenth and early sixteenth centuries' artists, such as Keyser, Martini, Pisanello, Taccola, Leonardo and Barbaro, may be considered as the precursors of this important phenomenon.[149] Renaissance technological culture may thus be considered as entering a new phase, in which practical knowledge tended to be fused with learned culture, beginning with the translation of Vitruvius into Italian and culminating in treatises originally written in the vernacular languages spoken by their authors. Chroniclers of the Late Middle Ages had not only referred subjectively to cities as objects of history, but their books were sensitive to innovations and technical achievements, for which the city became the 'theater'. These 'theaters' registered the technological improvements already achieved in the former centuries, in which an increase in the growth pace of water power utilization occurred, which had also been accelerated by the impetus given to the

148 For a discussion on the form and substance of these genre, see: Jan Lazardzig, "The Machine as Spectacle: Function and Admiration in Seventeenth Century Perspectives on Machines," in Helmar Schramm, Ludger Schwarte, Jan Lazardzig, (eds), *Instruments in Art and Science. On the Architectonics of Cultural Boundaries*, Berlin and New York: Walter de Gruyter, 2008, pp.153–175 and Wolfgang Lefebre, "Introduction," in Wolfgang Lefebre, (ed.), *Picturing Machines 1400–1700 (Transformations: Studies in the History of Science and Technology)*, Cambridge (MA): The MIT Press, pp. 1–12.

149 For a comprehensive overview of the main Italian artist-engineers and an analysis of their work, see Paolo Galluzzi, *Gli ingegneri del Rinascimento da Brunelleschi a Leonardo da Vinci*, Firenze: Giunti, 1996; Paolo Galluzzi, (ed.), *Prima di Leonardo: cultura delle macchine a Siena nel Rinascimento*, Milano: Electa, 1991; Paolo Galluzzi, "Portraits of machines in Fifteenth century Siena," in Renato G. Mazzolini, (ed.), *Non-verbal communication in science prior to 1900*, Firenze: L. S. Olschki, 1993, pp. 53–90 and "An interview with Paolo Galluzzi," towards the London inauguration of the Istituto e Museo di Storia della Scienza, Florence, Italy travelling exhibition *The Art of Invention: Leonardo and Renaissance Engineers*, that was on display at the Science Museum in London, England from October 15, 1999 to April 24, 2000. Website of the IMPA—Instituto Nacional de Matematica Pura e Aplicada (Brazil): http://w3.impa.br/~jair/einten1.html, accesed 18 June, 2012.

economy by maritime trade and colonial expansion. The authors excelled in the application of their mathematical and illustrating skills to new devices, convincing patrons to finance the publication of their work in print, attracting the attention of royal courts and municipal councils, striving to obtain exclusive rights to their inventions.[150]

Georgii Agricolae's *De re metallica* (1556)[151] and Jacques Besson's *Livre des instruments mathématiques et méchaniques* (1569)[152] were pioneer works in a genre that later included Jean Errard de Bar-le-Duc's *Le premier livre des instruments mathématiques méchaniques* (1584),[153] Agostino Ramelli's *Le diverse et artificiose machine* (1588),[154] Vittorio Zonca's *Teatro nuovo di machine et edificii* (1607),[155] Heinrich Zeising's *Theatri machinarum erster* (1613–1614),[156] Jacopo Strada's postumous work, *Künstliche Abriß allerhand Wasser- Wind- Roß- und*

150 Marcus Popplow, "Hydraulic Engines in Renaissance Privileges for Inventions and 'Theatres of Machines'," in Alessandra Fiocca, Daniela Lamberini and Cesare Maffioli, (eds), *Arte e Scienza delle Acque nel Rinascimento*, Venezia: Marsilio, 2003, pp. 73–83.

151 Georgii Agricolae, [Agricola], *De re metallica Libri XII: quibus officia, instrumenta, machinae, ac omnia denique ad Metallicam spectantia, non modò luculentissimè describuntur, sed et per effigies, suis locis insertas, adiunctis Latinis, Germanicisque appellationibus ita ob oculus ponuntur, Eiusdem De Animantibus subterraneis liber, ab autore recognitus: cum indicibus diversis…*, Basileae: Froben, 1561.

152 Jacques Besson, *Theatrum instrumentorum et machinarum*, originally written in 1569, Lugduni: Apud Barth, 1578.

153 Errard de Bar-le-Duc, *Le premier livre*.

154 Agostino Ramelli, *Le diverse et artificiose machine del Capitano Agostino Ramelli Dal Ponte Della Tresia Ingegniero del Christianissimo Re di Francia et di pollonia: nelle quali si contengono uarij et industriosi Mouimenti, degni digrandißima speculatione, per cauarne beneficio infinito in ogni sorte d' operatione*, Parigi: published by Agostino Ramelli, 1588. In this study we shall quote the English translation: Agostino Ramelli, *The various and ingenious machines of Agostino Ramelli: a classic sixteenth-century illustrated treatise on technology / translated from the Italian and French with a biographical study of the author by Martha Teach Gnudi; technical annotations and a pictorial glossary by Eugene S. Ferguson*, New York: Dover Publications, Aldershot (England): Scholar Press, 1987, for the texts and the Italian / French original for the drawings, unless indicated differently in the footnotes.

155 Vittorio Zonca, *Novo Teatro Di Machine Et Edificii / Per varie et sicure operationi / Con le loro figure tagliate in Rame / e la dichiaratione, e dimostratione / di ciascuna. / Opera necesaria ad Architetti, et a quelli, / chi di tale studio si dilettano*, Padova: Pietro Bertelli, 1607.

156 Heinrich Zeising, *Theatri machinarum erster* [*-letzter*] *Theill … auss italian; und franzoss: Sprach mit Fleiss transferirt durch Hieronymum Megiserum … in Druck geordnet und vorlegt von Henning Grossen dem Jüngern*, Leipzig: [s.n.], 1613–1614.

Handt Mühlen (1617),[157] Giovanni Branca, *Le machine,* (1629),[158] Isaac and Salomon de Caus' *New and Rare Inventions of Water-works Shewing the Easiest Waies to Raise Water Higher Then the Spring* (1644)[159] and Georg Andreas Böckler's *Theatrum machinarum novum* (1662).[160] In these technical treatises the number of water-lifting inventions almost surpasses all other types of machinery put together, although many of the innovations shown refer to gear transmissions rather than to basic technology.[161] Ramelli's bilingual descriptions (originally in French and Italian) are much more detailed than those found in previous and later illustrated treatises on technological devices.[162] The usefulness of these publications depended on the practical experience of their authors, about which we can learn from their biographies. Agricola, arguably the outstanding figure among these technological giants, exemplifies this link. Born in 1494, by 1527 he became town physician at Joachimsthal in the midst of the then most prolific metal-mining district of Central Europe, including active mining towns—Schneeberg, Geyer, Annaberg, Altenberg, Marienberg,

157 Jacobus de Strada and Octavius de Strada, (eds), *Künstliche Abrisz allerhand Wasser- Wind- Rosz- und Handt Mühlen: beneben schönen und nützlichen Pompen auch andern Machinen damit das Wasser in Höhe zuerheben, auch lustige Brunnen und Wasserwerck, dergleichen vor diesem nie gesehen worden ... Nunmehr ... durch den Truck publicirt ... durch,* (2 volumes), Vol. 1, Frankfurt am Main: Paulum Iacobi, 1617.

158 Giovanni Branca, *Le machine: volume nuovo et di molto artificio da fare effeta maravigliosi tanto spiritali quanto di Animale operatione arichito di bellissime figure con le dichiarationi a ciascuna di esse in lingua volgare et latina,* Roma: Iacomo Mascardi, 1629.

159 Isaac de Caus, *New and Rare Inventions of Water-works Shewing the Easiest Waies to Raise Water Higher Then the Spring: By which Invention the Perpetual Motion is Proposed: Many Hard Labours Performed: and Varieties of Notions and Sounds Produced: a Work Both Usefull Profitable and Delightfull for All Sorts of People,* translated into English from the French by John Leak, London: Joséph Moxon, 1659. The translation to English is from Isaac de Caus, *Nouvelle invention de lever l'eau plus haut que sa source avec quelques machines mouvantes par le moyen de l'eau, et un discours de la conduit d'icelle,* Londre: [s.n.], 1644. The text and plates of the 1644 French original were based on Salomon de Caus, *Les raisons des forces mouvantes, avec diverses machines tant utiles que plaisantes, auxquelles sont adjoints plusieurs dessings de grotes & fontaines,* Francfort: Jan Norten, 1615.

160 Georg Andreas Böckler, *Theatrum machinarum novum: Neu-vermehrter Schauplatz der mechanischen Künsten, handelt von allerhand Wasser-Wind-Ross-Gewicht- und Hand-Mühlen, wie dieselbige zu dem Frucht-Mahlen, Papyr- Pulver- Stampff-Segen- Bohren-Walcken-Mangen, und dergleichen anzuordnen,* Nürnberg: Verlegung Paulus Fürsten, 1661.

161 Marcus Popplow, "Hydraulic Engines," p. 82.

162 In Ramelli's book, out of 195 plates depicting machines, 116 are waterworks, used for raising water from wells and rivers, for drainage and irrigation. See Alexander G. Keller, "Renaissance Waterworks and Hydromechanics," *Endeavour,* volume XXV, (1966), pp. 141–145, p. 141.

Gottesgab, and Platten. Agricola read in the Greek and Latin authors all references to mining, spent all the time not required for his medical duties in visiting the mines and smelters, carefully recording practical tips from the most experienced among the miners.[163]

Some of the designs were primitive and almost unworkable, such as the Archimedean screws in Giuseppe Ceredi's *Tre discorsi sopra il modo d'alzar aque da' luoghi bassi*, published in Parma in 1567[164] (fig. 2.28). Ceredi obtained patents in Milan, Venice and Parma for the manufacture and installation of Archimedean screws for irrigation and the drainage of swamps, an application that, as we have seen, was known in antiquity, at least through Vitruvius's writings.[165] This would have been not unusual, as to get a patent it was not a necessary condition to be the inventor of a device, but only to be the first to possess it uniquely in any particular territory.[166] Alexander Keller, an authority on machines and technology of the Renaissance, indicates that the Archimedean screw was not usually used for irrigation in Europe during the Middle Ages, and as it has several configurations, even the prudent Italian authorities of the three cities were ready to grant the patents to Ceredi.[167] Another plausible justification could have been that Ceredi improved the efficiency of the device by changing the recommended ratio (16:1) given by Vitruvius for the length to central shaft diameter.[168]

163 Agricola, *De re Metallica*, (1950), p. vi.

164 Giuseppe Ceredi, *Tre discorsi sopra il modo d'alzar aque da' luoghi bassi*, Parma: Appresso Seth Viotti, 1567.

165 Stillman Drake, "An Agricultural Economist of the Late Renaissance," in Noel M. Swerdlow, Trevor Harvey Levere, (eds), *Essays on Galileo and the history and philosophy of science*, (3 volumes), Vol. 3, Toronto: University of Toronto Press, 1999, pp. 170–189, p. 172. Late sixteenth-century Venice was an attractive place for foreign experts, many of whom dealt with water-raising machines, as can be seen in: Roberto Berveglieri, *Inventori stranieri a Venezia, 1474–1788: importazione di tecnologia e circolazione di tecnici artigiani inventori: repertorio*, Volume 58 of *Memorie, Classe di scienze morali, lettere ed arti*, Venezia; Istituto veneto di scienze, lettere ed arti, 1995.

166 For an in-depth discussion on the beginnings of the quest to preserve the ownership of inventions, see Pamela O. Long, "Invention, Authorship, 'Intellectual Property,' and the Origin of Patents: Notes toward a Conceptual History," *Technology and Culture*, Vol. 32, No. 4, Special Issue: Patents and Invention (Oct., 1991), pp. 846–884. The concepts conveyed in that article would become the basis of Ms. Long's later book: *Openness, secrecy, authorship*.

167 Referred to in Drake, "An Agricultural Economist," p. 173.

168 Stephan R. Epstein, "Transferring Technical Knowledge and Innovating in Europe, c.1200–1800," part of the series: *Working Papers on The Nature of Evidence: How Well Do 'Facts' Travel?*, No. 01, London: Department of Economic History, London School of Economics, 2005, pp. 1–40, pp. 19–20.

FIGURE 2.28 *Giuseppe Ceredi's battery of Archimedean Screws (1567).*[169]

There are very few instances in which depictions represent devices actually used, such as in Vittorio Zonca's *tympanum*, located at Fusina, in the vicinity of Venice, to which the fresh water was transported by boat (fig. 2.29).[170] The contradiction between the ubiquitous representation of hydraulic contrivances in the popular Renaissance *theatrum machinorum* and the lack of utilitarian application "… reinforce the impression that this type of device, while becoming familiar to an educated few, was still considered new and exciting in the post-medieval period."[171]

Complex Water Lifting Configurations Relevant to the Present Study

Technology evolved into efficient machines, which were put to work for urban water supply in Europe during the Renaissance and the early modern period. The development of these devices progressed at diverse paces in different areas of the continent. Even if in some cities of Eastern Europe these enhanced

169 Giuseppe Ceredi, *Tre discorsi sopra il modo d'alzar aque da' luoghi bassi*, Parma: Appresso Seth Viotti, 1567, Plate 70, p. 90.

170 Popplow, "Hydraulic Engines in Renaissance," pp. 76–77.

171 Magnusson, "The Technologies of Water," p. 233.

FIGURE 2.29 *Vittorio Zonca,* tympanum *driven by the flow of a stream or by animal force (1607).*[172]

techniques were implemented, it was in Central and Western Europe where consolidation of the technologies and the first substantial commercial arrangements for urban water supply occurred.

The development of waterwheels driving water lifting plants brought about more efficient lifting systems and changed fundamentally the conditions under which European urbanization took place. Cities could not prosper without a stable water supply, and municipal authorities tried to cope with this need, not always successfully. As water supply became increasingly a factor of general interest, water developed also into an artistic medium to decorate cities and palaces. Visible indications are beautiful fountains that embellish old towns and palace gardens to this very day. The attention paid to urban technology projects went beyond the city walls and helped propagate their reputation, up to a point in which the practice of nicknaming notable machines was adopted.[173] The records of towns sometimes reflect the conflicts generated within a trade by introducing a new machine or a new process. Accounts of Cologne and Nuremberg register complaints against the nuisance caused by introducing bigger and louder machines in general, which lowered the value of

172 Zonca, *Novo Teatro Di Machine Et Edificii*, p. 61. Drawing description in pp. 62–63.

173 Mathieu Arnoux and Piere Monnet, (eds), "Introduction," in *Le technicien dans la cité en Europe occidentale, 1250–1650*, Rome: École française de Rome, 2004, p. 11. Nicknaming was especially popular when referring to siege weapons such as the *trebuchet*.

neighboring houses.[174] There is no doubt that the 'wailing' noise produced by water-lifting *norias* and other waterwheel types were a cause of anger as well.[175]

As will be discussed below, during the early modern period, most complex water-raising machines capable of reaching substantial heights seem to have been based either on Archimedean screws or on forcing pumps, driven by a waterwheel moved by the flow of a river or stream. These methods, which were contemporaneous to the waterworks on which the present study focuses, hereby presented according to their degree of complexity.

Initially, unlike the waterwheels driving mills, *norias* did not provide mechanical power to any other water lifting methods, but in the Late Middle Ages they became both the driving force and a component of much more complex water-raising machines, as we shall see was the case in the *Artificio* of Toledo. As with all water-lifting devices, it is Agostino Ramelli who managed to convey sophisticated applications of the scoop-wheel technology most clearly, as in fig. 2.30, which represents one of several variations on the same device in *Le diverse et artificiose machine*, suitable to lift water to moderate heights.

The earliest recorded complex designs based on Archimedean screws are from the first half of the sixteenth century, as can be seen in fig. 2.31, which depicts a special kind of water-raising system, built in Augsburg in 1538,[176] which under the name *Machina Augustana* has become a milestone in the history of technology. A detailed account of this complex machine was given by Girolamo Cardano, who passed through the city in 1550, and described in his *De Subtiliatate* its operating principle.[177]

Forcing pumps, used at the beginning of the sixteenth century for water extraction from flooded mining excavations, were described in Agricola's *De re metallica*.[178] The sixth book of this famous work contains designs of several

174 Arnoux and Monnet, "Introduction," p. 12.

175 See Mahmod Samman, "Rediscovering the Waterwheel—'Noria Al-Muhammadiyya,' an ASME International Historic Landmark," a presentation to the American Society of Mechanical Engineers, Northwest Houston meeting, August 2005. He describes the sound as a mixture of noise and musical notes that is often compared to organ music, its deepest notes in the range of 120–170 Hz.

176 Baur, *Brunnen: Quellen*, p. 119.

177 Cardani, *De Subtilitate*, (1580), p. 33. Description extracted from Wolf, *A History of Science*, p. 526.

178 Although sixteen years before Agricola another famous book dealing with mining was published—Vannoccio Biringuccio, *De la pirotechnia libri X dove ampiamente si tratta non solo di ogni sorte e diuersita di miniere, ma anchora quanto ricerca intorno à la prattica di quelle cose di quel che si appartiene à l'arte de la fusione ouer gitto de metalli come d'ogni altra cosa simile*, Veneto: V. Roffinello for C. Navo, 1540, it has many references to water power in the industry, but no detailed drawings of water raising devices.

FIGURE 2.30
Ramelli, complex scoop-wheel configuration (1588).[179]

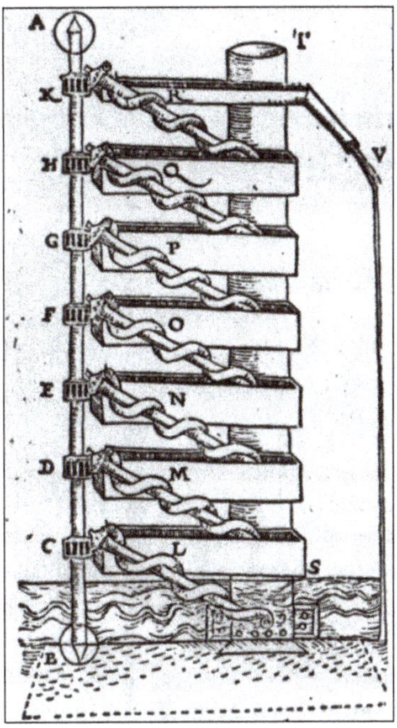

FIGURE 2.31
Cardano's illustration of the 'Machina Augustana.'[180] A vertical shaft AB is driven by a metal spur-wheel on the axle of the waterwheel, which supplies the motive power, and carries pinions driving the screws, which raise water in turn from each of a series of horizontal troughs to the next trough higher up. By this alternation of revolving screw and stationary trough the water is made to flow up to the top of the tower, from which the supply is taken.

179 Ramelli, *Le diverse et artificiose machine*, plate 41, p. 140.

180 Geronimo Cardano, Hieronymi Cardani Mediolanensis Medici De Svbtilitate libri XXI, Basilae: Per Ludovicum Lucium, 1554, p. 20. The first simplified description in French was already published in 1620, see: Ian Iaques von Wecker, *Les Secrets et merveilles de Nature*, Rouen: Claude le Villain, 1620, pp. 697–608. An almost identical drawing was submitted in an application for a privilege of invention in Venice in 1564 by Fedele Piccolomini—Archivio di Stato di Venezia [ASVE], Savi ed Esecutori alle Acque, Scritture 122; aut. min. 16/9/2013, n. 49/2003.

pumps driven by hydraulic wheels, such as the sophisticated one illustrated in fig. 2.32. Agricola qualifies this specific pump, which he notes had been invented ten years before the publication of his book in 1556, as ingenious, durable, useful and inexpensive to build.[181] The challenge of pumping technologies was being confronted in several regions almost simultaneously; from an entry in the minutes of the town council at Lausanne, Switzerland, on April 13, 1557, we learn: "Delivered to Master Jacques Besson, the engineer, two *écus*,[182] both for the water engine delivered by him to the said lords [of the council] and for a beginning of the award to him for the invention of fountains."[183] Judging by the small sum involved, it seems that Besson only produced the model of a pump, for a real one would have been much more expensive. Unfortunately, there are no further references to his activities at Lausanne or to any actual implementation of his 'invention of fountains.'[184]

Agostino Ramelli described in his treatise no less than 46 different force pumps configurations. One of the most complex, pertinent to the London and Paris projects investigated in this study, is shown in fig. 2.33. Four rocking arms driven by one waterwheel operate eight pumps through an intricate interlocking gear mechanism. All other authors of technical treatises in the sixteenth and seventeenth centuries included depictions of force pumps in their works, such as Heinrich Zeising,[185] Isaac de Caus,[186] Jacobus de Strada,[187] as well as in

---

181 In the same decade (1544) a huge water powered draining device was installed in a silver mine in Schwaz (Tyrol), which was operated by two men, performing the work previously made by 600 men; see Michael Mitterauer, *Why Europe?: The Medieval Origins of Its Special Path*, translated by Gerald Chapple, Chicago and London: The University of Chicago Press, 2003, p. 228.

182 Most probably the *écu au soleil*, a French gold coin. See John H. Munro, "Money and Coinage of the Age of Erasmus," in *The Correspondence of Erasmus*, (12 volumes), Vol. 1, (1484–1500), Toronto: University of Toronto Press, 1974, pp. 311–347, p. 315; Marie-Thérèse Boyer-Xambou, Ghislain Deleplace and Lucien Gillard, *Private Money and Public Currencies. The 16th Century Challange*, Armond (NY) and London: M. E. Sharpre, 1994, p. 73.

183 E. Chavannes, "Extraits des manuaux du Conseil de Lausanne, 1536–64," in *Mémoires et documents publiés par la Société d'histoire de la Suisse Romande*, 2d série, 1, Lausanne, 1887, p. 145. Extracted from Alex Keller, "The Missing Years of Jacques Besson, Inventor of Machines, Teacher of Mathematics, Distiller of Oils, and Huguenot Pastor," *Technology and Culture*, Vol. 14, No. 1 (1973), pp. 28–39, pp. 29–30.

184 Keller, "The Missing Years," pp. 28–29.

185 Heinrich Zeising, "Pump feeding fountain," (1613–1614), in *Theatri machinarum*, plate 20, page A-78-A.

186 Isaac de Caus, *New and Rare Inventions of Water-works*, plate xxvii, p. 36.

187 Strada, *Künstliche Abrisz allerhand*, Plate 73.

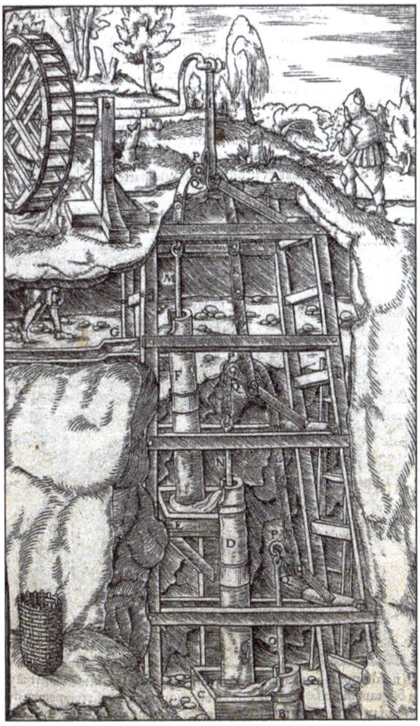

FIGURE 2.32 *Agricola's three stages water-raising pump.*[188]

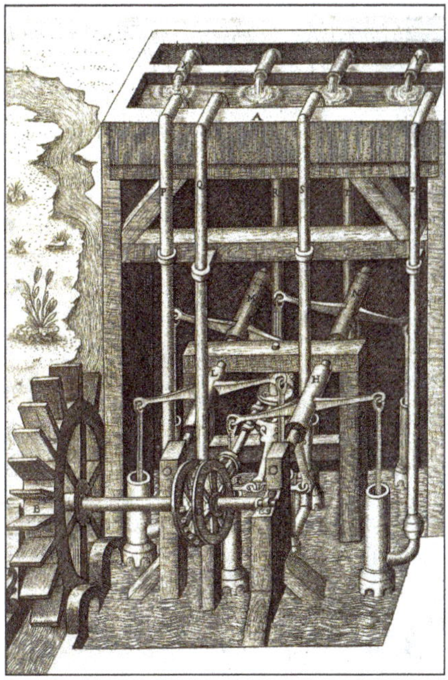

FIGURE 2.33 *Ramelli's complex eight force pumps device, driven by one waterwheel.*[189]

the surviving working drawings of Heinrich Schickhardt,[190] but none achieved either Agricola's practicality or Ramelli's clarity.

Water-Raising Machines at Atmospheric Pressure

Another radically different solution to deliver water at substantial heights utilizes water-raising machines in which no pressure is applied to the water. No record exists of this type of machines before the second half of the sixteenth century. These devices belong to two basic systems, which conceptually

188 Agricolae, *De re metallica Libri XI, I:* p. 145.

189 Ramelli, *The Various and Ingenious Machines*, plate 58, p. 175. See also a detailed description of the device, ibid., p. 176.

190 Heinrich Schickhardt, "Sketch of a waterwheel driving three pumps," one of several drawings concerning the renovation of the water supply of Montbéliard castle, dated 1607. The original manuscript is kept in Hauptstaatsarchiv Stuttgart (Germany).

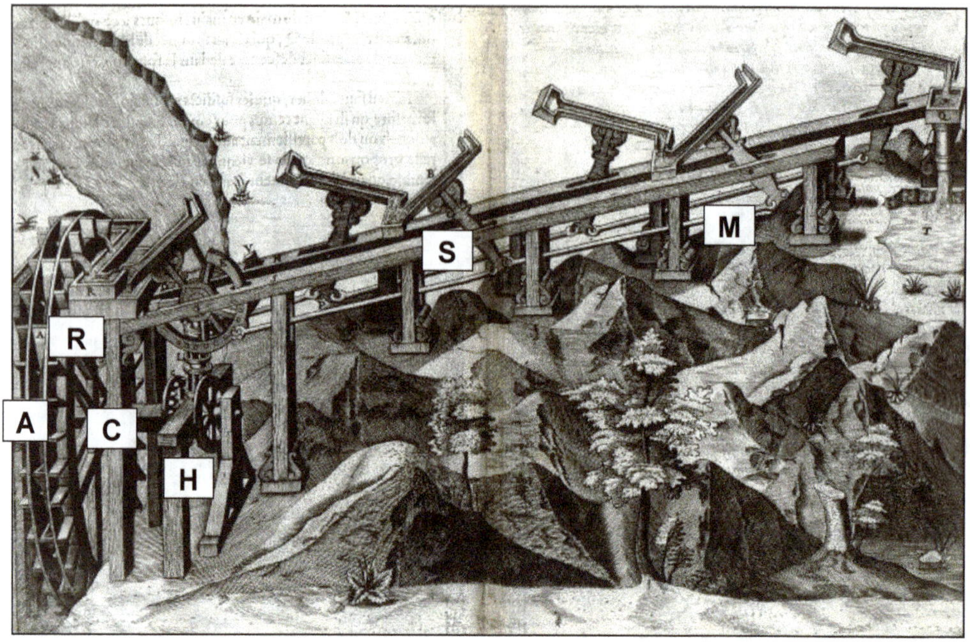

FIGURE 2.34 *Ramelli's 'inclined plane' water-raising machine (1588).*[191]
The large water-wheel A is turned by the force of the river flow. It takes water from the river and carries it into receptacle R. At the same time wheel A turns the two smaller toothed wheels C and H, which through a mechanism impart a reciprocating motion to bars M S. To these bars are attached levers driving scoops in their ends. A scoop would first descend to gather the water from the previous scoop and carry it until it spilled it into the following scoop. In this way the water would be gaining altitude.

are similar, though greatly differing in appearance: 'the inclined plane' (fig. 2.34) and 'the vertical tower' methods. Both of them are described in detail in Ramelli's *Le diverse et artificiose machine*; each of them is the basis for two of the three most accepted competing theories explaining the operation principle of the *Artificio* de Juanelo, one of the main subjects of the present study.

The 'inclined plane' is an extremely complex mechanism, which would require a high level of accuracy in its moving parts. Nevertheless, in typical Ramelli fashion, disregards these difficulties in the explanatory text: "This is another kind of machine with which the water of a river is very easily raised to

191 Ramelli, *Le diverse et artificiose machine*, plate 95, p. 148. Explanation taken from Ramelli, *The Various and Ingenious Machines*, p. 250.

the top of a mountain with the help of the river itself."[192] Ramelli's pioneering 'vertical water-raising machine', a somewhat simpler design than his 'inclined plane' concept, is depicted in Plate 96 of his book (fig. 2.35).

The analysis of both types of water-raising systems provides a good example of intellectual property theft in technical treatises. Ramelli was obsessed with keeping propriety rights over his inventions, complaining in his book that several of his assistants had stolen some of his designs, and subsequently mutilated and distorted them, while claiming they were theirs.[193] Out of fear that the detailed engravings in his theater of machines would be pirated, Ramelli also served as his own publisher. He did not differ from other engineers-inventors in a kind of technical schizophrenic attitude, displaying his knowledge to impress several constituencies, while at the same time concealing practical details whose lack would cause certain failure to a plagiarist competitor.[194] The end result was in each instance a single illustration depicting "... not real machines as much as idealized concepts, and lacked in visual perspective."[195] He did not live to see how others, who would become famous engineers, copied and 'embellished' his ideas and attributed them to themselves. Proof of up to what extent Ramelli's fears were not unfounded is provided by the actions of Georg Andreas Böckler (1644–1698), who in his *Theatrum machinarum novum* unashamedly copied 18 figures from Ramelli. The mirror-like images are the product of tracing the Ramelli originals to engrave the new copperplates according to their contour (when paper is pressed onto the new ink filled grooves in the copperplates, the resulting image is reversed).[196]

Heinrich Zeising had also openly copied Ramelli's Plate 95 depicting the "inclined plane machine" in his famous *Theatri machinarum*, published in 1608, twenty years after the publication of Ramelli's *Diverse et Artificiose Machine* (see fig. 2.36). A comparison with Ramelli's Plate 95 (fig. 2.34 above) shows the

192 Ramelli, *The Various and Ingenious Machines*, p. 250.

193 Ramelli, *Le diverse et artificiose machine*, second page [unnumbered] of the introductory letter of his book: "Alli benigni lettori. [To the kind readers]." In this respect, the advice to be extremely cautious about disclosing his innovations, given almost 150 years earlier by Bruneleschi to Taccola appears to be totally relevant, see Frank D. Prager, "A Manuscript of Taccola, Quoting Brunelleschi, on Problems of Inventors and Builders," *Proceedings of the American Philosophical Society*, Vol. 112, No. 3 (Jun. 21, 1968), pp. 131–149, pp. 140–142.

194 Sawday, *Engines of the Imagination*, p. 103.

195 Joel Mokyr, *The Gifts of Athena: Historical Origins of the Knowledge Economy*, Princeton: Princeton University Press, 2011, p. 62.

196 Ronald Brashear, "Ramelli machines: Original drawings of 16th century machines," *Website of the Smithsonian Institution Libraries*, http://www.sil.si.edu/ondisplay/ramelli/intro.htm, accessed 19 May, 2010.

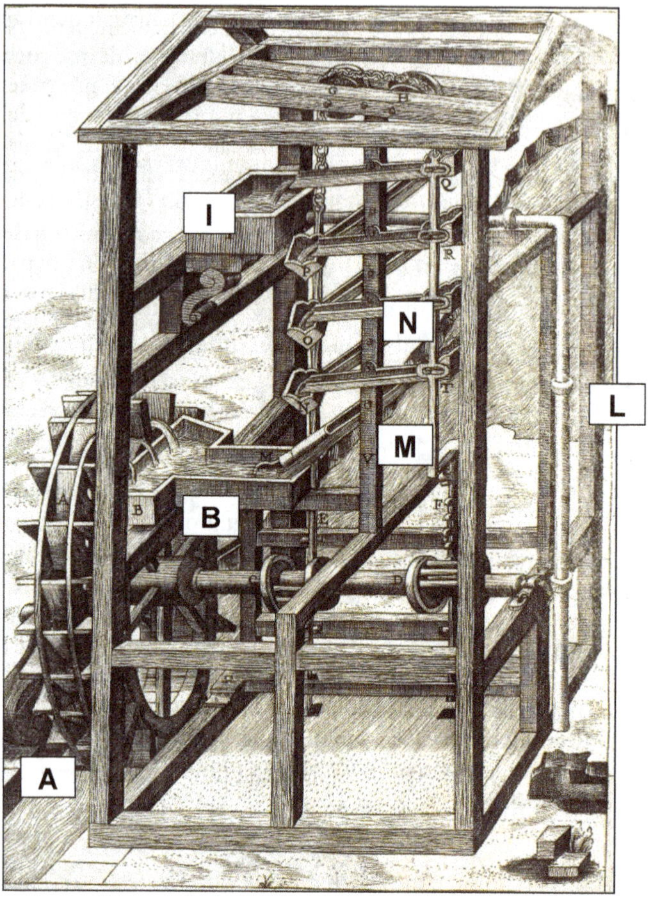

FIGURE 2.35 *Ramelli's vertical water-raising machine (1588).*[197]

The large water-wheel A is turned by the current of the river. It takes the water from the river with the scoops around its rim and carries it into the first receptacle B. At the same time and through a rocking mechanism it raises and lowers the dippers [ladles] with their carved channels. When the first dipper M is lowered, it takes the water from the first receptacle B and then carries it through its channel into the second dipper N. With the same movement the second dipper carries it into the third, and the third into the fourth box until it reaches the last receptacle I. Pipe L conducts the water to the place where it is needed.

197 Ramelli, *Le diverse et artificiose machine*, plate 96, p. 150. Explanation taken from Ramelli, *The Various and Ingenious Machines*, p. 256.

FIGURE 2.36 *H. Zeising, inclined plane water-raising machine (1608).*[198]

technical drawing to be identical, only the landscape in the background has been altered.

As with the 'inclined plane' blueprint, Ramelli's pioneering vertical conception was doomed to be copied as well. Its long lasting influence can be seen in the well-known work of Grollier de Servière *Recueil d'ouvrages curieux de mathematique et de mecanique* (1719), fig. 2.37, and in an identical drawing (with different captions) in Jacob Leupold's multi-volume set *Theatrum machinarum*, (1724–1739).[199]

But even Agostino Ramelli (1531–1600) himself, as well as other Italian engineers, such as Vittorio Zonca (1568–1603), Fausto Veranzio (1561–1617), Benedetto Castelli (1578–1643)[200] and Jacobus de Strada (1515–1588), had not a spotless record on what plagiarism is concerned: all of them plundered the writings of Francesco di Giorgio Martini (1439–1502).[201] Some historians have

198 Zeising, *Theatri machinarum*, plate 21, page A-84-A.

199 Jacob Leupold, *Theatri Machinarum Hydraulicarum, Tomus II: oder, Schau-Platz der Wasser-Künste, erster[-anderer] Theil ...: ein Werck, so nicht nur Künstlern, Künstmeistern, Berg-Leuthen und Künst-Steigern ... sondern auch Architectis, Ingenieurs, Commissarien, Beamten*, (2 volumes), Vol. 1, Leipzig: C. Zunkel, 1724, plate IX, Fig. 1, p. 22A.

200 Benedetto Castelli was the author of *Della Misura dell'acque correnti*, Bologna: HH del Dozza, 1660.

201 Ladislao Reti, "Francesco di Giorgio Martini's Treatise on Engineering and Its Plagiarists," *Technology and Culture*, Vol. 4, No. 3 (1963), pp. 287–298. For Martini's biography and an analysis of his works, see Gustina Scaglia, *Francesco di Giorgio: checklist and history*

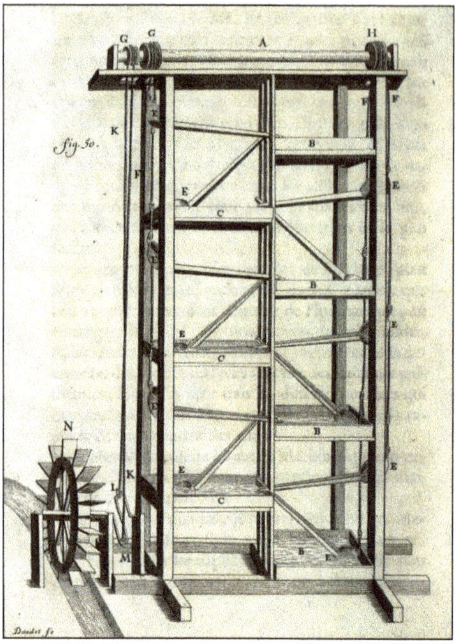

FIGURE 2.37
G. Servière, water vertical lifting machine (1719).[202]

adapted a radically opposite view on the issue of design appropriation, which they see as legitimate, maintaining that the common features in works by different authors were part of a contemporary "universal language of topological kinematic mechanisms."[203] It is not clear to what extent the drawings enabled competitors to make working machines based on them. Michael Oakeshott has maintained that, "... a printed or written page is not an independently generated beginning from which activity can spring ... [it] speaks only to those who know already the kind of thing to expect from it, and consequently how to in-

---

of manuscripts and drawings in autographs and copies from ca. 1470 to 1687 and renewed copies (1764–1839), Bethlehem (PA): Lehigh University Press and London: Associated University Presses, 1992.

202 Nicolas Grollier de Servière and Gaspard Grollier de Servière, Machine que l'on peut construire au bord d'une Rivière pour élever l'eau," *Recueil d'ouvrages curieux de mathematique et de mecanique, ou description du cabinet*, Paris: Chez David Forey, 1719, Plate XXII, fig. 50, described in p. 25 of the same book.

203 Francis C. Moon, *The Machines of Leonardo Da Vinci and Franz Reuleaux: Kinematics of Machines from the Renaissance to the 20th Century*, Dordrecht, (The Netherlands): Springer, 2007, p. 146.

terpret it."[204] This statement is reinforced by perusing technical treatises of the period, in which only the concept of a specific machine operation is shown. The scale of the drawings was intentionally distorted, they often contained technical details not actually applied in solving the problem, leaving out crucial practical 'tricks' in order to prevent the possibility of others copying it.

Transmission of Power through Distances

Plate 95 of Ramelli's book (fig. 2.34) shows two lines of articulated iron bars below and parallel with the main horizontal beams, transmitting work from a rotating waterwheel by moving back and forth, rocking the reciprocating troughs along and up to the end of the machine. Robert P. Multhauf has placed around 1550 the introduction of this mechanism into German mines, where it was used only to transmit reciprocating power.[205] The first known drawing of a primitive flat rod system or *Stangenkunst*, as the apparatus came to be known, is found in Jean Errard's *Premier livre des instruments mathématiques méchaniques*, published in 1584, four years before Ramelli's book (fig. 2.38). As will be later discussed, the mechanism was already used as an essencial component in the motion-transmission system of Juanelo's Artifice in Toledo, which started functioning in 1569, i.e. 15 years earlier than Errard's drawing. This invention would become of great importance from the seventeenth century onwards.

In 1659 the Englishman Robert D'Acres described the *Stangenkunst* (not by that name), in *The Art of Water Drawing*, as being a useful but as yet underdeveloped device. The further improvement of the mechanism was delayed by the English civil war, or in D'Acres' own words: "... these military discouraging times ..."[206]

---

204 Michael Joséph Oakeshott, *Political education: an inaugural lecture delivered at the London School of Economics and political Science on March 6, 1951*, Cambridge: Bowes and Bowes, 1951, p. 15, quoted in Cipolla, "The Diffusion of Innovations," pp. 47–48. See also: Kellenbenz, "Technology," p. 186.

205 Robert P. Multhauf, "Mine Pumping in Agricola's Time and Later," *U.S. National Museum Bulletin* 218, (1959), pp. 113–120, p. 118. The high degree of sophistication achieved by German mining experts in the use of *Stangenkunst* driven devices can be appreciated in a medal, coined in 1685, that shows the St. Anna mine, near Freiberg. An aqueduct is shown, one function of which is to feed a waterwheel in the house below, which in turn delivers power through the *Stangenkunst* to two open shafts. Brunswick Silver 4 Taler, Ernst August, 1685, *U. S. National Museum, Paul A. Straub coll.; Smithsonian photo 43334-A.*, Shown in Multhauf, "Mine Pumping," p. 116.

206 R. D'Acres', *The art of water drawing*, London: Henry Brome, 1660, pp, 16–17.

FIGURE 2.38 Stangenkunst, *Jean Errard (1584)*.[207]

One of the best-known representations can be found in Jacob Leupold's *Theatri machinarum hydraulicarum*, in his explanation of the power transmission system in the Marly waterworks, which will be discussed later in this book (fig, 2.39).

Waterwheels Sensitivity to Variations in the Level of the River Flow

It was well known in the sixteenth century that undershot waterwheels were very sensitive to the water level. To remedy this difficulty in water-driven mills, where water level varied, wheels were mounted on pontoons that were moored in the stream. Usually, the wheel was set between two boats, where the water velocity was the greatest. This 'floating' solution had already been used by the Roman General Belisarius to keep running mills when unable to do so with animals during the siege of Rome in the Gothic war in 537 CE[208] (fig. 2.40).

207 Bar-le-Duc, *Le premier livre*, planche 21.

208 Procopius, *History of the Wars*, Books 5–6 (Gothic War), (7 volumes) Volume III, translated by H. D. Dewing, New York: Cosimo, Inc., 2007, p. 191.

FIGURE 2.39
Stangenkünsten, *Leupold,* (1725).[209]

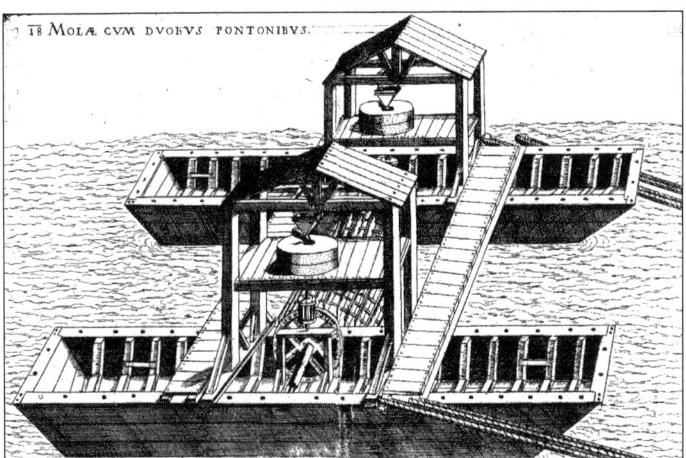

FIGURE 2.40 *Floating mill, keeping waterwheel at a constant depth.*[210]

209 Jacob Leupold, *Theatri Machinarum Hydraulicarum, Tomus II: oder, Schau-Platz der Wasser-Künste, erster[-anderer] Theil ...: ein Werck, so nicht nur Künstlern, Künstmeistern, Berg-Leuthen und Künst-Steigern ... sondern auch Architectis, Ingenieurs, Commissarien, Beamten,* (2 volumes), Vol. 2, Leipzig: C. Zunkel, 1725, plate XXVI, p. 44C.

210 Fausto Veranzio, *Machinae Novae: Cvm Declaratione Latina, Italica, Hispanica, Gallica Et Germanica,* Venetiis: [s.n.], 1617, Plate 18, between pp. 6–9.

FIGURE 2.41 *Watermills fixed to the Grand Pont, detail, Paris (1317).*[211]

In the following centuries the invention spreaded both to the East and to the West[212]—in Paris by the Late Middle Ages there were 55 boat mills on the river between Ile-Notre-Dame and the Pont-aux-Meuniers,[213] but due to the problems they caused to navigation, already by the twelfth century some had been replaced by watermills permanently fixed to the bridges across the Seine (fig. 2.41).[214] In the early sixteenth century a fixed mill operated under Pont Notre Dame, two more under Pont au Change and not less than 13 under Pont-aux-Meuniers.[215]

In *Ramelli's* book we find illustrations of typical mechanisms that achieved the same effect of keeping the depth of the wheel in the water constant for large waterwheel installations, such as the one illustrated in fig. 2.42. Ramelli even shows a complete floating scoop wheel, supplying water at variable

211 Yves de Saint-Denis, *Vie et martyre de saint Denis et de ses compagnons*, Detail of a French manuscript from 1317, (25.5 × 18 cm).

212 Forbes, *Studies in Ancient Technology*, p. 106.

213 Henry Heller, *Labour, Science and Technology in France, 1500–1620*, Cambridge: Cambridge University Press, 2002, p. 13.

214 Gies, *Cathedral, Forge and Waterwheel*, p. 113. Reynolds, *Stronger than a Hundred Men*, p. 59, provides a description of the evolution through time of boat mills into mills integrated into the superstructure of the bridges.

215 Jean Favier, *Paris au XV-e siècle, 1380–1500*, part of the series *Nouvelle histoire de Paris* (11 volumes), Vol. 3, Paris: Association pour la publication d'une histoire de Paris, Diffusion Hachette, 1974, pp. 28–29.

FIGURE 2.42
Waterwheel height regulating mechanism.[216]

levels.[217] As we shall see, solving this problem, present in all the cases treated in this study, required additional technical acumen.

Piping Networks[218]

In the period subject of our study, pipes were usually made of wood from some type of conifer, especially alder. Terra-cotta pipes, which had been used since Roman times, were fragile and costly, while wood was widely available and could be worked with simple tools, though it deteriorated rather quickly. The sources for wooden pipes were straight grown trees from which pieces of wood, preferable still covered with their bark, which contributed to better external protection, were cut. The maximum length of each tube was 4 to 5 meters. There were two basic methods to connect the pipes together: either the tube was tapered at one end and introduced into the following one, or

216 Ramelli, *Le diverse et artificiose machine*, plate 99, p. 154.

217 Ramelli, *Le diverse et artificiose machine*, plate 43, p. 66.

218 This discussion is largely based on Albrecht Hoffmann, "Zum Stand der Städtischen Wasserversorgung in Mitteleuropa vor dem Dreissigjährigen Krieg," in Frontinus-Gessellschaft e.V., (eds), *Die Wasserversorgung in der Renaissancezeit, Geschichte der Wasserversorgung* (5 vols.), Vol. 5, Mainz am Rhein: Verlag Philipp Von Zabern, 2000, pp. 101–144.

FIGURE 2.43
Drilling trunks, detail.[219]

with an iron or copper metal bushing, half of which inside the interior of two adjoining pipes. Water tightness was attained by a variety of means, such as the use of sealant made of mutton fat mixed with crushed bricks.[220] The dampness of the wood, once water flowed, caused the wood to swell, tightening the connections, thus reducing leaks. It was difficult to achieve effective diversions; forked branches of trees were the preferred method.[221] Pipes had an inside diameter which generally did not exceed 5 cm, so at low pressure conditions they were able to deliver limited quantities of water. When more water was needed, multiple lines were laid side by side. Shallow trenches were dug into which the piping was buried, protecting it from damage and frost.

Pipes were normally drilled using manual drills in a horizontal position along their length, Agricola shows work been performed in a vertical position

219 Agricolae, *De re metallica Libri XII*, p. 135.

220 George A. Antaki, *Piping and Pipeline Engineering. Design, Construction, Maintenance, Integrity, and Repair*, New York and Basel: CRC Press, 2003, p. 3.

221 Bertrand Gille, "The Fifteenth and Sixteenth Centuries in the Western World" in Maurice Daumas, (ed.), *A History of Technology and Invention*, translated from the French edition by Eileen B. Hennessy, (2 volumes), Vol. 2, New York: Crown, 1969, pp. 16–149, p. 122.

FIGURE 2.44 *Water-driven machine for trunks drilling.*[222]

as well (fig. 2.43). By the end of the sixteenth century, the expansion of urban distribution networks significantly increased the demand for pipes. This development brought about the establishment of *Bohrhäuser* [drilling houses]. These workshops were located near water courses, utilizing hydropower as the mover for their drilling machines, as illustrated in fig. 2.44.

Before laying a pipeline, it was critical to determine whether its angle of inclination from the source to the point of use, relative to a horizontal plane, was sufficiently large, but not more than necessary, to successfully bring water to the desired location. During the 1530s German mathematicians developed different methods to measure these angles, some of them based on astronomical instruments such as quadrants, as shown in fig 2.45.

The Diffusion of False Know-How

About the early 1500s several inventors designed unworkable perpetual motion devices based on force pumps or Archimedean screws (as an example, see fig. 2.46).

222 Isaac de Caus, *Nouvelle invention de lever l'eau plus haut que sa source avec quelques machines mouvantes par le moyen de l'eau, et un discours de la conduit d'icelle: avec beacoup de figures en taille douce / par Isaac de Caus, ingenyeur & architecte à Charles le Premier, Roy de la Grand Bretaigne*, London: Thomas Davies, 1657, Explanation in p. 24, figure in Plate XII.

FIGURE 2.45 *Gualtherius Hermenius Rivius, measuring heights with a quadrant* (*c.1549*).[223]

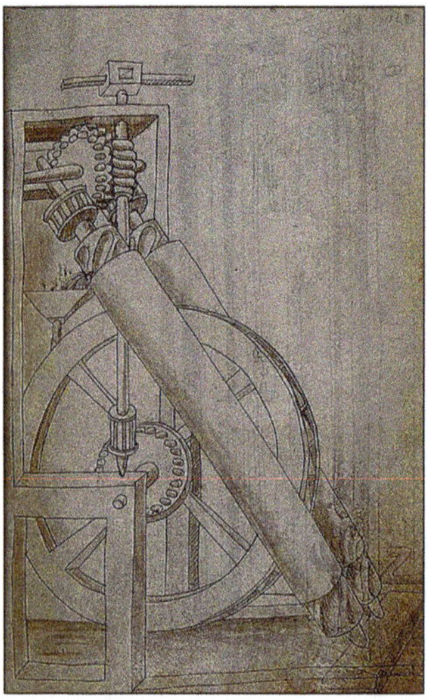

FIGURE 2.46
Two Archimedes' screws working a scoop-wheel.[224]

223 Walter Hermann Ryff (c.1500–1548), whose name was Latinized to Gualtherius Hermenius Rivius. "Vermessung der Höhe eines Gebäudes aus verschiedenen Distanzen mit einem Quadranten," 1547, Saxon State Library, Dresden, German Fotothek, Record 0000182.

224 Anonymous (after Taccola and Francesco di Giorgio), "Two Archimedes' screws working a scoop-wheel, which in turn pulls the screws," *Drawings of machines, Ms. Palatino 767*, Biblioteca Nazionale Centrale di Firenze, p. 111.

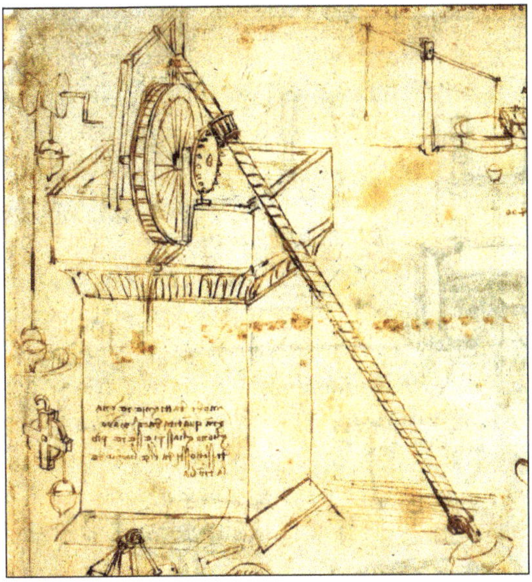

FIGURE 2.47
An Archimedian screw driven by a waterwheel, which lifts water to the top of the waterwheel.[225]

It was Leonardo da Vinci who put this quest in the correct perspective in the *Codex Forster*: "Oh speculators on perpetual motion, how many vain projects you have created in this search. Go and be the companions of the seekers for gold!"[226] and in the *Madrid Codex* as well: "… among the excessive and impossible beliefs of men can be found the quest for perpetual movement, called by some perpetual wheel…."[227] Notwithstanding his clear message, some of Leonardo's drawings are suspiciously remindful of his own experimenting with the quest for unattainable inexhaustible energy (fig. 2.47).

225 Leonardo da Vinci, "Machines for lifting and pumping water, towers and buildings on top of which water is lifted by Archimedean spirals," *Codex Atlanticus*, (12 Volumes), Vol. 3, fol. 1069, v., Pen and ink on paper, (61×44 cm), Biblioteca Ambrosiana, Milan. A sign that this drawing represents a try at perpetual motion is that the Archimedean screw turns moved only by the upper water wheel and no external source of energy is shown. The writing inside the drawing does not clarify Leonardo's intention.

226 *Codex Forster II*, fol. 92v., Extracted from Allan A. Mills, "Leonardo da Vinci and Perpetual Motion," *Leonardo*, Volume 41, Number 1, (February 2008), pp. 39–42, p. 42.

227 Ignacio González Tascón, *Fábricas Hidráulicas Españolas*, Madrid: Ministerio de Obras Públicas y Urbanismo, Centro de Publicaciones, 1987, p. 465.

The search for a water-driven *perpetuum mobile* device was widespread in Europe. From a report to the Doge on March 1605 by the Venetian Proveditors of Fortresses[228] we learn about two inventions of Mauritio Maggi, son of the Venetian Girolamo Maggi, a military engineer who, after serving the Republic in Famagusta, was put to death in Constantinople. The younger Maggi proposed to build a machine designed "… to raise water by perpetual motion by the strength of the water itself …" [… *per inalzar li acqui fermi a perpetuo moto colla forza delli istessi acqui*].[229] This would facilitate drainage and irrigation, essential for the vigorous development of agriculture in the Venetian Republic, which from the late sixteenth century through the seventeenth and eighteenth, promoted land reclamation in its mainland territories. The experts relied mainly on natural drainage by cutting trenches and canals, rather than on mechanical lifting water pumps, as the existing ones were not able to drain swamps.[230] On December 1605, the *proveditori* reported having received further explanations from Mauritio as to the working of his perpetual motion pump. They were convinced of its practicality and importance, but deferred the decision to the opinion of the Doge. But the machine was not built, for reasons that may appear obvious today.

The progress in the practical utilization of flowing water energy was not always accompanied by a parallel understanding of the laws of physics underlying the phenomena. Important figures, such as Vittorio Zonca, designed some water-raising machines that could not work. He even attempted to create a *perpetuum mobile* device, using the effect of the siphon, such as the one shown in fig. 2.48, by plagiarizing one of Giorgio Martini's ideas (an identical conceptual drawing by Giorgio Martini shown in fig. 2.49). The fact that siphons only

228 For the authority and responsibility of the proveditors, see: M. E. Mallett and J. R. Hale, *The Military Organisation of a Renaissance State: Venice c.1400 to 1617*, Cambridge: Cambridge University Press, 2006.

229 J. R. Hale, "Girolamo Maggi: a Renaissance scholar and military buff," *Italian Studies*, 40, (1985), pp. 31–50, p. 50.

230 Gaetano Cozzi, Michael Knapton and Giovanni Scarabello, "La Repubblica di Venezia nell'età moderna," in *Storia d'Italia*, (12 Volumes), Vol. 12–2, Turin: Unione tipografico-editrice torinese, 1992, p. 435. For an account of these undertakings, see: Salvatore Ciriacono, *Building on water: Venice, Holland, and the construction of the European Landscape in the Early Modern Times*, translated by Jeremy Scott, New York: Berghahn Books, 2006, chapter 2—"Irrigation and Land Drainage in the Venetian Republic during the Sixteenth and Seventeenth Centuries," pp. 62–101.

FIGURE 2.48
Zonca's unworkable water-raising device (1607).[231]

FIGURE 2.49
Di Giorgio Martini's unworkable water-raising device (c.1480).[232]

231 Zonca, *Novo Theatro di Machine*, p. 114.

232 Di Giorgio Martini, *Trattati di architettura*, fol. 36v., (c.1480).

work when the outlet opening is lower than the inlet was not understood at the time, which caused Zonca to surmise that his design would create a perpetual energy source—a typical case in which ignorance of the fundamental physic laws facilitated the diffusion of false knowledge and its naïve acceptance.

Art was also influenced by water supply technology, as exemplified by a wonderful painting by Carlo Saraceni, in which a contemporaneous water-lifting device is depicted in a scene taking place 2500 years earlier. The painting details were probably inspired by Ramelli's book.[233]

Before entering into the core subject of this book: the comparison of water supply undertakings in Toledo, London and Paris, and for the sake of perspective, it would be helpful to add a few remarks about the progress in other geographical areas. In the background, when analyzing a mosaic of similar water driven enterprises in other European cities, on the one hand Augsburg stands out as a reference on what technical progress is concerned, as the *Machina Augustana* testifies.[234] On the other hand, all Italian cities took totally different paths to fulfill their water supply needs.

233 Sawday, *Engines of the Imagination*, p. 15. The painting is in the National Gallery, London: Carlo Sareceni, "Moses defending the daughters of Jethro," 1609, Oil on copper, (28.5 × 35.3 cm). The closest drawing is in Ramelli, *Le diverse et artificiose machine*, Plate 76, p. 117.

234 The earliest book on the subject of the history of water supply to the city is Caspar Walter, *Hydraulica Augustana, Das ist: Ausführliche Beschreib- und Auslegung alles dessen was in des Heil. Röm. Reichs-Stadt Augspurg in den daselbst befindlichen Drey obern Haupt-Wasser-Thürnen so wohl, als in den Brunnen-Häusern, darinnen die Wasser-Druck-Wercke stehen, sammt den auf- und absteigenden Haupt-Wasser-Röhren, Röhr-Kästen oder Reservoirs, von welchen das Wasser durch einen weiten District der Stadt, nicht allein in die schöne publique Spring-Brunnen auf den Haupt-Plätzen, sondern auch in die privat-Häuser mittelst unter der Erden ligenden Canaelen oder Theihel geleitet wird, den Fremden, Passagiers und Liebhabern pflegt gezeiget zu werden: Woneben unterschiedliche nach dem verjüngten Maßstab verfertigte Modell von Wasser-Machinen, hölzernen Bruggen, Schnecken- oder Wändel-Treppen, gemahlte Taflen, worauf alles, was zu dem Brunnen- und Röhr-Wasser-Werck gehörig vorgestellet zu sehen*, Augsburg: Detleffsen, 1754. Other historians of the eighteenth and nineteenth centuries basically copied Walter: Paul von Stetten (der Jüngere), *Künst-, Gewerb- und Handwerksgeschichte der Reichsstadt Augsburg*, Augsburg: C. H. Stage, 1779 and Franz Joseph Kollmann and F. A. Oldenburg, *Die Wasserwerke von Augsburg*, Augsburg: Rieger, 1850.

In the late fifteenth and early sixteenth centuries Augsburg reached the zenith of its political and economic power. It took advantage of its privileged geographic position in the trade routes communicating the North with Italy, and by the 1490s surpassed Nuremberg as the financial capital of southern Germany.[235] The demands of the city's growing rich population required ever increasing quantities of water, but Augsburg's topographic situation did not allow its supply by aqueducts or pipelines from high natural sources, so water-raising systems with high reservoirs were necessary. An entrepreneur, Leopold Karg, who is considered the pioneer of the potable water supply to the city,[236] established already in 1412 the first waterworks, adjacent to the water flooded moat of the city.[237]

Augsburg is considered by German historians as a 'water know-how cluster' between the Late Middle Ages and the Industrial Revolution.[238] From the early Renaissance, the imperial city seems to have constituted the leading center of pumping plants in Central Europe. Not only did it have a series of waterworks, but it also possessed a large base of skilled workers in the construction and operation of pumping stations. The area of action of these professionals was not limited solely to Augsburg and its surroundings, but comprised other regions of Central Europe.[239] Thus, in early modern Germany, water-raising became a traditional craft with expertise concentrated in a few areas. For the construction and care of their pumping devices and pipelines, the cities needed adequate staff. The work was demanding and required know-how and experience which only years of training and practice could impart. Due to the short life expectancy in the early modern period not always cities succeeded in securing a strong base of skilled workers. An elite of urban

---

235 Kathy Stuart, *Defiled Trades and Social Outcasts: Honor and Ritual Pollution in Early Modern Germany*, Cambridge and New York: Cambridge University Press, 2006, p. 36.

236 Walter, *Hydraulica Augustana*, pp. 2–3; Von Stetten, *Künst-, Gewerb- und Handwerksgeschichte*, pp. 145–146, Kollman, *Die Wasserwerke*, pp. 72–73.

237 Kollmann, *Die Wasserwerke*, pp. 75–76.

238 "Hydraulic Engineering and Hydroelectric Power, Drinking Water and Decorative Fountains in Augsburg," application submitted to UNESCO for World Heritage status by the Augsburg Mayor Peter Grab on July 2012, p. 2.

239 Hoffmann, "Zum Stand," p. 129.

professionals developed, who passed the secrets of the craft from generation to generation.[240]

The Italian Strategy—Gravity Based Solutions

During the early Middle Ages the formidable water supply systems built by the Romans in Italy had decayed beyond repair, as for hundreds of years no investment was made to keep them up. The limitations of space and cost to build new big scale gravity fed pipelines and aqueducts became insurmountable, and even if the will and the resources to undertake such projects would have been available, it is doubtful if the know-how preserved would have been up to the task.

Scarce evidence, if at all, has been found in Medieval Italy about the use of complex hydraulic devices, such as water-screws, force pumps, and water-lifting wheels. This absence is curious, taking into account ancient Roman know-how and the Muslim earlier presence in the south. Paradoxically, the fifteenth-century drawings of Mariano Taccola, Pisanello, Martini and Leonardo show sophisticated hydraulic devices. These masters were followed by the likes of Ramelli and Zonca, Although these 'mechanical engineers' designed wonderful water lifting machines, which could have positively influenced water supply availability to the cities where they acted, it is arguable if their drawings were implemented even to the stage of working models. No records exist about Italian cities or towns having confronted their water supply needs at the time with full scale utilitarian devices driven by the flow of an adjacent river, on the banks of which they were usually built. Only in a few instances these technologies were put into practice to provide water to the private mansions of highly placed religious

240 Karl H. Fischer, "Die Wasserversorgung der Reichsstadt," in Stadtmagistrat Nürnberg, (ed.): *Die Wasserversorgung der Stadt Nürnberg von der reichsstädtischen Zeit bis zur Gegenwart. Festschrift zur Eröffnung der Wasserleitung von Ranna*, Nürnberg: W. Sebald, 1912, pp. 1–130, p. 121, extracted from Florian Ruhland, "Power, pleasure, and pollution: Water use in pre-industrial Nuremberg and Prague," *Klaudyán: Internet Journal of Historical Geography and Environmental History*, Volume 4, (2007), No. 2, pp. 5–18, p. 13.

authorities, but never for urban supply.[241] Instead, all of the main Italian cities tried, and usually succeeded, to solve their water problems either reconstructing old aqueducts or by other gravity based solutions.[242] It was during the Late Middle Ages and the early Renaissance that the rediscovery, renewal and improvement of ancient aqueducts technology know-how took place. Such was the case of Bologna,[243] Firenze,[244] Milano,[245] Verona,[246] and, above all, Rome itself.

Among the important engineering projects that were undertaken under papal patronage in Rome during the late-sixteenth and early-seventeenth centuries were proposals to restore and build three aqueducts.[247] The first

241 See, as an example, the device built for the residence of Cardinal Giovanni Ricci da Montepulciano (1498–1574), in Leonardo Lombardi, "Camillo Agrippa's Hydraulic Inventions on the Pincian Hill (1574–1578)," translated by Katherine W. Rinne, *The Waters Of Rome*, No. 5, (2008), pp. 1–10.

242 Roberta Magnusson and Paolo Squatriti, "The Technologies of Water in Medieval Italy," in Paolo Squatriti, (ed.), *Working with Water in Medieval Europe: Technology and Resource-use*, Leiden: Brill, 2000, pp. 217–266, p. 233.

243 Renato Drusiani, Giancarlo Leoni, Danilo Demaria and Nicoletta Lemb, "Water supply of Bologna (Italy) by Roman aqueduct: history, morphology and hydraulic, from ancient time to nowadays," *Water Science and Technology: Water Supply*, No. 10 (4), (2010), pp. 554–560.

244 Henry Coxe (pseud. [i.e. John Millard.]), *Picture of Italy; being a guide to the antiquities and curiosities of that classical and interesting country: containing sketches of manners, society and customs*, London: Sherwood, Neely & Jones, 1815, p. 394.

245 Giuliana Fantoni, *L'acqua a Milano*, Milano: Cappelli, 1990.

246 Paola Lanaro e Gian Maria Varanini, "I ponti 'costruiti': riflessioni sull'esempio veronese," in D. Calabi e Cl. Conforti, (eds), *I ponti delle capitali d'Europa dal Corno d'oro alla Senna*, Milano: Electa, 2002, pp. 58–74 and Gian Maria Varanini, "Energia idraulica e sviluppo urbano nella Verona comunale" in R. Comba, (ed.), *Paesaggi urbani dell'Italia padana nei sec. VIII–XIV*, Bologna: Cappelli, 1988, pp. 331–372.

247 A vast bibliography deals with these projects, from where this short summary draws. See David Karmon, "Restoring the Ancient Water Supply System in Renaissance Rome: The Popes, the Civic Administration, and the Acqua Vergine," *The Waters of Rome*, No. 3, (2005), pp. 1–13. Website of the University of Virginia:

http://www3.iath.virginia.edu/waters/Journal3KarmonNew.pdf, accessed 4 July 2016; Katherine Wentworth Rinne, "Water: the Currency of Cardinals in Late Renaissance Rome," in A. Calzona and D. Lamberini, (eds), *La Civilta` delle Acque tra Medioevo e Rinascimento*, Florence: Olshki, 2010, pp. 367–87; Katherine Wentworth Rinne, *The Waters of Rome: Aqueducts, Fountains, and the Birth of the Baroque City*, New Haven and London: Yale University Press, 2010; Katherine W. Rinne, "Between precedent and of these projects was a restoration of the ancient Aqua Virgo, by then called the Acqua Vergine, carried out between 1560 and 1570. This renovation was initiated by Pope Pius IV (1559–65) and completed under Pius V (1566–72). Previous popes had tried to rebuild the aqueduct—in 1453, Nicholas V became the first Pope to sponsor its restoration since the eighth century, with poor results, as the aqueduct required constant repairs. As water demand continued to grow, a comprehensive restoration was undertaken under Paul III. The history and construction techniques used in these efforts were described by Pirro Ligorio in volume 16 of his *Libri delle antichità*, composed between 1550 and 1565.[248] Past water sources, recoverable materials and still standing structures were used whenever possible. Distributing the water was even more difficult, as almost none of the previous within walls infrastructures remained. Thus, the need to redevelop such systems, though through a failure ridden learning curve, ultimately created a new breed of hydraulic experts. The result was the construction of numerous innovative conduits and fountains, instrumental in contributing to the physical transformation of the city that was considered crucial for its urban and spiritual renaissance as the centre of the Christian world during the Counter Reformation.

The Acqua Vergine provided the technical and administrative framework upon which the Acqua Felice and the Acqua Paola projects were built. Pope Sixtus V (1585–90) sponsored the Acqua Felice between 1585 and 1587. The Acqua Paola, built between 1607 and 1612 under Pope Paul V (1605–21), followed along the course of the ancient Acqua Traiana. The scale of these schemes was enormous—the success of these major enterprises caused a profound effect on the city, by creating a comprehensive water distribution system, composed of the three individual networks that together served the entire city for the next three hundred years.

experiment: restoring the Acqua Vergine in Rome (1560–70)," in L. Roberts, S. Shaffer and P. Dear, (eds), *The Mindful Hand: Inquiry and Invention from the Late Renaissance to Early Industrialization Editors*, Amsterdam: Edita, 2007, pp. 94–115; and Katherine Rinne, "Fluid Precision: Giacomo della Porta and the Acqua Vergine fountains of Rome", in Jan Birksted, (ed.), *Landscapes of Memory and Experience*, London: Taylor and Francis, 2000, pp. 183–201.

248 For details on Ligorio's manuscript, see Karmon, "Restoring the Ancient Water Supply System," footnote #89.

The outstanding water-raising project inaugurated in Toledo in 1569, which is one of the three projects on which we shall now focus, was built contemporaneously to most of these undertakings. As will be shown, it greatly surpassed the height to which water was raised in all former European waterworks (and future ones for 100 years) that used flowing water as the energy source, and therefore also necessitated new technical solutions.

TOLETVM
Septentrio
Oriens
El Jugenio
Tagus fluuis
Templum Archiepiscopat. Toletani
Palatium Regium Toletanum

CHAPTER 3

Toledo

Spain and Toledo in the 1560s

The population of the Spanish monarchy in the late sixteenth century is estimated to have been between 6.5 and 9 million people, much less than in France, Germany and Italy, but much more than in England or Portugal. Eighty percent of the population was concentrated in the Kingdom of Castile. The population of the kingdom had steadily increased since the mid-fifteenth century, an expansion that continued into the late sixteenth century, until the plague of 1598–1600, which caused the death of half a million people, temporarily reversed the trend.[1]

The two crowns of Castile and Aragon, which had been united by the Catholic kings, were not on equal footing.[2] The Spanish expansion in Europe

---

Opposite page—Toledo in 1598: Joris Hoefnagel, "Toletum," in Georg Braun and Franz Hogenberg, *Civitates Orbis Terrarum*, first Latin edition of volume V, Cologne, 1598, incorrectly marked 1566, Plate 15.

1 The lower population estimate is given by James Casey, *Early Modern Spain: A Social History*, London: Routledge, 1999, p. 23; the higher estimation by Fermín Caballero, *Manual geográfico-administrativo de la monarquía española: Obra útil para empleados, hombres de negocios, viajeros y curiosos, con 1.750 artículos en orden alfabético*, Madrid: Imprenta de A. Yenes, 1844, p. 132.

2 The discussion on Toledo governance and Toledan society draws from several sources: María Asenjo González, "Perfíl socioeconómico de la ciudad de Toledo en el siglo XV a través de sus ordenanzas," *Cuadernos de historia de España*, Nº 77, (2001-2002), pp. 109–144; Ángel Santos Vaquero, "Alcalde mayor de alzadas en el ayuntamiento de Toledo," *Espacio, Tiempo y Forma, Serie IV, Historia Moderna*, tomo 16, (2003), pp. 11–29; Francisco José Aranda Pérez, *Poder y poderes en la ciudad de Toledo: gobierno, sociedad y oligarquías en la Edad Moderna*, Madrid: Universidad de Castilla La Mancha, 1999; Linda Martz, "Implementation of Pure-blood Statutes in Sixteenth Century Toledo," in Bernard Dov Cooperman, *In Iberia and Beyond: Hispanic Jews Between cultures: Proceedings of a Symposium to Mark the 500th. Anniversary of the Expulsion of Spanish Jewry*, Cranbury, New Jersey and London: University of Delaware Press, 1998, pp. 245–272; Mariano García Ruiperez, "El AMT y la investigación histórica sobre la Edad Moderna," *Cuaderno de historia Moderna*, No. 22, (1999), pp. 201–227; Enrique Lorente Toledo, "Carlos I y su relación con Toledo: de la rebelión a la Corte," *Archivo secreto: revista cultural de Toledo*, No. 1, (2002), pp. 210–227; Julián Montemayor, *Tolède entre fortune et déclin (1530–1640)*, Limoges: Presses Univ. Limoges, 1996; Antonio Martín Gamero, *Historia de la Ciudad de Toledo: Sus claros varones y monumentos*, Toledo: Severiano López Fando, 1862 and

 | DOI 10.1163/9789004312425_005

and the rest of the world during the sixteenth century was a Castilian enterprise—soldiers, diplomats, theologians, statesmen and resources and money came out of Castile more than from the other parts of the Peninsula. Secondly, the population of Castile as a whole seemed to have been reluctant to accept innovative policies and was more subservient to its kings. The third factor lies in geography: Castile occupies a central place in the Peninsula. It is there where kings resided more frequently; from Toledo or Madrid and from the Escorial it was easier to address the problems posed by all other territories. The superiority of Castile was symbolized by its linguistic dominance. Castilian became the language of the empire, spoken in the army, in Spain's overseas territories and in foreign courts where Spaniards lived. It became the preferred language used by Spanish authors. From the sixteenth century the aristocracy and educated people adopted it as the language of Spanish culture.

Throughout its history, Toledo, the capital of Castile, had been a place of importance due to its location and its strategic position, standing on a steep rock surrounded almost completely by the Tajo River. From the late eleventh century, its central role in the Christian defense system was enhanced by its significance as an ecclesiastical center and its ties to royalty. By the early fifteenth century it was a great Spanish city, safe within its walls and aware of its leading role based on its religious primacy.[3]

The workload of the King in ruling an empire in which "the sun never sets"[4] could be borne, at least partly, by the monarch himself, as was often the case under Felipe II, who spent half his life among documents (hence his nickname '*El Rey Papelero*' [The Paperwork King]),[5] or could be delegated to a *válido*—the monarch's favorite, who was the highest unofficial officer at Court.[6] The use of *válidos* for carrying out the daily management of affairs led to the

---

Francisco de Pisa, *Descripción de la imperial ciudad de Toledo i historia de sus antiguedades i sus grandezas, i cosas memorables, los reies que la an señoreado, o gouernado, i sus Arzobispos más celebrados*, (Dos partes: primera y apuntamientos para la segunda parte), primera parte, Toledo: Diego Rodriguez, 1617.

3 For a general history of the city, see: Antonio Martín Gamero, *Historia de la Ciudad de Toledo: Sus claros varones y monumentos*, Toledo: Severiano López Fando, 1862.

4 For a concise account of the Spanish Empire world impact during the 16th century, see: Sibille Stocker and Christian Windler, *Instituciones y desarrollo socio-economico en España e Hispanoamerica desde la epoca colonial*, Bogota (Colombia): Fundes, 1994.

5 Manuel Castillo Martos, "De mano e imprenta: textos científicos y tecnológicos, (siglos XVI y XVII)," in Antonio Acosta Rodríguez, Adolfo Luis González Rodríguez, Enriqueta Vila Vilar, (eds), *La Casa de la Contratación y la navegación entre España y las Indias*, Sevilla: Universidad de Sevilla, Consejo Superior de Investigaciones Científicas, 2003, pp. 567–604.

6 Nicolás García Tapia, *Técnica y poder en Castilla durante los siglos XVI y XVII*, Salamanca: Junta de Castilla y León, Consejería de Cultura y Bienestar Social, 1989, p. 175.

formation of a class of *letrados* [lawyers] that allowed upwards social mobility from unprivileged positions or, more commonly, from the ranks of the lower nobility. This alteration of the traditional social hierarchy caused envy and jealousy from the old aristocrats, whose advice had been frequently rendered to kings. The monarchs strove to have them serving the court in diplomatic or military missions, away from positions where they could interfere with the royal authority. At the same time loyalty to kings was guaranteed by those who owed their positions directly to the monarch and had no ambition other than to retain the king's favor. In a society where family background, not merit or work, had always been the justification of social status, some of these *válidos* achieved a rank to which they could not aspire to obtain by their lineage. A few attained high confidence from kings, such as Martin Gaztelu with Carlos I,[7] Juan de Idiáquez, Mateo Vázquez de Leca and Antonio Perez with Felipe II[8] and later the Duke of Lerma under Felipe III.[9]

The extraordinary complexity of the administrative authorities structure of Toledo in the sixteenth century is a key issue in the quest to unravel the motivation behind the actions that drastically affected—for the worse—all public works, including water supply. Throughout Spain, the municipalities preserved their autonomy in the legal, economic, social and political spheres. They relied on their own militias, the judicial power of the mayors, their own financial and economic resources (obtained from manufactories, services, fines, fees, taxes, etc.), the limited trade contacts and the difficulty for kings to control an area beyond a short radius from the court, due to limited means of communications. In the absence of a strong intermediate level of authority in the territory, the municipal institution, inheritance of the Roman *municipum*,[10] had an extraordinary vitality, reinforced by the repopulation that followed the *reconquista* in the Middle Ages. This process brought about a freedom unparalleled in other parts of Europe. Governance evolved, especially from the thirteenth century onwards, from the open council and general assembly of all neighbors, to a system in which real power was concentrated in the hands of a few *hombres buenos* (citizens with full civil rights and socioeconomic

7 William Stirling-Maxwell, *The cloister life of the emperor Charles the fifth*, Oxford: Oxford University Press, 1857, p. 52.

8 F. Chacón Jiménez and Nuno Gonzalo F. Monteiro, (eds), *Poder Y Movilidad Social: Cortesanos, Religiosos y Oligarquías en la Península Ibérica (Siglo XV–XIX)*, Murcia (Spain): Editum, 2006, p. 145.

9 For an extensive biography, see: Patrick Williams, *The Great Favourite: the Duke of Lerma and the court and government of Philip III of Spain, 1598–1621*, Manchester: Manchester University Press, 2006.

10 A Roman '*municipum*' was a city whose citizens had the privileges of Roman citizens but was governed by its own laws.

standing) or *caballeros* (members of the nobility). The latter were organized in powerful corporations—*ayuntamientos*[11] [councils] *de ciudades o villas* [towns or villages], which enjoyed charters that gave them jurisdiction over wide rural communities and uninhabited areas.[12]

King Juan II of Castile, by decree of March 10, 1421, divided the *ayuntamiento* of the city into two main bodies: *cabildos de regidores y de jurados* [councils of aldermen and of judges]. The former included twenty-four seats divided between *caballeros* (two thirds) and *hombres buenos* (one third). Though it was a deliberative body whose decisions were taken by majority vote, in practical terms it was dominated by the representatives of the nobility. The second body—the *cabildo de jurados*—belonged to the common people and was formed in the beginning by forty-two members, originally elected by the parishioners. They had voice but no vote in municipal decisions, supervising and auditing the performance of the *cabildo de regidores*, often raising protests when decisions were against their common good, and when necessary, appealing to the king. The Toledan historian Francisco de Pisa notes that "the posts of *jurados* were at an earlier time provided by the votes of the parishioners,"[13] which indicates that as had happened with the *cabildo de regidores*, even the most democratic municipal institution abandoned elections for its members and chose a perpetual and hereditary link with the influential families of the city. The officials nominated by the Crown were the *corregidor*, who exercised the royal jurisdiction over a town and its district (a title created by the Catholic Monarchs in 1477) and another three officials of lower rank. All these positions were systematically allocated to members of distinguished families, but the real work was performed in their name by *letrados*.[14]

The involvement of royal authority in the municipal control in general and in Toledo in particular increased in the Late Middle Ages. The *corregidor* acted as a governor with powers in justice, police, finance, war, commerce, and public works. He convened the assemblies of the two *cabildos* comprising the *ayuntamiento*, chaired them, moderating the discussions, reserving his casting vote in case of tie or disagreement among *cabildo* members. The *corregidor* was usually a nobleman, a high ranking military officer or a person of renowned intellectual capacity, normally not from the city itself. As the

11 The name '*ayuntamiento*' is derived from the act of assembling, namely that functionaries 'se ayuntarán' (will assemble), De Pisa, *Descripción de la imperial ciudad de Toledo*, primera parte, p. 36.

12 For an account of this process, see: Aranda Pérez, *Poder y poderes*, pp. 48–55.

13 De Pisa, *Descripción de la imperial ciudad*, primera parte, p. 36.

14 Aranda Pérez, *Poder y poderes*, p. 86.

King's representative he had absolute power over all aspects of urban life, often causing conflicts with the city's *regidores*, who often tried to curtail his power, demanding his respect for the central interests of the city. In Toledo, as in all other Spanish cities, the tension between the *regidores* and the *corregidor* is reflected in contemporaneous texts written by figures who also served in these capacities.[15] Nevertheless, in most cases the *corregidores* were well received by the members of the *Ayuntamiento*, who were glad to have prestigious figures, who acted practically as heads of their institution.[16] The *corregidores* strove to be close to the Court, to be able to enjoy the opportunities offered by the vast Spanish empire. Their lust for power and control was satisfied by the social and economic benefits of their office. In return, they promoted the king's agenda for city management.

During the period treated here, the position of *regidor* was considered personal property, and it could be sold or transmitted to one's heirs. It was therefore considered to be a good long-term investment. Prospective purchasers, who sought the best value for their money, were not only seduced by the prospect of nobility but were also attracted by tax exemption. In the sixteenth century the Crown had the right to approve the transmission of these offices, but it was rare that the wishes of an officeholder were not respected.[17] From the Crown's point of view, the sale of municipal offices to the men of trade and finance afforded an opportunity to tap the resources of these prosperous citizens for their constantly starved coffers,[18] disregarding the active resistance of other well-established elites, such as the *caballeros*—who had been

---

15 See Jerónimo Castillo de Bobadilla, *Política para corregidores y señores de vasallos en tiempo de paz y de guerra y para Iuezes eclesiásticos y seglares y de sacas, aduanas y de residencias y sus Oficiales y para Regidores y Abogados y del valor de los corregimientos y Goviernos Realengos y de las Órdenes*, (2 volumes), Vol. 2, Madrid: Luis Sánchez, 1597, p. 162 and Juan de Castilla y de Aguayo, *El Perfecto Regidor*, Salamanca: Cornelio Bonardo, 1586, fol. 75 v.—fol. 86 v. Ronald W. Truman, *Spanish Treatises on Government, Society and Religion in the Times of Philip II*, Leiden, Boston and Koln Leiden: Brill, 1999, pp. 117–184, uses these last books as two of the three ones he presents in his analysis of treatises composed in Spain.

16 Castillo de Bobadilla, *Política para corregidores*, p. 154.

17 Ibid., p. 250.

18 I. A. A. Thompson, "The Purchase of Nobility in Castile, 1552–1700," in *War and Society in Habsburg Spain*, Aldershot (UK): Variorum, pp. 313–378, p. 313. For an in-depth discussion on the financial troubles that plagued Spain in the sixteenth century, see: Carla Rahn Phillips, "Time and Duration: A Model for the Economy of Early Modern Spain," *The American Historical Review*, Vol. 92, No. 3 (June 1987), pp. 531–562 and I. A. A. Thompson, *War and Government in Habsburg Spain, 1560 - 1620*, London: Athlone Press, 1976, specially chapter III: "The Problem of Financing," pp. 67 and ff.

traditionally at the top of the social hierarchy.[19] The buyer, as a *regidor*, would be able to take part in and possibly influence local politics, as well as increase his income by participating in the schemes organized by the city council. And even if the officeholder preferred not to attend the council meetings, the title of *regidor* made him a member of the local ruling oligarchy, which conferred a fair degree of social status and prestige within the community.[20]

The number of offices available in the Toledo city council varied with time.[21] Their number increased during the second half of the sixteenth century: Carlos I added a total of fifteen new *regidurías* in Toledo; Felipe II added more, and in 1581 there were thirty-six *regidores*, twenty-five sitting on the nobles' bench and eleven among the citizens.[22] In addition, six officials had the right to sit and vote in the municipal council. Thus, a total of forty-two individuals comprised the city council, nearly double the number of members established by the Catholic monarchs, who preferred a smaller deliberating body in order to increase administrative efficiency.[23]

Early Attempts to Supply Water to Toledo

Each power that dominated Toledo took advantage of the knowledge accumulated by the previous rulers. Rome devastated the former society of the Iberian Peninsula, but contributed advanced technical skills. It undertook great engineering works that gained the admiration of the subject populations. The construction of a complex water supply system to solve the chronic dearth in the city of Toledo in the second century CE is a good example of that successful policy.[24]

19 Francisco José Aranda Pérez, "Los mercaderes de Toledo en el seiscientos, bases económicas y status sociopolítico," *Investigaciones históricas: Época moderna y contemporánea*, No. 12, 1992, págs. 71–96, p. 92.

20 Martz, "Implementation of Pure Blood," p. 249. According to Enrique Soria Mesa, "Los estudios sobre las oligarquías municipales en la Castilla moderna. Un balance en claroscuro," *Manuscrits* Nº 18, (2000), pp. 185–197, p. 191, about 40% of the *regidores* were of *judeoconverso* [converts from Judaism] origin in the XVIth century.

21 See: Pedro de Alcocer, *Hystoria, o descripción de la Imperial cibdad de Toledo: con todas las cosas acontecidas en ella desde su principio, y fundación*, Toledo: Iuan Ferrer, 1554, Capítulo xliij, fol. cxxij.

22 Martz, "Implementation of Pure Blood," pp. 249–250.

23 Martz, *Poverty and Welfare*, p. 94.

24 Lidia Arenillas, Miguel Arenillas, Carmen Diaz-Guerra and José María Macías, "El abastecimiento de agua a Toledo en época Romana," in José María Macías and Cristina I.

The historical urban nucleus rose to almost one hundred meters above the Tajo River. The Roman engineers decided to take the water not from the Tajo itself, but from a small river, affluent of the left margin of the Tajo, at a location that lay at the appropriate height allowing the conveying of water by an aqueduct. A dam was built near the village of Mazarambroz, at a distance of about 40 km. from Toledo, creating a reservoir from which water was conveyed through an open aqueduct to the highest parts of Toledo, lying at about 170 m. below the height of the far-away reservoir. Along the aqueduct there were towers for the regulation of the water level.[25]

The Roman engineers had to solve the problem of crossing the Tajo gorge, which was too deep and wide for a bridge. This difficulty was overcome by building a siphon. The aqueduct water ran into a distribution tank from which a row of parallel lead pipes came out from the other side. They descended along one side of the gorge, crossed the bottom on a 'venter' [belly] bridge and climbed uphill to the receiving basin from which the water continued to its destination in the urban area (fig. 3.1). The siphon was designed at the optimal

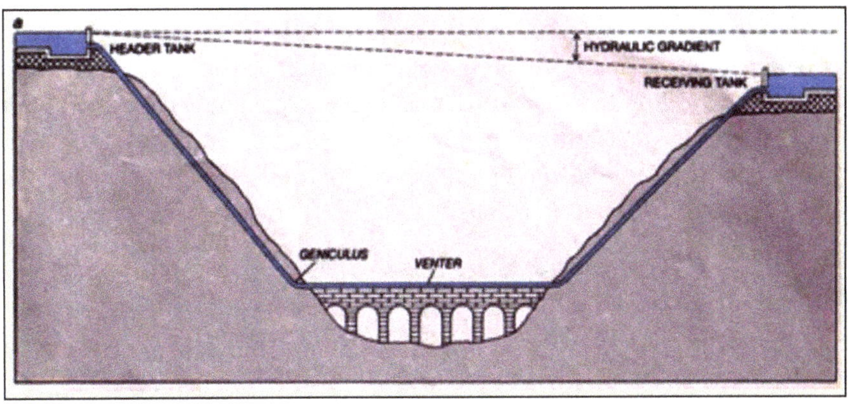

FIGURE 3.1 *Sketch of the 'venter' bridge over the Tajo.*[26]

Segura, (eds), *Historia del abastecimiento y usos del agua en la ciudad de Toledo*, Madrid: Confederación Hidrográfica del Tajo 1999, pp. 35–48.

25 Arenillas, "El abastecimiento de agua, pp. 44–45. For a map of the course of the aqueduct, see: Fernando Aranda Alonso, *El Sistema Hidráulico Romano de Abastecimiento a Toledo*, Toledo: Diputación Provincial, IPIET, 1997, p. 127.

26 A. Trevor Hodge, "Engineering Works," in Örjan Wikander, (ed.), *Handbook of Ancient Water Technology*, Technology and Change in History, (ongoing series), Vol. 2, Leiden: Brill (2000), pp. 67–94, p. 79.

depth required to keep the pressure sufficiently low, in order not to endanger the lead and ceramic pipes available in that period.[27] This accomplishment was well-known in later periods—a sketch by Mariano di Iacopo (Taccola) from the beginning of the fifteenth century conceptually describes the Roman siphon.[28]

In the fifth century, Barbarian tribes began settling in the Iberian Peninsula, and among them the Visigoths. They preferred to establish their communities in the Central Iberian plateau, where Roman presence was less prominent than in the Ebro valley. They transformed Toledo into their capital city, and since the Visigoth kings were highly adapted to the Roman customs and practices, they generally kept alive institutions and utilized the infrastructures left over by the Romans. No records have been left or, at least, discovered, referring to the water supply to the city under their rule, and it is almost impossible to identify which changes and improvements the Visigoths introduced, if at all.

The Muslims took over Toledo in 711 CE, under the leadership of Táriq ibn Ziyad. At an undetermined date, Islamic engineers built a device to raise water from the river, though it is not clear if it reached the upper area of the city.[29] The only extant textual evidence was left by Al-Idrisi in the middle of the twelfth century. According to his short testimony, it consisted of a huge water wheel, or *noria*, located close to the Alcántara Bridge. Although the description is not precise, it is clear that the dimensions were gigantic, as even 90 *codos* (*codos* instead of *estadales* in the almost certainly mistaken translation of the original text), are about 40 meters.[30] The text is confusing, as it means that

27 Elena Sánchez López and Javier Martínez Jiménez, *Los acueductos de Hispania: construcción y abandono*, Madrid : Fundación Juanelo Turriano, 2016, p. 124.

28 Mariano di Iacopo (detto il Taccola), "Bellows operated by a man; sketch of the siphon for the aqueduct of Toledo," in *Codex Latinus Monacensis 197 II*, Bayerische Staatsbibliothek München, c. 72v, in Paolo Galluzzi, "Art and Artifice in the Depiction of Renaissance Machines," in Wolfgang Lefèvre, Jiirgen Renn and Urs Schoepflinm, (eds), *The Power of Images in Early Modern Science*, Basel, Boston and Berlin: Birkhäuser, 2003, 1989, pp. 47–64, p. 50.

29 For a highly speculative discussiones on the dating and the alternative configurations of the device, see: Ana Isabel Cerrada, Juan Carlos de Miguel and Cristina Segura, "Un Pasado Oscuro. Visigodos y Musulmanes," in Macías, *Historia del abastecimiento*, pp. 49–90.

30 It is almost certain that the correct translation from Arabic is the length unit *codo*. This is supported by the translation to the French by Amédée Jaubert, *Geographie d'Edrisi*, Paris: La Societé de Geographie, 1840, cited by Luis Moreno Nieto and Angel Moreno Santiago,

at least part of the Roman aqueduct was still standing in the twelfth century, while it seems to have been destroyed centuries before.[31] Thomas Glick adopts Al-Idrisi's version of water flowing in an aqueduct as the driving force of the great *noria*. Glick describes the *noria* of Islamic Toledo as a compartmented wheel with a rim of pots, moved by the force of the water alone, which lifted water from the aqueduct. "This type of wheel," he writes, "although mechanically simple, was typically very large in size and was associated with public works such as dams, royal gardens, or major irrigation installations ..."[32]

The Spanish scholar Basilio Pavón Maldonado has interpreted Al-Idrisi's account in a sketch (shown in Fig. 3.2). Based upon the dimensions of the Alcántara Bridge, built in Roman times, Maldonado estimates that water was lifted by the *noria* to a height of about 30 meters and subsequently raised to the

FIGURE 3.2 *Hypothetical reconstruction of the* Noria *of Toledo.*[33]

Juanelo y su Artificio, Antologia, Toledo: d. b. ediciones, 2006, p. 56. Even accepting *codo* as the unit of measurement, from the text it is impossible to define the exact dimensions. One *codo* could signify either 41 cm. (*codo ma'muni*) or 58 cm. (*codo rasasi*), and consequently the *noria* diameter could have been between approx. 37 m. and 52 m.

31 For an in-depth discussion of this issue see: Basilio Pavon Maldonado, "En torno al acueducto y la rueda hidráulica árabe de Toledo según Idrisi," *Al-Andalus Magreb: Estudios árabes e islámicos*, No. 5, (1997), pp. 273–294.

32 Thomas F. Glick, *Islamic and Christian Spain in the early Middle Ages*, Princeton: Princeton University Press, 1979, p. 236.

33 Basilio Pavón Maldonado, *Tratado de arquitectura hispanomusulmana, I. Agua (aljibes, puentes, qanats, acueductos, jardines, ruedas hidráulicas, baños, corachas)*, (4 volumes), Vol. 1, Madrid: Consejo Superior de Investigaciones Cientificas, 1990, fig. 300, p. 280.

Alcázar level by "an artifice of unknown characteristics."[34] Testimonies by Arab writers and European travelers refer to numerous *norias* in the lower parts of the city, used to irrigate orchards and gardens through a network of crisscrossing ditches.[35] But it is particularly the unique huge *noria*, which supplied water to the higher part of the city, which has become a myth in the history of water supply to the city.

Toledo was conquered in 1085 by Alfonso VI, who incorporated to his kingdom a large part of the territory of the Toledo *Taifa* (an independent Muslim-ruled principality) territory. Until after the end of the reign of Fernando and Isabel, the Catholic Kings, three complementing methods were used to solve the water supply problem: *azacanes*,[36] who were water carriers who transported water on the back of mules from the river to the highest part of the city; *aljibes*, or underground cisterns in which rain water collected from slanted tiled roofs was stored[37] and *aguadores*, who brought, with the help of beasts of burden, the clearest drinking water from springs and wells in the other side of the Tajo. The spring waters obviously fetched the highest price, which only the rich could afford.[38] Water from the river was drawn from holes made by the *azacanes* in its sandy banks (fig. 3.3), thus allowing for some type of filtration to take place, a method still encountered in nineteenth-century Toledo.[39]

---

34 Pavon Maldonado, Basilio, "En torno al acueducto," p. 280.

35 Leopoldo Torres Balbas, "La Albolafia de Córdoba y la gran noria toledana," *Al-Andalus*, No. 7, 1942, pp. 465–466; Micer Andrés Navajero, "Viaje por España," in Antonio María Fabié, (ed. and translator) *Viajes por España de Jorge de Einghen: del Barón León de Rosmithal de Blatna, de Francisco Guicciardini y de Andrés Navajero, con una introducción*, Madrid: Librería de los Bibliófilos Fernando Fe, 1879, pp. 231–352, pp. 253–254.

36 Sebastian de Covarrubias Orozco, *Tesoro de la Lengua Castellana o Española*, original written in 1611, edited by Felipe C. R. Maldonado, revised by Manuel Camarero, Madrid: Editorial Castalia, 1995, p. 145. Also quoted in: Miguel de Cervantes Saavedra, *El Ingenioso Hidalgo Don Quijote de la Mancha*, edited by Florencio Sevilla Arroyo and Antonio Rey Hazas, Madrid: Centro de Estudios Cervantinos, 1994, footnote 46, p. 205. Many of the *azacanes* were 'French'—see: "Discursos leídos ante la Real Academia de la Historia en la recepción pública al Ilmo. Señor D. Jerónimo López Ayala y Álvarez de Toledo el día 23 de Junio de 1901," in Jerónimo López Ayala y Álvarez de Toledo, *Toledo en el siglo XVI, después del vencimiento de las comunidades*, Madrid: Hernández, 1901, p. 150.

37 Toledo has two raining seasons, one in spring and the other one in autumn. Summers are hot and dry, winters cold and dry. For an in-depth account of the different architectural aspects of the *aljibes*, see: Pavón Maldonado, *Tratado de arquitectura*, pp. 13–90.

38 Ana Isabel Cerrada, Juan Carlos de Miguel y Cristina Segura, "De la conquista Cristiana a la modernidad," in Macías, *Historia del abastecimiento*, pp. 70–90, p. 74.

39 Ana Isabel Cerrada, José María Macías, Juan Carlos de Miguel and Cristina Segura, "Renacimiento e ilustración, el abastecimiento industrial," in Macías, *Historia del abastecimiento*, pp. 93–114, p. 103.

FIGURE 3.3
Azacán, *nineteenth century.*[40]

Part of the population obtained water in meager quantities from wells, mostly contaminated and salty, situated outside the town in *La Granja, La Cruz and Caño Quebrado.*[41] Water prices varied seasonally, increasing significantly in summer,[42] and abuses were constantly reported. The city therefore regulated the volume of the unit to be used when selling water to be 5 1/4 *azumbres*,[43] measured in marked containers.[44]

40 Photograph obtained, during a personal visit, from the Centro de Interpretación del Toledo Histórico—Centro Cultural San Marcos, in the San Marcos Church, nowadays Centro Cultural San Marcos, Toledo, in 2006.

41 Luis Hurtado de Toledo, "Memorial de algunas cosas memorables que tiene la imperial ciudad de Toledo dirigido á la CRM del rey D. Phelipe de Austria, monarca de las Españas," in Carmelo Viñas y Ramón Paz, (eds), *Relaciones de los Pueblos de España Ordenadas por Felipe II, Reino de Toledo*, tercera parte, Madrid: Instituto Balmes de Sociología, Instituto Juan Sebastián Elcano de Geografía, Consejo Superior de Investigaciones Científicas, 1963, p. 504.

42 José Cristóbal Sánchez Mayendia, "El Artificio de Juanelo en la Literatura Española," *Cuadernos Hispanoamericanos* 103, (1948), pp. 73–92, p. 74.

43 One azumbre is equivalent to 2.05 liters. See: Yolanda Congosto, *Aportación a la historia lingüística de las hablas andaluzas (siglo XVII): Descripción de una sincronía, Sevilla:* Universidad de Sevilla, 2002, p. 183.

44 Julián Montemayor, *Tolède entre fortune et déclin (1530–1640)*, Limoges: Presses Univ. Limoges, 1996, p. 58.

Water Supply to Toledo in the Sixteenth Century[45]

Toledo reached the zenith of its urban growth under Carlos I, who also built the monumental Bisagra Gate and had the old medieval *Alcázar* transformed into a majestic Renaissance palace.

When Felipe II decided to establish the Court in Madrid in 1561, the city extended through an area of approximately 80 hectares and a population of about 20,000 people, and was thought to be one of the most healthful cities in Castile, with drinking water sources much better than those in Toledo.[46] Many historians and writers reported so in their books, describing the city in the late sixteenth century as having a temperate climate, mild winds and low clouds, with plentiful provided water to its fertile land, stressing its beautiful natural surroundings.[47] These literary efforts to find justification to the decision to make Madrid the seat of the court try to convey an idyllic reality, despite the many inconveniences and demographic, social, urban, sanitary and public order problems for the old medieval village.

Toledo, although it entered into a period of urban decline, remained a monumental city and maintained its prominent position. Nevertheless, in the middle of the sixteenth century Toledo still lacked a public system of water supply, and as it had been for hundreds of years, water shortage was constant. When compared to Granada, which was annexed to Castile only in 1492, the water supply and sewerage facilities of Toledo, which by then had already been under Christian domination for 450 years, were significantly backward or just plainly inexistent.

The interest in a system of public water supply in Toledo was not only related to the needs of a large population, but was also considered necessary to satisfy the king's will to irrigate the gardens of his palace—the *Alcázar*. The biggest difficulty was the height difference—more than ninety meters—between the river and the foot of the northeast tower of the *Alcázar*, in Zodocover, Toledo's central square. This place was considered to be an appropriate location for the construction of the water reservoirs for the distribution to the city core—

45 This discussion is largely based on the first chapter of Luis de la Escosura y Morrogh, *El Artificio de Juanelo y el Puente de Julio César*, Memoria publicada por la Real Academia de Ciencias Exactas, Físicas y Naturales de Madrid, Madrid: Imprenta de Don Luis Aguado, 1888.

46 James Casey, *Early Modern Spain: A Social History*, London and New York: Routledge, 2002, p. 153.

47 Instituto de Estudios Madrileños, *Madrid en el siglo XVI*, Madrid: Instituto de Estudios Madrileños (Consejo Superior de Investigaciones Científicas), 1962, p. 63.

the few blocks between the square and the cathedral that were the center of the economic life of the city, where every Tuesday an agricultural market took place.[48] An ironic popular saying those days reflected the belief that the realization of such a project was impossible: "… como subir el agua a Zocodover" "… like lifting water [from the river] to Zocodover".[49]

Although the Roman aqueduct in Segovia had been repaired in 1481,[50] curiously no records exist about the intention to examine the possibility of repairing the Roman aqueduct to Toledo as well. However, a document dated April 18, 1485, describes the first known attempt to raise water to the city by mechanical means. It is a letter sent from Cordoba by the Catholic kings to the "*Corregidor, alcaldes, alguacil, regidores, jurados, caballeros, escuderos* [squires], *oficiales* [public officials] and good men from the most noble and loyal city of Toledo," about a plan to construct a device to supply water to the Zodocover square, just besides the *Alcázar*.[51] This document has great significance, as it shows that already by the end of the fifteenth century Spanish rulers were keen in adopting unproven technologies for a most ambitious undertaking. The only historian who has referred to the role of the Catholic Kings in this enterprise is Francisco de Pisa, who in 1605 noted that they procured a good method or artifice to raise water from the Tajo to the highest point in the city.[52]

Sometime after his inheritance of the Spanish Crown in 1516, Carlos I received a plan to rebuild the great water wheel of the times of the Islamic domination, as recorded by the Venetian Ambassador to the Spanish Court, Andrés Navagero, who travelled throughout Spain in 1524.[53] But no action was taken, and it appears that in the end, reproducing the huge water wheel was not considered a viable option for addressing the lack of water in the city. Nevertheless, different attempts to find a solution were made, based on the

48 Lorente Toledo, "Carlos I y su relación," p. 220.

49 Bartolomé Villalba y Estaña, *El pelegrino curioso y grandezas de España*, originally published in 1577, edited by Pascual de Gayángos, Madrid: Imprenta de M. Ginesta, 1886, p. 194.

50 Bertrand Gille, *Histoire des Techniques*, (2 volumes), Vol. 1, translated by P. Southgate and T. Williamson, New York: Gordon and Breach, 1986, p. 574.

51 AMT, *Archivo Secreto*, cajón 6, leg. 2, No. 4, pieza 1, extracted from María Isabel del Val Valdivieso, "Un exponente del buen gobierno urbano: el abastecimiento de agua en la Castilla medieval," in María Isabel del Val Valdivieso y Olatz Villanueva Zubizarreta, (eds), *Musulmanes y cristianos frente al agua en las ciudades medievales*, Santander (Spain): Universidad de Castilla-La Mancha, 2008, pp. 359–380, pp. 365–378.

52 De Pisa, *Descripción de la imperial ciudad*, p. 23.

53 Navagero, "Viaje por España," p. 254.

technology available at the time.[54] In 1526 Enrique III of Nassau, who had become Marquis de Cenete by marrying the Marchioness of Cenete in 1524 and was grand chamberlain to the King, contracted German technicians to build a device, with an operating principle similar to the one used in that period for water extraction from flooded mines. The importance and power of this personality can be appreciated by the fact that he had been nominated one of two godfathers of Felipe II.[55] The planned undertaking was recorded in 1526 in the *Libro de Recepciones* [Register of Arrivals] of the *Monasterio de la Concepción Francisca*.[56] Waterwheels in the Tajo drove the pump pistons, which moved in cast iron cylinders. Two major difficulties impaired the workings of the device: floods, which damaged the tower where the main driving wheel was located,[57] and, especially, the excessive pressure of the water, which caused the pipes to burst.[58] The water was supposed to be raised at least 100 meters, more than twice the water raising height that had been attempted until then in other European cities, such as Augsburg.[59]

In 1550 several patents were granted to Antón Ruiz Canalejo, from Córdoba, for machines curiously described as follows: one 'capable of moving water peacefully, but able of overcoming summits of mountains,' another 'full of prudent tricks', yet another 'encouraging still waters' and finally a 'scale of boxes capable of raising water to unlimited heights'. The last description may suggest that it was a precursor of Juanelo's *Artificio*, which will be discussed later. The

54 For a description of some of those failed attempts, rich in details about budget and taxation, but with no reference to the relevant technical methods used, see: Almudena Sánchez Palencia, "En torno a Juanelo Turriano," *Anales Toledanos XIX*, Toledo, (1984), pp. 70–91.

55 Aurelio Espinosa, *The empire of the cities: Emperor Charles V, the comunero revolt, and the transformation of the Spanish system*, Leiden (The Netherlands): Brill, 2009, p. 201.

56 Quoted in Sisto Ramón Parro, *Toledo En La Mano, Ó: Descripción Histórico-Artística De La Magnífica Catedral Y De Los Demas Célebres Monumentos Y Cosas Notables Que Encierra Esta Famosa Ciudad*, (2 volumes), Volume 2, Toledo: Imprenta y Librería de Severiano López Fando, 1857, p. 659.

57 Torres Balbas, "La Albolafia de Córdoba, p. 465.

58 Antonio Ponz, *Viage de España*, (1776), (18 volumes), Vol. 1, Madrid: Viuda de Ibarra, Hijos y Compañía, 1787, p. 156 and De Pisa, *Descripción de la imperial ciudad*, p. 24.

59 Estevan de Garibay y Camalloa, *Compendio Historial de las Chronicas y Vniversal Historia de Todos los Reynos de España, Donde se Poden en Suma los Condes, Señores de Aragon, con los Reyes del Mesmo Reyno: y Condes de Barcelona, y Reyes de Napóles y Sicilia*, (4 volumes), Vol. 4, Barcelona: Sebastian de Cormellas, 1628, p. 259.

patents were granted for 15 years, and soon extended for a further 25 years.[60] In the Municipal Archives of Toledo, a document from the then Prince Felipe is kept, dated 1553, authorizing a certain Sebastián Navarro to raise water from the Tajo.[61] The same Felipe, shortly after his accession to the throne, granted in Brussels in November 1556 a patent for ten years to Fernando de Leiva for a device to raise water for irrigation and other uses adapted from existing pumps that had been used for drainage of moats. These machines were to be driven either by animal force or by the flow of a river.[62] Nothing is known about implementation of any of these devices.

The situation reached a critical stage in the early 1560s, when Toledo was the most populated city in Castile—according to some estimates, numbering about 80,000 inhabitants.[63] The authorities of the city evidently believed that water provision by machines capable of lifting water from the abundant flow of the Tajo to the highest part of the city was the preferred choice. A conceptual drawing, dated 1561, which may be related to these discussions, illustrates this idea (fig. 3.4).

60 Nicolás García Tapia, "Descubrimientos e invención técnica. La actividad de los inventores españoles," in Ana María Carabias Torres, (ed.), *Las relaciones entre Portugal y Castilla en la época de los descubrimientos y la expansión colonial* (Papers submitted to the Congreso Hispano-Portugués that took place in Salamanca, 1992), Salamanca, 1994, pp. 165–180, p.173–174.

61 AMT, *Archivo Secreto*, Cajón 6, leg. 2, Núm. 4, exhibited during the *Muestra Documental sobre Toledo durante el reinado de Carlos V*, Ayuntamiento de Toledo, 2000. For an account of the contribution of the AMT to historic research, see: Mariano García Ruiperez, "El Archivo Municipal de Toledo y la investigación histórica sobre la Edad Moderna," *Cuaderno de historia Moderna*, No. 22, (1999), pp. 201–227.

62 AGS, *Cámara de Castilla*, Libro de Cédulas No. 126, fol. 267, extracted from Nicolás García Tapia, *Patentes de Invención Españolas en el siglo de oro*, Madrid: Oficina Española de Patentes y Marcas, 1994, p. 40.

63 Jerónimo López de Ayala Álvarez de Toledo y del Hierro, Conde de Cedillo, "Discursos leídos ante la Real Academia de la Historia en la recepción pública al Ilmo. Señor D. Jerónimo López Ayala y Álvarez de Toledo el día 23 de Junio de 1901," in *Toledo en el siglo XVI, después del vencimiento de las comunidades*, Madrid: Imprenta de los hijos de M. G. Hernández, 1901, p. 54. Linda Martz, "Toledo y los Toledanos en 1561," *Volume 5 of Monografías* (*Instituto Provincial de Investigaciones y Estudios Toledanos*), Toledo, 1974, p. 13, presents an in-depth discussion on Toledo's population movement in the second half of the 16th century. For another thorough analysis of the issue see: Richard L. Kagan, "Contando vecinos. El censo toledano de 1569," *Studia Histórica. Historia moderna*, No. 12, 1994, pp. 115–136.

FIGURE 3.4
Conceptual device to raise water from the Tajo to the Alcázar (*1561*).[64]

In August 1561, with the Imperial capital already in Madrid, the city representatives, under the presidency of the powerful *corregidor* Don Gastón de Peralta, Marqués de Falces, Conde de Santiesteban de Larín (at the time the *alcalde mayor* was Iñigo de Tolosa),[65] met once again to discuss potential solutions for the water supply problem.[66] The conclusion reached in that meeting was that in order to prevent the bursting of the pipes, several stages would be needed in the water-raising device to be built. As a result of this decision a new commission was granted in 1562 to the Flemish engineers Juan de Coten and Jorge Ulrique,[67] who tried to use the waterwheels of a rented mill located on the

64 Anonymous drawing, AGS, *Mapas, Planos y Dibujos*, XXVII-3. España. Ministerio de Educación, Cultura y Deporte. Archivo General del Simancas. MPD,027,003.

65 Extracted from a list of the *corregidores y alcaldes de la ciudad de Toledo*, included in Francisco José Aranda Pérez, "Nobles, discretos varones que gobernáis a Toledo. Una guía prosopográfica de los componentes del poder municipal en Toledo durante la Edad Moderna (corregidores, dignidades y regidores)," in Francisco José Aranda, (ed.), *Poderes intermedios, poderes interpuestos. Sociedad y oligarquías en la España Moderna*, Cuenca: Ediciones de la Universidad de Castilla-La Mancha, 1999, pp. 227–310, p. 236.

66 Peralta was a scion of an aristocratic family, serving loyally the Emperor since being a young man. When Felipe II was crowned, he served him as well, and the king named him *Corregidor* of Toledo when the city became the residence of the court and later (in 1566) appointed him Viceroy of Mexico. See Ignacio José Rubio Mañé, *El Virreinato, Orígenes y jurisdicciones y dinámica social de los virreyes*, (4 volumes), Vol. 1, Mexico: Universidad Nacional Autonoma de Mexico, 1983, p. 230.

67 The Flemish technicians are mentioned in a letter: AGS, *C. & S. R.*, leg. 271, fol. 79.

Tajo, the 'Barranchuelo,' to work elevating pumps, but they too failed in their attempt.[68] Subsequently, the French clockmaker and engineer Louis de Foix[69] was granted in 1563 a patent for a device to collect and lift water to high places through the use of "subjection" (sic.), a term that is open for interpretation.[70] In any case, by order of Felipe II, De Foix was paid in September 1564 the sum of 300 *ducados*, a considerable amount of money in those days.[71] According to the text of the *cédula* ordering this payment, De Foix's commission was to build models and *ingenios* related to the raising of water to Toledo. No additional details are provided, but if a new invention was in fact proposed as a model, no attempt was made to implement it.

It was soon after that Giovanni (Juanelo) Turriano, who had been brought from Italy by the emperor Carlos I, was elected as the designer and ultimately became the entrepreneur who built the *Artificio* or *Ingenio* that bears his name.[72]

Juanelo Turriano

Turriano was born about 1500 in a village near the city of Cremona, in Lombardy, a region that came under the reign of the Spanish monarchy after the imperial victory in Pavía, in 1525. His given name turned in Spain to Juanelo,[73] the nickname under which he became famous. He was an individual formed outside the academy, whose knowledge had to do more with practical learning than with theory and scholarly activities. Most likely, according to the manner of the time, he learned in his father's workshop to construct and repair mechanical instruments. After this first domestic formation, he became an apprentice

68 Instructions by Felipe II to pay for the rent of the Barranchuelo Mills, AGP, *C. R.*, tomo 3, fol. 210 v.

69 Sometimes spelled De Fois or De Fox. For De Foix's biography see: Claude Grenet-Delisle, *Louis de Foix: horloger, ingénieur, architecte de quatre rois*, Bordeaux: Fédération Historique du Sud-Ouest, 1998.

70 AGS, *Cámara de Castilla*, Libro de Cédulas No. 131, fol. 38, extracted from García Tapia, *Patentes*, p. 45.

71 Instructions by Felipe II to pay Luis de Fois 330 *ducados*. AGS, *C. R.*, tomo 2, fols. 407 v., 408.

72 In Covarrubias's *Tesoro*, p. 504, the term '*Ingenio*' is explained by referring to the example of Juanelo's *Artificio*. The modern meaning is: an ingenious complex device or mechanical artifice. Sometimes written *yngenio*, the modern Spanish word is spelled *ingenio*, which is used in the text, except in quotes from the period.

73 His name was spelled in different forms, such as Joanelo, Gianello, etc. In the text Juanelo is used, except in quotations, which are presented as written in the original.

FIGURE 3.5
Portrait of Juanelo Turriano.[74]

in a jewelry workshop in Cremona, and later passed to Milan, where he developed into a masterful clockmaker and mechanic.[75]

The Emperor Charles V (Carlos I of Spain) was crowned in Bologna in 1530, where, through fortuitous circumstances, he became acquainted with Juanelo. Ferrante Gonzaga (1507–1557), at the time commander-in-chief of the Imperial

74 Anonymus, "Portrait of Juanelo Turriano," oil on canvas, (68.5×54.3 cm), Civic Museum 'Ala Ponzone', Cremona (Italy). Angel del Campo y Francés, *Juanelo Turriano, semblanza iconográfica*, Fundación Juanelo Turriano, Madrid 1997, contains an extensive research on Juanelo Turriano's physical appearance.

75 Juanelo's biographical data is based on José Antonio García-Diego, *Juanelo Turriano, Charles V's Clockmaker. The man and his legend*, translated into English by Charles David Ley, Madrid: Editorial Castalia, 1986, chapter I; Bernardo Revuelta Pol y Dolores Romero Muñoz, "Juanelo Turriano. relojero e ingeniero cremonés" in *Realismo y espiritualidad. Campi, Anguissola, Caravaggio y otros artistas cremonenses y españoles en los siglos XVI–XVIII*, Valencia: Ajuntament Alanquas, 2007, pp. 73–81 and Marino Viganò, "'Parente et alieuo del già messer Janello'—Primeras notas sobre Bernardo y Leonardo Turriano," in Alicia Cámara, Rafael Moreira y Marino Viganò, *Leonardo Turriano ingeniero del rey*, Madrid: Fundación Juanelo Turriano, 2010, pp. 203–227. In this last paper a profusion of data extracted from Italian sources is presented for the first time. For an account of the social context and relationships in which Juanelo's career developed, see Cristiano Zanetti, *Juanelo Turriano. De Cremona a la Corte: formación y red social de un ingeniero del Renacimiento*, Madrid: Fundación Juanelo Turriano, 2015. The extent of his fame while alive is described in Daniel Crespo Delgado, "Juanelo Turriano: ingenio y fama", en Alicia Cámara y Bernardo Revuelta (coord.), *Ingenieros del Renacimiento*, Madrid, Fundación Juanelo Turriano, 2014, pp. 9–25.

army in Italy and later governor of Milan, had the intention to present as a gift to the Emperor the 'Astrarium,' the famous astronomical clock constructed in 1364 by Giovanni Dondi (1330–1388), which was considered a masterpiece of medieval technology.[76] It included the celestial motions of sun, moon, and five planets, a perpetual calendar of all religious feasts, both fixed and movable,[77] and even took account of the apparently irregular orbits of the Moon and Mercury (as predicted by the Ptolemaic system) and the observed irregularities in the orbit of Venus.[78] Not surprisingly, no clockmaker was able to put it back in operation. After examining it, and verifying its poor state of conservation, eroded by corrosion, Turriano reached the conclusion that though he could not repair it, he was capable of building an improved version. He dedicated twenty years to design a wonderful astronomical clock for the emperor, who was known to be captivated by clocks. He spent an additional three years building it, developing mechanical tools that enabled him to manufacture wheels and gears speedily and accurately. The clock, which made him famous, "… showed hours, minutes, solar and moon hours, planetary movement and the signs of the Zodiac. It was built with more than 1500 parts, three springs that produced the movement and eight planetary spheres."[79]

In 1554, the king and emperor brought Juanelo from Milan to Brussels, where he remained at his service up to the monarch's death in 1558. Following Carlos I's renouncement of the throne, Juanelo was put in charge of the clocks and planetariums collected in Yuste, in his position as clockmaker at the service of Felipe II.[80] The new King did not inherit from his father the passion for precision machinery,[81] but Juanelo was equally able to advise him on the construction of great engineering works, hydraulic ones in particular. In fact, he had already taken part in works of such nature in Italy, building two hydraulic machines required by Pope Pius V in 1563 and 1567.[82] He was also

76 For an in-depth discussion about this clock, see: Silvio A. Bedini and Francis Maddison, "Mechanical Universe. The Astrarium of Giovanni de' Dondi," *Transactions of the American Philosophical Society, New Series*, vol. lvi, part 5, (October 1966), pp. 1–69.

77 Lynn Townsend White, "Medical Astrologers an Later Medieval Technology," in *Medieval Religion and Technology: Collected Essays*, Los Angeles: University of California Press, 1978, pp. 293–315, p. 303.

78 Herbert Alan Lloyd, *Giovanni de' Dondi's horological masterpiece, 1364*, Oxted (Surrey): [s.n.], 1956, extracted from White, *Medieval Technology*, p. 126.

79 Bautista, "Juanelo," p. 100.

80 AGP, *C. R.*, tomo 2, fol. 224.

81 García-Diego, *Juanelo Turriano, Charles*, p. 117.

82 Bernardo Revuelta Pol y Dolores Romero Muñoz, "Juanelo Turriano, relojero e ingeniero cremonés" in "Realismo y espiritualidad: Campi, Anguissola, Caravaggio y otros artistas cremoneses y españoles en los siglos XVI–XVIII," catalogue of the exhibition that took

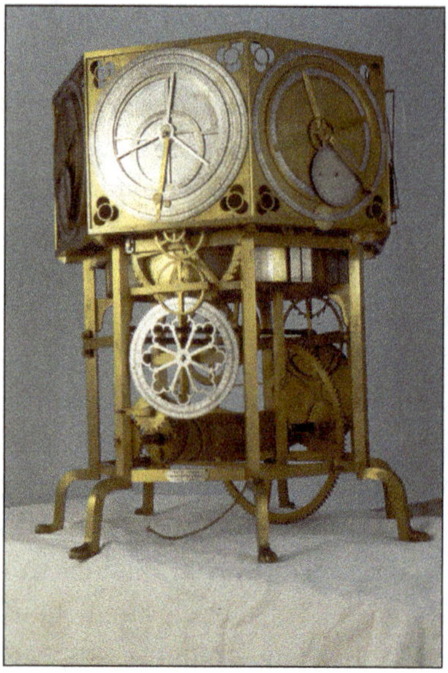

FIGURE 3.6
De Dondi's Astrarium (*Reprod.*).[83]

awarded a patent by the same Pope for two machines to collect water from fountains, rivers and lakes.[84] He later served as an adviser in several projects in Spain, such as the channel in the Madrid neighboring village of Colmenar and the dam built on the Tibi marsh, and designed devices to drain flooded mines in Andalucía.[85] From his arrival to Spain in 1558 until 1565 he resided and worked in Madrid, playing an active role in the manufacturing of the bells for the Escorial Monastery.[86] In later years he served as an adviser,

place in the Castillo de Alaquàs (March–July, 2007), Valencia: Ajuntament d'Alaquàs, 2007, pp. 73–81, p. 78. We learn from a letter from Rome, sent by the Comendador Mayor de Alcántara, Luis Dávila y Zúniga, to Felipe II, dated 12 and 13 May, 1563, that the Pope requested Felipe to 'lend' him Juanelo for two years, for certain [sic] work he wanted Juanelo to perform. Luis Cervera Vera, *Documentos biográficos de Juanelo Turriano*, Madrid: Fundación Juanelo Turriano, 1996, p. 23.

83 Luigi Pippa (costruttore), "Reconstruction of the astrarium," Museo della Scienza e della Tecnologia, Milan (Built 1961–1963).

84 Felice Zanoni, "Un brevetto pontificio d'invenzioni del '500. Janello Torriano e un documento dell'Archivo segreto vaticano," *Bolletino storico cremonese*, 1940, pp. 145–154, pp. 147–148.

85 Enrique López Castellón, *Historia de Castilla y León*, (10 volumes), Vol. 5, Valladolid: Reno, 1984, p. 368.

86 An interesting document tentatively dated 1578 demonstrates Juanelo's know-how on the manufacturing of bells and his musical knowledge. AGS, *C. & S. R*, leg. 261, fol. 170,

simultaneously with scholars from the universities of Alcala and Salamanca, to the commission charged by Pope Gregory XII to reform the calendar in order to correct the misalignment between the astronomical time and the Christian calendar.[87] Part of the Spanish contribution to this matter is a manuscript signed by Juanelo in 1579 and addressed to Felipe II, who transferred the document to the Pope.[88] In addition, several myths have Juanelo as protagonist, such as the building of a wooden automat (*hombre de palo*), almost two meters high, which walked gracefully from Juanelo's house to the Archbishop's, to ask for food for Juanelo, and returned to his house carrying it.[89]

The *Artificio*

Although hard evidence is not available, everything indicates that both the King and the Toledo authorities were simultaneously engaging more than one engineer, letting them compete to find the best solution to the water supply needs of the city. Thus, while Louis de Foix was receiving his own patent and building models for a water-raising device, at an unknown date, most probably earlier than or in the first half of 1562, Juanelo submitted a proposal to the King and the city to build a machine that would lift water from the river up to the *Alcázar*.[90] The issue of what contribution the same De Foix had, if at all, in the construction of the *Artificio*, has become a matter of controversy. According to

extracted from José Antonio García-Diego, "Cinco documentos relativos a Juanelo Turriano," *Toletum, Boletín de la Real Academia de Bellas Artes y Ciencias Históricas de Toledo*, No, 8, (1977), p. 264. García-Diego was the first historian who published this document.

87 For a detailed account of the work performed by Juanelo in this respect, see: Ángel Fernández Collado, "Juanelo Turriano y la aportación Española a la Reforma del Calendario Gregoriano," *Toletum* 23, (1990), pp. 151–159.

88 Juanelo Turriano, *Breve discurso a S.M. el Rey Católico en torno a la reducción del año y reforma del calendario*, edited by José Antonio García-Diego y J. M. González Aboin, Madrid: Fundación Juanelo Turriano, 1990.

89 José Amador de los Rios, *Toledo Pintoresca: Ó Descripción de sus Mas Célebres Monumentos*, Madrid: Imprenta y Libreria de Don Ignacio Boix, 1845, p. 202.

90 Supporting the estimation for the date of his proposal is a *Cédula* granted on July 16th, 1562, increasing his anual assignation from two hundred to four hundred *ducados*. Juanelo was required to reside in the Court, and "... make clocks and other things that his profession capacitated him ..."—AGP, *C. R.*, t. 2, fol. 224. Another document extended by the King on August 25, 1563, reinforces the argument—it allows Juanelo to be either in Madrid or Toledo to "... do certain things we have ordered him to do ...," AGP, *C. R.*, t. 2, fol. 378 v.

Eugenio Llaguno, who wrote in the early decades of the nineteenth century, the French, disregarding any Italian contribution, attributed the *Artificio* design to De Foix, on the basis of a letter by Christophe de Thou (1508–1582), President of the Parlement de Paris, in which De Foix is referred to as the architect of the Escorial and the inventor of the *Artificio*.[91] This contention was refuted almost immediately by Spanish writers, dismissing it as ridiculous.[92] Compromising, Llaguno suggested the hypothesis that the French position could have derived from the possibility that De Foix was a disciple of Juanelo and in that capacity could have made the models that demonstrated the feasibility of the Italian's concept.[93] Finally, a French historian claimed in 1864 that the original letter by Christophe de Thou had been misunderstood, and that it referred to only for the models De Foix had made, and not to the full scale of the *Artificio*, which he acknowledged was built by Turriano.[94]

It is, however, highly unlikely that the two projects were related. Should Juanelo's endeavors in the water-raising scheme had been public knowledge, the wording in the documents granting Juanelo an increase in his annual salary[95] and his freedom to be in Toledo[96] would have been different, whereas the language used seems intentionally obscure, concealing the nature of 'the other things' he was expected to produce, which obviously included the *Artificio*.

Though we do not know with certainty what drove Juanelo to take this initiative, obviously he felt able to succeed where several other technicians had failed and must have hoped to earn a lot of money in the process.[97] Ambrosio de Morales, a close friend and admirer of Juanelo,[98] provides a clue to Turriano's motivation when referring to his happiness and pride whenever other people

---

91 Llaguno y Amirola, *Noticias de los arquitectos*, p. 76. The content of the letter is detailed in a book by Christophe de Thou's son: Jacques Auguste de Thou, *Jaobi Augusti Thuani Historiarum sui temporis tomus primus* [*-septimus*], (first published in 1604), Londini: excudi curavit Samuel Buckley, 1733, p. 634.

92 Pedro Salazar de Mendoza, *Origen de las Dignidades Seglares de Castilla y Leon: con relacion Sumanaria de los Reyes de estos Reynos*, Toledo: D. Rodriguez de Valdivielso, 1618, p. 167 and Diego de Colmenares, *Historia de la insigne ciudad de Segovia, y Conpendio de las Historias de Castilla. Segunda inpresion*, Madrid: Diego Diez, 1640, pp. 541–542.

93 Llaguno y Amirola, *Noticias de los arquitectos*, p. 77.

94 Philippe Tamizey de Larroque, *Louis de Foix et la Tour de Cordouan*, Bordeaux (France): Chaumas, 1864, p. 12.

95 AGP, *C. R.*, t. 2, fol. 224.

96 AGP, *C. R.*, t. 2, fol. 378 v.

97 García-Diego, *Juanelo Turriano*, p. 6.

98 Enrique Redel: *Ambrosio de Morales. Estudio biográfico*, Córdoba: Imprenta del Diario, 1909, p. 129.

showed amazement at his technical prowess.[99] The solution to the water-raising project, which Juanelo presented by means of a "model in very small size form,"[100] was probably based on an idea that had matured during the long period in which he was serving Carlos I, after hearing a high-ranking Spanish official, the Marqués del Vasto, Don Alonso de Ávalos (1502–1546), complain about the lack of water in the royal city, and the difficulties encountered in solving this problem.[101]

Finally, on 16 April 1565, at the office of the notary Gaspar Testa, Turriano signed a contract with the King's representative, Doctor de Lagasca, the City of Toledo, whose *corregidor* was then Fernando Carrillo de Mendoza, two *regidores* (Gurierre de Guevara and Francisco de Rojas) and the *jurado* Alonso Dávalos.[102] He committed himself to construct a machine that would raise water from the river in a continuous motion up to the esplanade of the palace, delivering with each movement 16 *libras* of water, equivalent to 12,400 liters daily.[103]

The record of previous failures in the quest for a proper water supply system for Toledo was such, that neither the city nor the king were ready to cover the costs of yet another attempt. Thus, it was agreed that the construction of the work would be done at the expense of Juanelo and that neither the king nor the city would be liable for any payment before the water ran through the device to their full satisfaction. From the water raised, a seventh part was to be retained for the service of the *Alcázar*, the other six parts were to be available for the city.[104] If the effort would not succeed, all costs would be "to Juanelo's damage and loss." If, however, the water raised by the device would meet the specifications as far as quantity and delivery location were concerned, the city would proceed to complete the relevant civil works, Turriano would receive from the king 8,000 *ducados* 15 days after the completion of the device, and from the city an annual payment of 1900 *ducados* for him and his heirs. This payment, however, obliged Juanelo and his successors, to provide the

99 Ambrosio de Morales, *Las antigüedades de las ciudades de España*, Madrid: Oficina de Benito Cano, 1792, (First edition: Alcalá de Henares: Casa de Iuan Iñiguez de Lequeríca, 1575), p. 334.

100 Ibid., p. 331.

101 Ibid.

102 AHP, Gaspar Testa, *Protocolo* (notarial protocol) 262, fols. 248–252 v.

103 One *libra* is equivalent to 460 grams, see: German Lafont Mateo, *Pampliega, Torrepadierne y Santiuste. Mil años de Historia. Siglos VII al XVII*, Salamanca: German Lafont Mateo, 2010, p. 143. The equivalent daily volume of Juanelo's commitment was extracted from Reti, "El Artificio," p. 16.

104 AHP, Gaspar Testa, *Protocolo* (notarial protocol) 262, fol. 252 r.

maintenance and repair of the building and works between the river and the *Alcázar*. These were extraordinarily tough requirements.

When the contract was signed Juanelo was sixty-five years old and already had a glorious past. He was known, admired and sought by princes and kings who valued his service and advice. It is difficult to understand how a man in his condition, at an age when others seek peace and security, was willing to risk everything in an enterprise of such magnitude under the drastic conditions of such a contract. His unconditional faith in the word of the sovereign may, in part, justify this attitude, but most probably his motivation resulted from his desire to see implemented in practice an idea that was revolutionary for his times. The actual construction of the device has left very little documentary evidence, and the chronicles are also silent in this regard, which is in some way natural, as the building and operation would have been on Juanelo's exclusive account, without official administrative controls, which are precisely those that leave their mark in the archives.

The *Artificio* was built within a relatively short period of three years and was inaugurated on 23 February 1569.[105] Months later, on the 13th of May of the same year, the representatives of the crown and the city officially controlled the volume of water supplied by the machine, which turned out to be about fifty percent superior to the one stipulated in the contract (about 17,000 liters per day[106] instead of the 12,400 liters required by the contract). Juanelo, the people of Toledo and the King himself were seized by enthusiasm. An important document later drafted attests to Turriano's boundless optimism: it specifies the "... things needed, such as metal and other materials for the *ingenios* your Majesty will command to construct for the water for Toledo."[107] According to the same document, the intention was to add at least three more similar devices: the first, attached to the existing one, the second in San Juan de los Reyes, and the third, from the Pedro Lopez's mill to the Puerta de Bisagra. Only the first of these projects came to fruition, as we shall see presently.

The cost of the works was high and Juanelo, deeply indebted, claimed the payment of the first *Artificio*, which the City refused to make, notwithstanding Juanelo's full compliance with his contractual commitments. The City alleged that all the water raised by the first *Artificio* was turned towards the palace,

105 Reti, "El Artificio," p. 16.

106 Equivalent to 'one thousand six hundred pitchers, each one containing four *azumbres*' ['mil seiscientos cántaros de a cuatro azumbres de agua'], Llaguno y Amirola, *Noticias de los arquitectos*, p. 103.

107 AGS, *C & S. R*, leg. 271, fol. 229, (undated, but later than 1569).

without any of it left for urban needs.[108] Meanwhile, the *Artificio* raised water day and night, carefully maintained by and at the expense of Juanelo, who feared losing everything if he neglected the device, which required his constant attention due to the rapid wear of its elements. Juanelo wrote to the king asking for justice and expressed his bitterness and resentment against the rulers of the city. Felipe II tried to help Juanelo and forcefully commanded in December 1573 the *Corregidor* and the *Ayuntamiento* to fulfill their commitments towards Juanelo,[109] but this instruction only gave birth to additional meetings of the *Ayuntamiento*, and no payment was forthcoming to relieve the plight in which the king's faithful servant found himself. By 1575, Juanelo was financially ruined. He had neither been paid the sum due to him for the construction of the *Artificio* nor the expenses of six years of maintenance. Juanelo drafted another letter, begging to receive the monies due to him for six years, as he had to repay huge debts to his creditors. He stressed the injustice of driving the builder of such a magnificent work to an abject poverty.[110] Turriano's desperate economic situation worsened, and in view of such problems and his extreme necessity, Felipe II decided to take action, demanding an explanation from the city for its hostility. Luis Gaytán de Ayala was empowered by the city to negotiate with the King's ministers, and protracted discussions followed. Juanelo, who was in poor condition and in bad health, was represented by Antonio Fassole, his granddaughter's husband.

Did the city of Toledo and the King have good reasons not to pay what was promised to Juanelo? Was the device not working properly? Was the project uneconomical or its water dirty? The same year, in another letter to the King, Juanelo insisted on the proper functioning of his device.[111] The city delayed answers, giving neither specific reasons, nor facts or figures. It only declared that "Juanelo not only did not stand by his commitments, but the city did not enjoy any fruit or benefit."[112] This was not entirely false, but it was not Juanelo's fault. The water reached the Zodocover square, but was used exclusively by the palace.

108 AMT, *Libro de actas 12*, records of 24 and 29 October, 1574, letter from the City to the king dated 11 november, 1574 (AGS, *C. & S. R*, leg. 271, fol. 232) and letter from the City to the king dated 18 January, 1575 (AGS, *C. & S. R*, leg. 271, fol. 236).

109 AGP, *C. R.*, tomo 4, fol. 59.

110 AGS, *C. & S. R*, leg. 271, fol. 233, the document is not dated but it is estimated to be from the beginning of 1575.

111 AGS, *C. & S. R*, leg. 271, fol. 238.

112 AGS, *C. & S. R*, leg. 271, fol. 236.

The final solution adopted by the King for the city and agreed by Juanelo was to build another device.[113] Thus, on 20 March 1575, an agreement was reached, the main terms of which established that the water of the first machine would be completely reserved for the *Alcázar*, while a second machine was ordered, whose cost was estimated at 8–10,000 *ducados*, to be paid entirely by the King.[114] The water provided by the second machine would be left for Turriano and his heirs, who would have the exclusive right to sell it to the city. The new device was to be completed within five years and was not to be sold, in whole or in part, without permission of the King, who received the option to buy it at the highest price that another potential buyer would offer. The public land required for the construction was to be donated to Juanelo by the city, which promised to pay Juanelo a lump sum of 6,000 *ducados*. An extant protocol of an internal discussion between the King and his advisers, which took place two weeks before the signature of the agreement, shows that the King had added a crucial note, in his own handwriting: "if I will need the water, I would be able to take it."[115] The agreement was ratified by the King on 21 March 1575.[116]

Following the agreement, Turriano seemed to be satisfied, and work was renewed at a fast pace. However, conflicts resurfaced soon. Perusing the copious corpus of documents in the Archivo de Simancas and the Archivo General de Palacio, which include letters to and from the King to the *Ayuntamiento* authorities, letters from the officers in charge of the construction, etc., it is evident that the Toledo municipal authorities, headed and inspired by the *corregidor*, did what they could in order to cause Juanelo to fail in his undertaking, or at least impair his ability to conclude the *Ingenio* in time. Materials were denied or delayed, commitments to put guard on the construction site were not kept, false rumors were spread purposely about the water quality being compromised by the use of metal tubes, and above all, monies due to be paid to Juanelo according to the new agreement were not forthcoming. When Juanelo requested the city to buy an existing water mill, indispensable for putting in motion all the additional machinery, the City officials did not yield and the mill had to be ultimately acquired by the King. Turriano, who had attached

113 A detailed account of the complicated saga that finally led to the new contract is given in: Daniel Damler, "The Modern Wonder and its Enemies: Courtly Innovations in the Spanish Renaissance," in Claus Zittel, Gisela Engel, Romano Nanni and Nicole C. Karafyllis, (eds), *Philosophies of technology: Francis Bacon and his contemporaries*, 2 volumes, Vol. 1, Leiden: Brill, 2008, pp. 429–456, see pages 438–442.

114 AGP, *C. R.*, tomo 4, fol. 168.

115 AGS, *C & S. R*, leg. 271, fol. 240.

116 AGP, *C. R., tomo* (volume) 4, fol. 172.

the second machine to a wall separating both devices, built at his expense, demanded a refund for that addition, which was denied to him under the pretext that he had been already paid previously. The king intervened on Juanelo's behalf several times, severely admonishing the city, but even with all his power, the bureaucratic *Ayuntamiento* machine managed to raise obstacles at every possible instance, thus delaying progress in the construction.[117]

As if all that was enough, encouraged by Juanelo's technical success, several entrepreneurs proposed to make even better, bigger or less costly machines. Captain Baltasar Tristán commited himself in October 1577 to build a device to raise water from the Tajo, at a daily amount of 500 *botas de Nápoles*, containing 36 *cántaros de Toledo* per *bota* (equivalent to a total of 25,000 liters).[118] He planned to install fountains in the *Alcázar* yard, in the Zodocover square, in a site adjacent to the Iglesia Mayor, and wherever else it would be needed. As claimed, the device would be less complicated than Juanelo's, and would be cheaper by 6,000 *ducados*. Tristán also pretended to possess secrets "for the war," which he did not reveal.[119] In similar terms, a 1578 memorandum from Captain Cristóbal Suazo proposed to lift water to three fountains in the city and the *Alcázar*, at his own expense.[120] Another was the case of the attempts made in 1584 by the Italian Cardinal Deza in Rome, behind Juanelo's back, on behalf of the Spanish Cardinal Granvela, a minister of Felipe II, to engage the services of the Brescian engineer Valente Valenti.[121] It was said that Valenti had succeeded in raising water in Rome to a level of almost 800 *palmos*[122] to the

117 Damler, "The Modern Wonder," pp. 442–444.

118 Each *cántaro de Toledo* contained 34 pounds of water from the Tajo river, so 1 *bota de Nápoles* was equivalent to about (36x34x.46)=536 liters. Wyndham Beawes, *A Civil, Commercial, Political, And Literary History Of Spain And Portugal: In Two volumes*, (2 volumes), Vol. 1, London: Faulder, 1793, p. 294.

119 García Tapia, *Ingenieria y arquitectura*, p. 275.

120 Ibid.

121 The story is detailed in Raffaele Marchetti, *Sulle acque di Roma antiche e moderne*, 2 volumes, Vol. 1, Roma: Sinimberghi, 1886, extracted from: Benito Ruano, Eloy, "Un competidor de Juanelo Turriano (y otro proyecto de *Artificio* para Toledo)," *Separata of Studia Histórica et Philológica in honorem M. Batllori*, Roma: Instituto Español de Cultura, 1984, pp. 83–88. The same article includes a copy of the draft contract to be signed with Valenti. It is not clear if the Medici cardinal was Ferdinando, son of Cosimo I de Medici or Alessandro Ottaviano de Medici, who at the time were both members of the College of Cardinals.

122 Most probably it refers to *palmus*, a Roman unit of length measurement of the with of a hand, equivalent to 7.4 cm, resulting in a total height of 60 m. The Spanish *palmo* is equivalent to 21 cm, which would represent an unachievable height of 188 m.

orchard of Cardinal Medici. Valenti had also been part of the 'L'Acqua Felice' aqueduct project, whose scope was to bring water from Palestrina to the Villa Medici in Pincio.[123] Nevertheless, neither Valenti's nor the others' initiatives came to fruition, and there are no surviving records indicating whether the construction of any of these proposed machines was even started.[124]

Notwithstanding the hurdles, Juanelo's second *Artificio* was concluded satisfactorily in 1581, with a final cost to the crown of 11,000 *ducados*, but the city did not comply with the financial commitments it took upon itself in the agreement, and Juanelo, too old and sick, had no strength to keep fighting. Not surprisingly, the King appropriated to himself all the water of the second machine, as the first one, which had meanwhile deteriorated, provided less water than it had before. Juanelo intended to repair it at his expense, but even so the City did not receive a single drop of water, because everything was reserved for the palace.

Juanelo's letters to the King, dated 1584–1585, reveal a desperate state of mind and disastrous economic conditions. The King tried to help his old servant. But times were bad and no funds were available, even to the King. More time was wasted in bureaucratic formulas: before making a decision, the water flow was to be measured once again, which was done in April 1585 with results exceeding the specified quantities—about 16.000 liters in 24 hours. The report notes that the rocking motion of the new device could be increased by changing the relation between the gears, an improvement over the first *Artificio*.[125]

It is not clear who was behind the resistance to the success of the *Ingenio* and what were the motivations for such a resistance. It is obvious that the water carriers must have been against it, as the success of the *Ingenio* would deprive them of their income. The profitability of their business was once proverbial in Spain "… where [in Toledo] usually the *aguadorers* are *gabachos*,[126] and they become very rich with only one or two mules."[127] But there may have been a deeper reason for the lack of acceptance of technological innovation by the municipal authorities. Their sentiment, probably expressed behind closed doors, seemed to be: 'if the king wants to install expensive water games, he should do it exclusively at his own expense.' In fact, from the very beginning Felipe was the driving force and the *Ayuntamiento* the reluctant companion.[128]

123 Antonino Bertolotti, *Artisti lombardi a Roma nei secoli XV, XVI, XVII*, (2 volumes), Vol. 1, Forni (Italy): U. Hoepli, 1881, p. 94.

124 García Tapia, *Ingenieria y arquitectura*, p. 275.

125 AGS, *C & S. R.*, leg. 271, fol. 243 and fol. 244 (17 April, 1585).

126 In Spain, it is used since the 1500s as a pejorative for French people.

127 Covarrubias, *Tesoro*, p. 145.

128 For a discussion of these reasons, see: Damler, "The Modern Wonder," pp. 444–450.

Tired and ill, Juanelo sent to the King request after request, complaining about his extreme misery, brought by the huge debts he contracted to finance the construction. Juanelo's last letter to the King, a few days before his death, reflects his utter despair. The letter reached the King too late. On June 11, 1585, Juanelo dictated his will, a pathetic document stored in the archives of Toledo.[129] Two days later he died, leaving a septuagenarian and widowed daughter, Barbara Medea, to whom a year later the King ordered to pay 6,000 *ducados* for the transfer of the second *Ingenio* to his property. This payment was delayed and curtailed, and she became so poor that in 1601 that, in consideration of services rendered by her father, the King granted her a life pension of four *reales* a day.[130] She left a daughter, Doña María Turriano, to whom the King allocated in 1603 two *reales*[131] per day, as she lived in abject poverty. Another son of Barbara Medea, named Juanelo as his grandfather, was vested with the responsibility for the maintenance of the *Artificio*, in return for an annual fee of 100 *ducados*, as well as a daily payment of 4 *reales*.[132] He died in 1597, leaving his family in misery. His son, Gabriel Joanelo, after inheriting his father's job under the same conditions, joined the army and was killed in Sicily in 1616. We learn from a *cédula real* dated 27 September, 1625, that king Felipe IV ordered payment to Gabriel's only heir, his sister Catalina, of long overdue fees for the maintenance of the *Artificio*, which had not been forthcoming since 1605 and up to Gabriel's death in 1616.[133] That was the end of Turriano's family.

Juanelo was buried without great honors in the church of the Monastery of El Carmen in Toledo.[134] One of his friends, Esteban de Garibay, who was present in the ceremony, wrote: "There was not the due attendance merited by one who was a famous prince in all things to which he applied his most shining intelligence and skills."[135] Juanelo's descendants continued protesting for years about the disaster that had befallen on them, indirectly caused by the construction of one of the more admirable engineering undertakings of its time. The triumph of Juanelo was ephemeral, as even though the devices were

129 Archivo Histórico Provincial de Toledo, *Protocolo* (notarial protocol) de Juan Sánchez de Canales, libro 1594, fols. 10 to 12. Quoted in Cervera Vera, *Documentos biográficos*, pp. 220–222.

130 AGP, *C. R.*, tomo 6, fol. 528.

131 The *real* was a silver coin of 34 *maravedies*, which was the Spanish smallest unit of account.

132 AGP, *C. R.*, tomo 8, fol. 386, v.

133 AGP, *C. R.*, tomo 12, fol. 252, v.

134 Del Campo y Francés, *Juanelo Turriano*, p. 48.

135 Esteban de Garibay y Zamalloa, "Memorias de Garibay," in *Memorial Histórico español, Colección de documentos, opúsculos y antigüedades, que publica La Real Academia de la Historia*, (11 tomos), tomo 7, Madrid: Imprenta José Rodríguez, 1854, p. 420.

used, albeit intermittently, for nearly half a century, they afforded him none of the material fruits which he justly deserved.

Even admirers of the *artificios* admitted that it was obvious that Juanelo's devices did not solve the critical problem of water scarcity in Toledo. The 'pilgrim' [pelegrino] referred to in Villalba y Estaña in his 1570s travelogue is quoted telling an old man whom he met in the city "… you lack water, [but you have] too many water carriers …"[136] There were outright critics of Juanelo's invention as well. The cosmographer Andrés de Céspedes, who saw the device at the end of the sixteenth century, wrote: "the machine is ingenious, but very violent [sic.] and of little use, since it is constantly in need of repairs."[137] Another testimony is given by Jehan Lhermite (1560–1622), a Flemish noble who joined Felipe II's court in Madrid in 1587, as a teacher of mathematics and French of the future king—Felipe III.[138] He visited Toledo for the first time in 1596, and was amazed by the *artificios*,[139] but following his second visit, in 1600, he complained about being unable to obtain an explanation of the operation principle of the devices, despite having pressured the king's chief architect to obtain such information. Probably influenced by this failure, he noted that even had he been able to describe in detail Juanelo's invention, nobody would have attempted to imitate him, as the investment and the running costs were excessive: "… each water pitcher cost the king not less than one *real*."[140] Furthermore, he remarked that since their construction the machines were idle on Sundays and holidays and at night, because somebody was constantly required to attend to its functioning. Francisco de Pisa (1534–1616), a historian of Toledo, noted that the device was in need of constant and expensive repairs, though "in 1604 another artifice was built, which was better and easier to maintain."[141] He most probably referred to Juan de Fernández Castillo's *Ingenio*, to which we shall refer later. The sculptor who made a bust of Juanelo, kept nowadays in the Museum of Santa Cruz in Toledo, described the *artificios* as a "… work of greatness but

136 "… faltaros agua, sobrar aguadores …," Villalba y Estaña, *El pelegrino curioso*, p. 198.

137 Andrés de Céspedes, *Libro de instrumentos nuevos de geometría, muy necessarios para medir distancias y alturas, sin que interuengan números, como se demuestra en la práctica: demás desto se ponen otros tratados, como es uno, de conduzir aguas, y otro una question de artillería, en donde se ponen algunas demostraciones curiosas*, Madrid: Juan de la Cuesta, 1606, fol. 40, v.

138 José García Mercadal, *España vista por los extranjeros*, (2 volumes), Vol. 2, Madrid: Biblioteca Nueva, 1920, p. 203,

139 Jehan Lhermite, *Le Passetemps*, written between 1597–1602, (2 volumes), Vol. 1, Antwerpen: Cr. Ruelens, 1890, p. 288.

140 Ibid., pp. 277–278.

141 De Pisa, *Descripción de la imperial ciudad*, p. 23.

without substance ..."[142] Antonio Ponz (1725–1792) had a lucid view of the huge expense involved in maintaining the *Artificio* in good running order, due to the large number of bars, wheels, pipes, buckets and other parts. He estimated that it was more expensive than having horses carry water to every house in the city. Maintenance was indeed an issue: Ladislao Reti found in the Spanish archives two items of heavy expense related to the *Artificio*: lubricating tallow for the moving elements of the *Artificio* and charcoal for the forges used to repair the parts which frequently wore out or broke down.[143]

Iconographic Representations and Exact Location of the *Artificio* Building[144]

Contemporary iconographic representations of technological devices can sometimes be extremely helpful for tracking down their historical development and understanding the methods used for their operation. However, there are several discrepancies among the few maps and drawings that have remained from the period in which the *Artificio* operated. As was customary in early modern cityscapes, landmarks which symbolized religious and governmental institutions or sometimes even technological achievements are sometimes depicted exaggeratedly larger than other monuments.[145] Such was the case with Toledo's cathedral and *Alcázar*.[146] Even the dating of iconographic testimonies may sometimes be misleading, particularly when noted according to the year of publication, rather than that of their drawing. All these drawbacks are apparent when studying the representations of the *Artificio*, some of which are hereby presented chronologically.

A drawing dated 1566 (fig. 3.7), probably carried out by the Flemish manuscript illuminator, Joris Hoefnagel, during the early 1560s and published in 1572, shows the area of the city where the *Artificio* was being built, with no reference to the ongoing works. The *Alcázar* is surmounted by a tower, which is not

142 Quoted in Nicolas García Tapia, "Nuevos datos técnicos sobre los artificios de Juanelo," *Anales Toledanos* XXIV, Diputación Provincial, Toledo 1987, pp. 141–159, p. 143.

143 Ladislao Reti, (ed.), *The Unknown Leonardo*, London: Hutchinson, 1974, p. 281.

144 For an extensive gallery of images showing the Artificio, its ruins and the surrounding area, see: Moreno Nieto, *Juanelo y su Artificio*, pp. 183–219.

145 For an illustrative example (a view of Paris surveyed in 1652), see: John Brian Harley, "Deconstructing the Map," *Cartographica: The International Journal for Geographic Information and Geovisualization*, Volume 26, Number 2, Summer 1989, pp. 1–20, pp. 6–7.

146 Julián Montemayor, *Tolède entre fortune et déclin (1530–1640)*, Limoges: Presses Univ. Limoges, 1996, p. 18.

FIGURE 3.7 *Detail of upper right corner of View of Toledo—drawn in 1566, published 1572.*[147]

FIGURE 3.8 *Detail of upper right corner of View of Toledo, incorrectly dated 1566, drawn in 1580, first published in 1598, showing the* Ingenio.[148]

---

147 Georg Braun and Frans Hogenberg, *Civitates Orbis Terrarum*, (6 volumes), Vol. I, 1572 (Ausgabe *Beschreibung vnd Contrafactur der vornembster Stät der Welt*, Köln: Kempen, 1582, plate 4v. Depicted by Joris Hoefnagel.

148 Joris Hoefnagel, "Toletum, in Georg Braun and Franz Hogenberg, *Civitates Orbis Terrarum*, first Latin edition of volume v, Cologne, 1598, incorrectly marked 1566, Plate 15. The whole drawing is shown in the page opposite the beginning of this chapter.

represented in later images. The earliest pictorial record in which the *Artificio* clearly appears is a spectacular bird's eye view of the city of Toledo drawn by the same artist (fig. 3.8). This engraving includes only three inscriptions: the *Ingenio*, the city geographic situation—*Oriens* and *Septentrio*—and the Tagus *Fluvius* [the Tajo River]. The fact that the *Ingenio* is the only structure noted, and that the word is written in Spanish in the context of a Latin map, can be interpreted as a sign of admiration for the machine. Obviously the author had new information, as other changes are apparent, such as the *Alcázar*, similar now to later depictions by El Greco.

Only the upper and third section of the *Artificio* is shown in this view, while the lower two sections remain hidden from this perspective. The building that housed the waterwheels and the impelling machinery, just beside the river, fits accurately the extant photographs of its ruins, taken in the nineteenth century, such as the one by Eugène Sevaistre, showing the ruins of the *Artificio* in 1857 (fig. 3.16).[149]

In the 1580s, the Italian engraver Ambrogio Brambilla produced a map (fig. 3.9) with an accompanying legend that mentions the "*Artificio de lagua*" [The water artifice]. The *Artificio* is depicted as a long aqueduct of uniform height, running from the Alcántara Bridge to the *Alcázar*. Its caption is referred to as the *Artificio* in the list of notable buildings of the city. No special construction

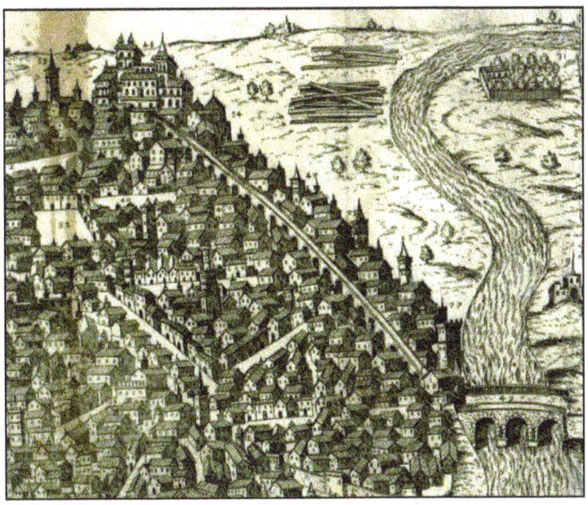

FIGURE 3.9
Brambilla's Plan (1585). Upper right side detail showing the Artificio.[150]

149 Eugène Sevaistre, "Ruins of the Artificio," photograph (1857), kept in the Real Biblioteca del Patrimonio Nacional de España, extracted from Moreno Nieto, *Juanelo y su Artificio*, p. 206.

150 Ambrogio Brambilla, "Toledo," engraving, (48x72.5 cm), Romae: Joannes Orlandi formis, 1602 (original map depicted in 1585).

for the machinery is shown on the river bank. Although this drawing does not offer any clue to the nature of the device, it obviously shows the importance attributed to it by whoever described or represented the city.

By the last quarter of the sixteenth century, Domenikos Theotokopoulos, known as 'El Greco' (1541–1614), settled in the city and painted many works of art, including two remarkable pictorial testimonies of the *Artificio*, in the context of two very different landscapes. The first one is his "View of Toledo," painted about 1597, in which we can see a series of towers placed in a row in the mountain slope, exactly where the *Artificio* was located (fig. 3.10).

The second picture, "Vista y Plano de Toledo," painted about 1610, includes in its foreground the figure of a young man holding a map of the city, which is considered to be Toledo's first relatively accurate map (fig. 3.11). This fact has led some historians to attribute the map to El Greco's son, Jorge Manuel, who was a gifted architect, though a prominent Spanish historian has declared that neither the father nor the son had depicted it.[151] These two paintings, which show the city through the eyes of a careful observer, will serve us later, in analyzing the yet unknown operation principles of the machine.

FIGURE 3.10
Detail of View of Toledo by El Greco—enlarged (1597).[152]

151 José Botella Llusiá and Antonio Fernández Molina, (eds), *Marañón en Toledo: Sobre "Elogio y Nostalgia de Toledo,"* Cuenca: Univ de Castilla-La Mancha, 1999, p. 86.

152 El Greco, (Domenikos Theotokopoulos), "View of Toledo," oil on canvas, (121.3×108.6 cm), 1597, Havemeyer Collection, The Metropolitan Museum of New York.

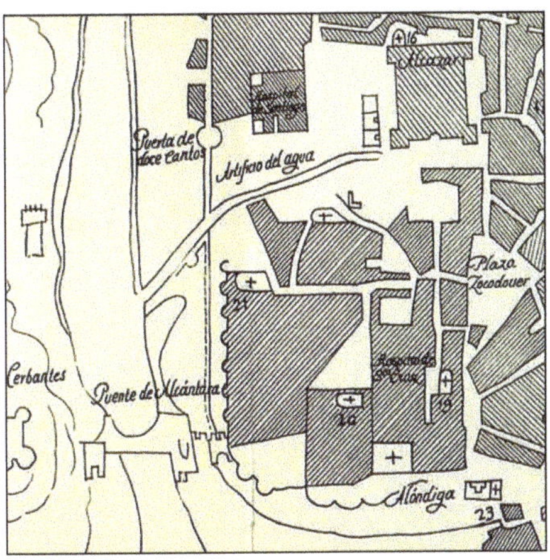

FIGURE 3.11
Top, "Vista y Plano de Toledo," El Greco (1610);[153] *below, enlarged detail, highlighting the* Artificio *alley.*[154]

An engraving dated around 1650 shows the machinery building by the river, roofless, though all the other buildings protecting the *Artificio* which are depicted are covered with slanted roofs. The upper section, entering the North East tower of the *Alcázar*, is also clearly discernible (fig. 3.12).

153 El Greco, "Vista y Plano de Toledo," oil on canvas, (132x228 cm), 1610, Museo El Greco, Toledo.

154 Detail of "Vista y Plano de Toledo," redrawn by Alfonso Bacheti, 1967, from the original. Julio Porres Martin-Cleto, "El Artificio de Juanelo," *Temas Toledanos*, No. 47, IPIET, Diputación Provincial, Toledo, (1987), p. 17.

FIGURE 3.12 *The* Ingenio *and the Alcántara Bridge, detail, c.1650.*[155]

FIGURE 3.13
The Artificio *in the Colonels' plan (1668).*[156]

155 Juan de Colmenar, *Les Délices de l'Espagne et du Portugal*, (5 volumes), Vol. 2, Leiden: chez Pierre Vander Aa, 1707, p. 321.

156 Carlos y Fernando de Grunenbergh, "Perspectiva de la Ciudad de Toledo vista de la otra parte de la Puente del camino de Sevilla," (the Colonels' plan), in *Memorial que los Coroneles Carlos y Fernando de Grunenbergh han dado a la Reina Maríana de Austria, sobre la proposición realizada de hacer navegable el río Manzanares desde el Pardo hasta Toledo: se manifiestan los motivos que tuvieron para haber hecho dicha proposición, los fundamentos y razones que hay para la ejecución y facilidad de su fábrica, las utilidades que de ella resultan, las condiciones para su realización y la demostración que han hecho de las corrientes del dicho río y del Jarama*, Madrid: [s.n.], 1668, p. 27. For a in-depth research on the Grunenbergh's Project, see: Dolores Romero Muñoz (coordinación, Daniel Crespo

In 1668 two military entrepreneurs, Carlos y Fernando de Grunenbergh, colonels in the Spanish army, proposed to the Court to dredge and widen the Manzanares River running between Madrid to Toledo, in order to make it navigable, as they took for granted that the Tajo would soon be made navigable as well between Toledo to Lisbon. Among the documents that they submitted was a perspective plan of the city, in which details of the *Artificio* building and a sketch of a waterwheel can be observed (fig. 3.13).

Ambrosio de Morales noted in 1575 that Juanelo built a bridge, "like the one Caesar built during the siege of Marseille,"[157] so as to enable the raised water to cross a wide street towards the *Alcázar*. However, most historians, including Escosura, estimated, by using Caesar's description, that Morales mistakenly confused the bridge with the one Caesar's engineers built to transfer his army across the Rhine, in order to fight the Sugambri.[158]

A plan of the city, drafted by Arroyo Palomeque at the end of the seventeenth century or the beginning of the eighteenth, showed for the first time a

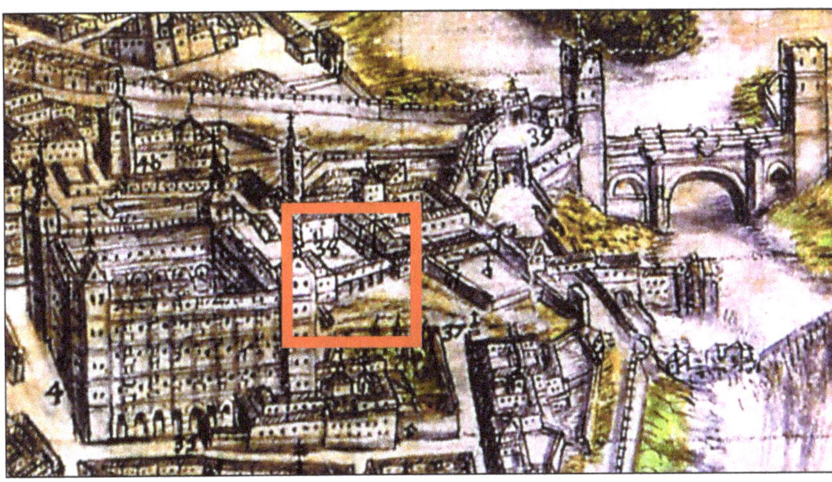

FIGURE 3.14 *Detail of map in "View of Toledo by Arroyo Palomeque," showing the machinery building, the stepped stations and the 'Julius Caesar' bridge (inset).*[159]

Delgado), *La navegación del Manzanares: el proyecto Grunenbergh*, Madrid: Fundación Juanelo Turriano, 2015.

157 Morales, *Las antigüedades*, p. 335.

158 Theodor Mommsen, *The history of Rome*, translated by William Purdie Dickson, (4 volumes), Vol. 4, New York: Scribner, 1871, p. 311.

159 José de Arroyo Palomeque, "Panoramic view of Toledo," c.1720, extracted from Julio Porres Martin-Cleto, *Toledo a traves de sus planos*, Toledo: Diputación Provincial, IPIET, 1989, fig. 10. An extensive discussion about this view is found in Julio Porres Martín-Cleto, Rafael Juan del Cerro Malagón y José Luis Isabel Sánchez, *Panorámica de Toledo de Arroyo Palomeque*, Toledo: IPIET, 1992.

depiction of the 'Julius Caesar' bridge across the wide street (detail in Fig. 3.14). The shape and height of the *Artificio* building housing the driving machinery is relatively well documented in several nineteenth-century engravings, drawings and photographs that have survived, of which two have been chosen for their clarity, and are shown in figs. 3.15 and 3.16. As will be discussed below, estimating on the basis of those photographs, the height of the building housing the waterwheels, adjacent to the river can provide clues to the functional parameters of the *Artificio,* and represents an argument to refute one of the hypotheses about its working system.

FIGURE 3.15 *The* Artificio *ruins, lithography (1853).*[160]

FIGURE 3.16 *Photograph by Eugène Sevaistre, Ruins of the* Artificio *(1857).*[161]

160 Francisco Javier Parcerisa, "Puente de Alcántara," Litography by J. Donon, in J. M. Quadrado, *Recuerdos y bellezas de España—Castilla La Nueva* (2 volumes), Vol. 1, Madrid: Imprenta de D. José Repullés, 1853, image between pp. 278–279.

161 Eugène Sevaistre, "Ruins of the Artificio," photograph (1857), kept in the Real Biblioteca del Patrimonio Nacional de España, extracted from Moreno Nieto, *Juanelo y su Artificio*, p. 206.

The *Artificio*'s Technology—The Operational Principle

The fact that no drawings of the *Artificio* mechanism, with the only exception of a very crude sketch of its concept, were made or survived, and that the device was dismantled about 50 years after its inauguration, have prevented a systematic and accurate reconstruction of its technology. This riddle has motivated several researchers to try and solve it by suggesting different theories, summarized in the following paragraphs.

Escosura y Beck—The Inclined Plane Theory

A Spanish mining engineer, Luis de Escosura y Morrogh, made the first serious study on the *Artificio* technology, after receiving in 1860 a commission to propose alternative solutions to the water supply problem of Toledo, not satisfactorily solved even by then. Intrigued by Juanelo's approach, Escosura published his own findings in 1888.[162] His reconstruction was mainly based on the text of Ambrosio de Morales, in the *Antigüedades de las Ciudades de España*, which we quote here extensively for its importance:

> The essence of this invention is to join or to hinge small wooden bars crossed in their middle and at their ends, as in Roberto Valturio's machine intended to raise a man to heights, though this [invention] of Juanelo has additional advantages and subtleties. With all the bars thus articulated, when moving the wooden bars closest to the river, the motion is transmitted to all the others up to the *Alcázar*, with great calmness and smoothness, as required from a machine built for perpetuity. But what is more wonderful is to have fitted and hinged to the end of the wooden bars long brass tubes, almost one *braza*[163] in length, with two cups of the same metal in their ends, which raise and descend with the movement of the wooden bars: when moving down one goes full and the other empty, and when both meet by their sides, they remain motionless until the full one spills its charge to the empty one. Then, the full one rises to pour the water through the tube into the empty one, and the one that becomes empty lowers to join itself with the full one at the back, which

162 Luis de la Escosura y Morrogh, *El Artificio de Juanelo y el Puente de Julio César*, Memoria publicada por la Real Academia de Ciencias Exactas, Físicas y Naturales de Madrid, Madrid: Imprenta de Don Luis Aguado, 1888.

163 Equivalent to 1.82 meters. José Luis Galán García, *Sistemas de unidades físicas*, Barcelona: Reverté, 1987, p. 19.

> also descends in order to fill it. Thus both cups of a tube are sometimes empty, alternatively changing, so that the one that had a full cup is left completely empty, and the empty one becomes full ...[164]

It was almost impossible to interpret Morales' description without any illustration, and it was only by chance, according to Escosura, that when he was about to give up on his quest, while perusing Agostino Ramelli's book on ingenious machines, first published in 1588, he came across Plate 95,[165] which he basically recognized as fitting Morales' description. Looking into the matter in depth, Escosura understood that there were discrepancies between Ramelli's conception and Morales' narrative portrayal. He consequently introduced several modifications in order to reconcile them.

The great stumbling block for Escosura's theory was to interpret the "*... small wooden bars, crossed in their middle and at their ends, as Roberto Valturio's machine intended to raise a man to heights ...*," as Morales description began.[166] Escosura acknowledged that only in the realm of pure speculation could he try to complete the explanation. Just for the specific purpose of complying with the description, he devised an extremely complex kinematic mechanism, based on the appearance of the Valturio's scale (fig. 3.17), though it is highly improbable that Juanelo would unnecessarily complicate his machine in such a way.[167]

A copy of Escosura's book reached the hands of the famous engineer Theodor Beck, who translated it into German, and added his own observations.[168] Beck differed with Escosura only on the kinematic structure of the machine, since he maintained that transmission of movement to such a distance as the one between the river and the *Alcázar* would progressively accumulate irregularities and loose fits, impairing the mechanism's operation. Yet Beck's even more complex approach does not improve the feasibility of his model representing Juanelo's device.[169]

---

164 Morales, *Las antigüedades*, pp. 332–333. Another contemporaneous description (1577) can be found in Villalba y Estaña, *El pelegrino curioso*, pp. 194–195, though a careful comparison shows it is a summary of Ambrosio de Morales' account.

165 Ramelli, *Le diverse et artificiose machine*, plate 95, p. 148. Shown in this study in chapter II, fig. 2.34.

166 Morales, *Las antigüedades*, p. 332.

167 Escosura y Morrogh, *El Artificio*, pp. 40–45.

168 Theodor Beck, "Juanelo Turriano (1500–1585)," in *Beiträge zur Geschichte des Maschinenbaues*, Berlín: Springer, 1899, pp. 365–390.

169 Ibid., pp. 383–387.

FIGURE 3.17
Valturio's scale.[170]

The Vertical Solution, Proposed by Ladislao Reti

The Escosura/Beck's concept was accepted for about 100 years and the engraving in Plate 95 of Ramelli's book was considered to be a faithful depiction of a mechanism similar to the water-raising machine built by Juanelo. After an exhaustive study by the historian of technology and Leonardo da Vinci scholar, Ladislao Reti (1901–1973), in the course of which he tramped over the hills of Toledo and inspected surviving traces of structures that were probably part of the sixteenth-century waterworks, he put forward an alternative, more convincing theory. It was based on a new interpretation of the documents referring to the *Artificio* kept in the Archivo General de Simancas and other archives, a different analysis of the descriptions given by Ambrosio de Morales and another account by another contemporaneous witness—Sir Kennelm Digby (1603–1665), the English physicist, naval commander and diplomat, who visited Toledo in the early seventeenth century and described the *Artificio* in his monumental philosophical book—*Two Treatises*.[171]

Reti's exhaustive study led him to conclude that the actual Toledo machine rather resembled another Ramelli drawing, but with important modifications,

170 Robertum Valturium, *De Re Militari, Libris XII*, Parisiis: Apud Christianum Wechelum, fub infigni fcuti Bafilienfis, 1535, p. 259.

171 Sir Kennelm Digby, *Two Treatises*, Paris: Gilles Blazot, 1644.

as expounded in his presentation at a conference in Toledo in June 1967.[172] His vertical mechanism is based on the machine depicted in Plate 96 of Ramelli's book.[173] The principle of operation is illustrated in Fig. 3.18, which represents one of the towers of oscillating arms of the device. In each tower a semicircular wheel swings around a pivot fixed to a column with an up and down movement, driven by a connecting rod-crank based mechanism moved by a second waterwheel. The semicircles drive slide chains that alternatively lift and lower a series of braces to which tubes with metallic cups or spoons, are attached at their ends. The cups receive the water, and on the other end, the tubes, which are bent in an elbow shape, spill it in the following cup. The first cup (#1) collects water from the lower basin. Subsequently the rising cup #1 passes the water to cup #2. By alternative movements of elevation and descent, the structure of braces causes the water to be transferred successively from one cup to another, finally arriving to the top basin, raised all the way at atmospheric pressure. Reti also established the number of oscillating channels and cups, as well as the stages in which they were placed, based on an inventory of the *Artificio* parts (kept in the Archivo General de Simancas),[174] taken in March 1639 as a result of legal proceedings about brass thefts that took place in February of the same year, to which we shall refer later.

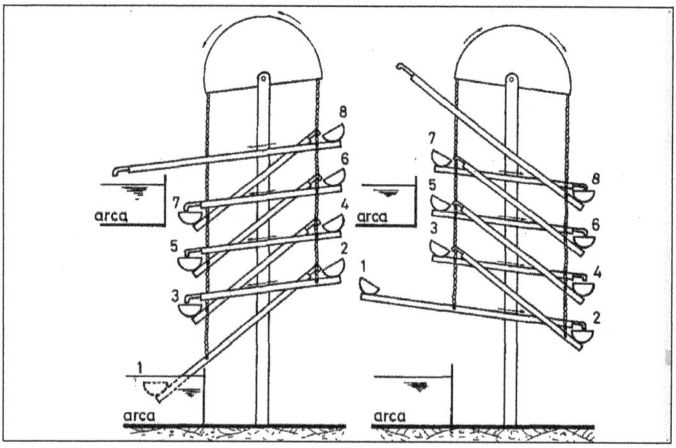

FIGURE 3.18 *Ladislao Reti's vertical system principle of operation.*[175]

172 The complete text of the conference was published as Reti, "El Artificio."

173 Ramelli, *The Various and Ingenious Machines*, p. 257. Shown in this study in chapter II, fig. 2.35.

174 AGS, *C & S. R.*, leg. 272, fol. 20 r.

175 Porres, *El Artificio*, p. 17.

The illustration in Fig. 3.19 shows how, according to Reti, the river flow drove the *Artificio* by means of two waterwheels. A first stage of elevation used a chain of buckets driven by the primary waterwheel, which raised the water fourteen meters up, to an intermediate basin. From that height, Reti reconstructed a series of 24 towers, each one containing 8 oscillating cups (altogether 192 cups) and two intermediate basins (for a total of 48 basins),[176] which Reti virtually located in the slope leading to the *Alcázar* in accordance with documents in Simancas.[177] He calculated that the water rose up 45 centimeters in each step, totaling 86 meters for the 192 steps. However, there are discrepancies between his theory and contemporaneous writings, such as to the description by the sixteenth century painter, Federico Zuccaro: "… and there are an infinite number of such cups, eight of which are always full and eight empty, in each order that are many …"[178] Reti himself acknowledged that the number of cups in each tower could not have been always eight, and that their distribution must have varied according to contour of the slope, consequently the number of towers could have been different as well.

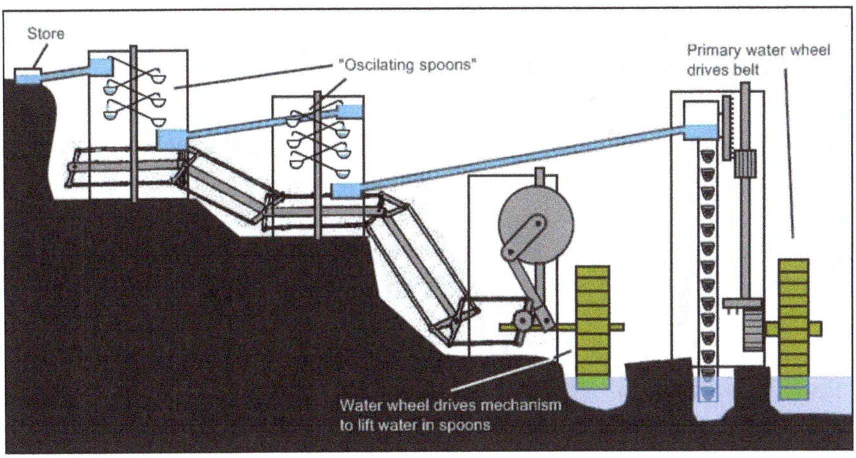

FIGURE 3.19 *Reti's schematic representation of the operation principle of the* Artificio.[179]

176 Reti, "El Artificio," p. 22.

177 Reti, "El Artificio," p. 23.

178 Gibbs, "Federico Zuccaro," p. 50, quoting a letter written by Zuccaro in 1586, extracted from J. Dominguez Bordona, "Federico Zuccaro en España," *Archivo Español de Arte y Arqueología*, No. 7, 1927, p. 77–89, p. 77.

179 Porres, *El Artificio*, p. 32. For the sake of clarity the representation includes only two towers and six cups per tower.

Reti found support for his theory in Digby's account of his impressions while visiting the *Artificio* in the first years of the seventeenth century: "… In the bottom, there was an indented wheel, which turning round with the stream, gave motion at the same time to the whole engine …"[180] Digby's lengthy description of the device can be readily understood as very similar to Morales' as far as the technical part is concerned, adding more detail and allowing for a better visualization of the *Artificio*'s function: "The Artificio consisted of a multitude of little troughs or square ladles set one over another, from the bottom to the top and upon two several divided frames of timber …" The philosopher Digby's lengthy description of the device is taken as a metaphor of man: "And thus, the two sides of the machine were like two legs that by turns treaded the water; as in the vintage, men press grapes in a vat …"[181] Reti saw in the allegory of the two legs alternatively pressing grapes in a vat a figure similar to the motions described above in his representation of the device.

Curiously, Reti did not mention an earlier first-hand testimony that sustains his ideas, contained in the "*Memoria de algunas cosas notables que tiene la Imperial ciudad de Toledo*,"[182] [Report on some remarkable things which the Imperial City of Toledo possesses] dedicated in 1576 to Felipe II by the Toledan Luis Hurtado, Rector of the San Vicente parish. In Hurtado description we can find support to Reti's theory: "… with eight stages of metal tubes, four in each ladder, which laboriously throw into the *Alcázar* two pipes, with a diameter similar to a *Real de a ocho*,[183] and these work day and night, as their mover is the same river with some wheels and in a supernatural way …"[184]

Years later, after Reti's death,[185] a text was discovered, which, because of its 'guaranteed objectivity,' became fundamental to the general positive acceptance of his theory. Manuel Severim de Faria, (1583–1655) who was *chantre* (precentor) of the See of the city Évora in Portugal, wrote a travelogue, published as *Itinerario Hispánico del Chantre de Évora of 1604*, where he described

180 Digby, *Two Treatises*, p. 205.

181 Ibid., p. 206.

182 Hurtado de Toledo, "Memorial de algunas cosas," p. 501.

183 The *Real de a ocho*, a silver coin minted in Castile since 1407, had a diameter of 38 mm. Patrick Richard Carstens and Timothy L. Sanford, *The Republic of Canada Almost*, Bloomington (Indiana): Xlibris Corporation, 2013, p. 269.

184 Hurtado de Toledo, "Memorial de algunas cosas," p. 501.

185 Julio Porres Martin-Cleto, "Consideraciones y nuevos datos sobre el Artificio de Juanelo," *Boletín de la Real Academia de Bellas Artes y Ciencias Históricas de Toledo*, 40, 1999, p. 129 tells about Reti's reservations concerning his own theory, saying he hoped that additional documents to be found in the future would confirm his hipothesis.

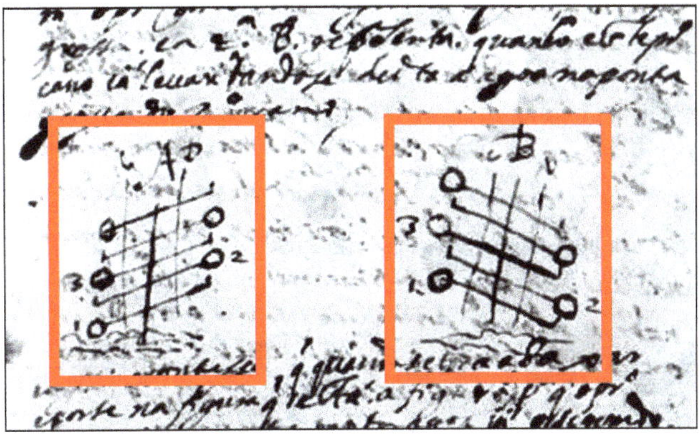

FIGURE 3.20 *Operation sketch in Severim text.*[186]

his extensive trip in Spain, including also Toledo.[187] His conceptual description of the *Artificio* fits Reti's explanations. Moreover, the sketches that appear in the travelogue—the only known contemporaneous depiction of the operation principle of the *Artificio*, constitute arguably the almost definite confirmation of Reti's representation (fig. 3.20).

Juan Luis Peces Ventas' Improvement of the Inclined Plane Theory

A new, alternative explanation of the *Artificio* mechanism was proposed by Juan Luis Peces Ventas, in an article published in 1997,[188] in which he argued that Reti's vertical solution is not consistent with all the documentation related to the *Artificio*. His solution followed the line of Escosura and Beck—it was also based on the Valturio's ladder analogy—but replaced the straight segments with right-angle ones. With this modification two types of rhombus are created, causing the stretching and elevation of one when the contiguous one is flattened.

186 Severim de Faria, "Itinerario Hispánico," p. 159.

187 Manuel Severim de Faria, "Itinerario Hispánico del Chantre de Évora en 1604," *Separata, Revista de Estudios Extremeños*, Tomo XLII, No. 1, (Enero–Abril, 1986). The original is in Lisboa: Manuel Severim de Faria, *Peregrinaçao de Balthazar de Faria Severim, Chantre de Évora, ao mosterio de Guadalupe, no anno de 1604*, Biblioteca Nacional de Portugal, Fundo Geral, Cód. 7642, ff. 118r–119v.

188 Juan Luis Peces Ventas, "El Artificio de Juanelo Turriano," *Revista de Informacion de la Junta de Comunidades de Castilla-La Mancha*, No. 23, 1997. He adapted the text to his website: http://home.worldonline.es/agallar/ (no longer available) on October, 1998.

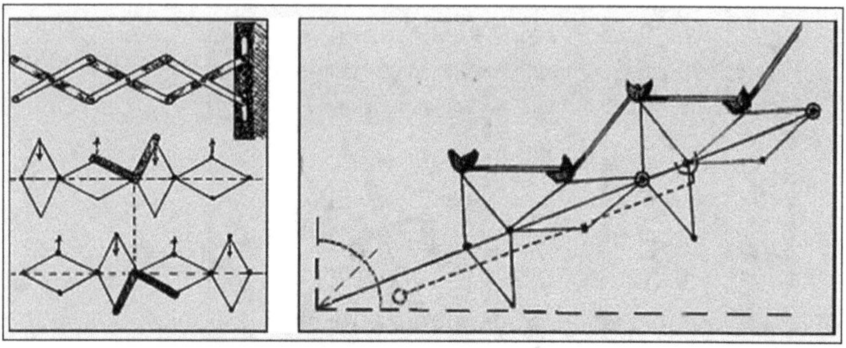

FIGURE 3.21 *Peces Ventas' modification to the inclined plane system, left-top: Conventional Valturio scale, below and right: modified Valturio scale.*[189]

If cups are placed in the upper vertices of the rhombus, tilting the whole mechanism when setting it up in motion, the key principle of operation is revealed, the two positions (A) and (B) can be seen in Fig. 3.21 (left) and the resulting movement of the cups in Fig. 3.21 (right).

I had the opportunity of personally interviewing for several hours Mr. Peces Ventas in Toledo in February 2007, in the Centro de Interpretación del Toledo Histórico—Centro Cultural San Marcos, in the San Marcos Church.[190] Although he tried, the hardest task for Peces Ventas was to detract from the value of the sketches drawn by the chantre of Évora that are generally considered to be 'the proof' of the correctness of the vertical solution. He indicated that in the *chantre*'s sketch the river water is represented by undulating lines in the lower part, immediately below the drawing. Quoting from de Severim de Faria's description of the sketch, according to which "… so that the first ('A') takes the water with the widest part …," Peces Ventas claimed that the observer obviously did not see the gigantic chain of buckets that elevated the water up to the basin where Juanelo's special device started its operation. According to him, the *chantre*'s description also ignores the waterwheel which drove the whole system. Furthermore, Peces Ventas remarked that it is not certain that the second *Artificio* was in working condition in 1604, as it was damaged in 1598, and the budget for its repair was allocated only in 1612, whereas the first one had already been dismantled. Therefore, he claimed, it is doubtful that

189 Peces Ventas' website: http://home.worldonline.es/agallar/discussion, accessed July 12, 2012. Not available anymore, presently shown in: https://juanellotoriano.wikispaces.com/Juan+Luis+Peces+Ventas, accessed 12.12.2016.

190 Mr. Peces Ventas passed away in 2012 in Toledo.

such an eyewitness actually saw the *Artificio* in operation. Moreover, in his zeal to refute the vertical solution, Peces Ventas also stated that in the Spanish version of Severim de Faria's description, which he presented to me, different measurement units were used, such as 'a hand' and 'fingers', and at a certain point 'centimeters' are used as well, although the decimal system became accepted only two hundred years later. However, a careful look at the original Portuguese version shows that no mention was made of centimeters in the passage indicated by Peces Ventas, but the term "... a small *patacón*[191] size ..." is the expression used. So this apparently conclusive argument is totally invalid.

Peces Ventas maintained that the *Artificio* was obviously based upon one type of solution only, and he felt he amply demonstrated that it could not have been the vertical one, even if that was (and still is) the prevailing version accepted by the historical community. In order to prove his thesis, he decided to build a scale model, motivated by his conviction that the inclined plane system was the true method used by Juanelo, and that such a model also suits the narration of Morales and the documentation found in the archive at Simancas. His conceptual model is exhibited in the same institution where I interviewed him, where in my presence he operated it to show me its feasibility.[192] It must be admitted that it reproduced the required motions in an impressive way. The model allows studying and appreciating the mechanisms of traction, those of temporary stoppage for pouring water from one cup to the following one, cancellation of tolerances, verification of the state of indifferent equilibrium of the whole device, constant operation and uniform effort on the driving mechanism coming from the waterwheel, all in accordance with Morales' description. There is no doubt in our mind that his system would work if replicated in full scale, with a driving water wheel delivering enough energy. Yet all that does not necessarily prove that it was the principle actually used by Juanelo.

Nicolás García Tapia's Proposition

Nicolás García Tapia's contribution is based on Reti's solution, but with a series of adjustments. Digby's precise description indicates that these cups were placed one above the other, in two rows, one facing the other. It is thus clear that the 16 cups were arranged in both sides of the machine, so that each side

191 Ancient Spanish silver coin. Kris E. Lane, *Pillaging the Empire: Piracy in the Americas, 1500–1750*, New York: M.E. Sharpe, 1998, p. 31.

192 Due to the delicate mechanism of the model, he was the only person allowed to do so.

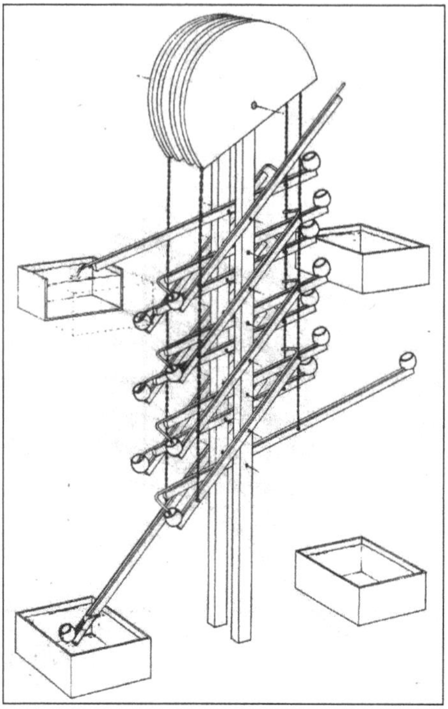

FIGURE 3.22
Perspective of García Tapia's reconstruction.[193]

had two rows of 4 cups each, each row having its scooping water cup, which alternatively was introduced into its basin, in order to elevate the water. In that way, both sides of the machine would have resembled two feet that, alternatively, stepped into the water. Thus eight cups were full and eight empty, as Zuccaro said. Their lifting height corresponded one to a simple set of 8 cups—3.6 m., being the true altitudes of the elevation divisible by such a dimension. With this set up, the lifting heights of each one of the stages is kept, and some of the towers would have 16 cups and others 8, according to the topography and the slope of the terrain.

193 García Tapia, "Nuevos datos técnicos," p. 158. A wonderful animation of the elevating towers was submitted by Ricardo Reinoso Delgado, *Estudio y Simulacion del Artificio de Juanelo Turriano*, proyecto fin de carrera, Universidad de Sevilla, 2003. The animation can be seen in the website: http://juanellotoriano.wikispaces.com/Modelo+de+GT%2C+animaci%C3%B3n. Lucas González Conde, in *Reconstrucción Virtual del Artificio de Juanelo Turriano para elevar agua del Rio Tajo a Toledo*, final thesis, Universidad de Valladolid, 2005, includes detailed computer designs of the components and sub-assemblies of the Artificio, in accordance with García Tapia's conception. The thesis contents

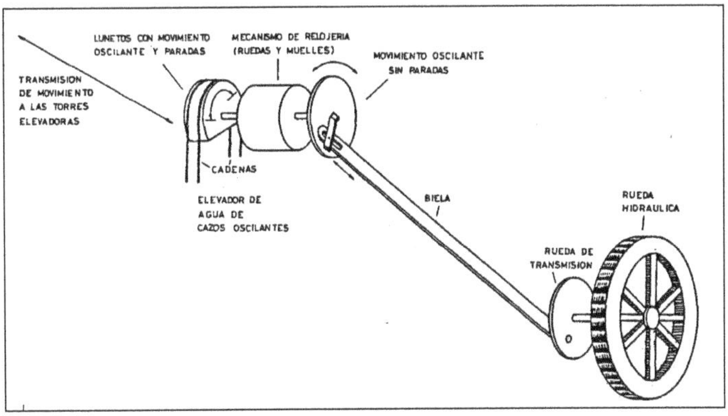

FIGURE 3.23 *García Tapia's transmission mechanism.*[194]

Summarizing, Nicolás García Tapia's proposal is basically identical with Reti's, with the addition of a second structure, which moves alternatively to the first one, driven by the chains hanging from the half wheels, as seen in Fig. 3.22. The basins of the different towers communicated to each other by means of brass channels, when there was a level section between them. In this way, by successive stages, the water reached the last tower, attached to the *Alcázar*, where the final water deposit was located and from where the water was distributed. García Tapia's reconstruction shows that there were two input and two output open basins, situated on either side of the tower. When the first cup was introduced into the basin to take water, the opposite one was raised, passing the water on to the next, and each basin was full when the one opposite it was empty. The water was thus made to flow continuously with a coordinated and precise movement. This configuration represents the only theory that fully fulfills the reference in all documents to "two alleys of water in the *Ingenio.*"

In another article, written in 1987, García Tapia proposed two additional improvements to Reti's theory.[195] The first refers to the motion transmission from the waterwheels worked on the river bank up to the half moon pulleys that rock the cup arms in the towers. Based on a document drafted in April 1626

were submitted by Lucas González Conde and his supervisors—Miguel Bermejo Herrero, María Gloria del Rio Cidoncha, Juan Martínez Palacios in their presentation: "Reconstrucción Virtual del Artificio de Juanelo Turriano para Elevar Agua del Río Tajo a Toledo. De la Tradición al Futuro," *Congreso Internacional Conjunto XVII INGEGRAF—Congreso Internacional de Ingeniería Gráfica*, Sevilla, España, 2005, pp. 1–9.

194 García Tapia, "Nuevos datos técnicos," p. 159.

195 Nicolás García Tapia, "Nuevos datos técnicos," *Anales toledanos.*

by a group of officers of the *Alcázar* who inspected the *Artificio*, in which they described what was in essence a connecting rod-crank mechanism,[196] García Tapia sustains that the hydraulic wheels transmitted the circular motion into an oscillating one through a connecting rod, as shown in fig. 3.23. The second observation consists of an alternative solution to comply with Morales' description: "… facing each other by their sides, they remain motionless for the time necessary to allow the full one to spill the water into the empty one …"[197] Considering Juanelo's fame as an accomplished clockmaker, García Tapia deduced that he must have used a simple clock mechanism, as schematically represented in in the same figure. He found further indirect evidence in another Simancas document, which specifies that it was necessary to lubricate "*… axles, springs*[198] *and other moving parts …*"[199] No springs are needed in Reti's model, but would be needed in a clock-like apparatus.

Francesc Xavier Jufre García's Concept: Operating Valturio Scales

A gallant effort to put forward a revolutionary theory about the way Juanelo's device worked has been presented by a young Catalan industrial engineer—Francesc Xavier Jufre García. He has exposed his concept in depth in a paper submitted to the *II Congreso Internacional de Patrimonio e Historia de la Ingeniería*, which took place in June 2007 in Las Palmas de Gran Canaria,[200] in a book published in Madrid in 2008,[201] as well as in an impressive website.[202] His proposal is totally based on adapting Valturio's vertical scales to lift a pair of connected cups attached to horizontal arms. The scales, located on an inclined plane, would move alternatively up and down, in such a way that when a scale is completely extended and its cups in their uppermost position, water is poured into the cups at the top of the following scale, as the latter is folded. The system requires transferring siphons in each cup and fixed elements to prime the siphons, a mechanism to keep the cups unmoved while water is

196 AGS, *C & S. R.*, leg. 332, fol. 428.

197 Morales, *Las antigüedades*, p. 332.

198 "… ejes, resortes …," which are standard items in clock mechanisms.

199 AGS, *C & S. R.*, leg. 271, fol. 190.

200 Francesc Xavier Jufre García has kindly sent me a summary of his presentation to the *II Congreso Internacional de Patrimonio e Historia de la Ingeniería*, which took place in June 2007 in Las Palmas de Gran Canaria.

201 Francesc Xavier Jufre García, *El artificio de Juanelo Turriano para elevar agua al Alcázar de Toledo* (s. *XVI*). *Modelo con escaleras de Valturio*, Lleida (Spain): Editorial Milenio, 2008.

202 Francesc Xavier Jufre García's website: http://www.artificiodejuanelo.org.

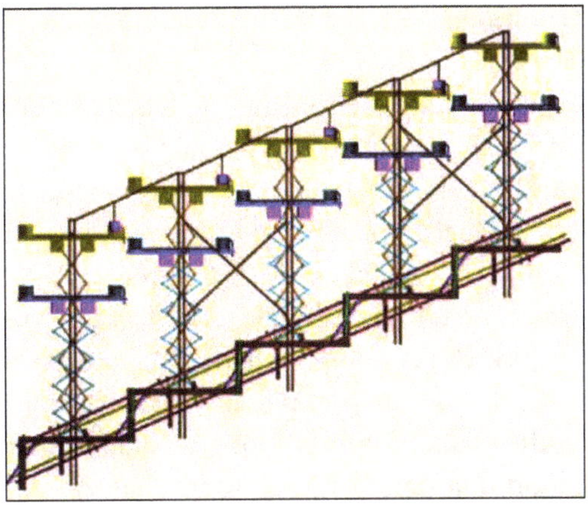

FIGURE 3.24
Jufre's 'Valturio' scale with water lifting cups.[203]

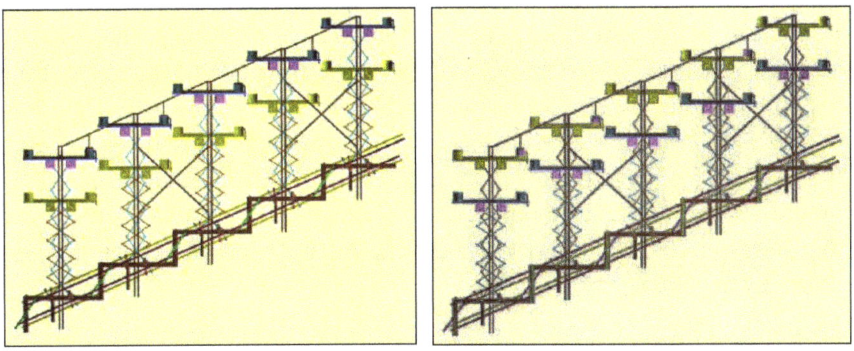

FIGURE 3.25 *Jufre's conceptual design—Valturio vertical scales alternate movement.*[204]

transferred and guide rails to maintain a smooth motion of the whole device (see figs. 3.24 and 3.25).

In such a way water would be transferred from scale to scale in many steps, overcoming the difference in height. A system of power transmission driven by a waterwheel would impart motion to the scales, similarly to the other theories. According to Jufre García, the method presented in his paper and book, fills all the gaps encountered in the other existing theories, while being consistent with the literal and visual texts.

203 Extracted from Jufre García's presentation to the *II Congreso Internacional de Patrimonio e Historia de la Ingeniería* (see footnote No. 200).

204 Ibid.

Unanswered Technical Questions

An important question that was hardly addressed in the models described above is the existence or non existence of a mechanism that could keep the device working in draught periods in which the level of the river was too low, or alternatively, too high due to floods.[205] Zuccaro's description of 1586, in which he noted that "… *the waterwheels are made with such ability that the when the river rises it does not affect them, which is truly an achievement of ingenuity and art …*,"[206] has served the historian Jack Gibbs to argue that the wheels were mounted on a floating platform, with the addition of a pulley and a counterweight to maintain the tension in the chain of buckets. José Antonio García-Diego concurred with Zuccaro's report that the *Artificio* had a mechanism that allowed for the regulation of the height of the driving waterwheels in accordance with the water level in the river, though noted that no clue exists as to how it was done.[207] García Tapia, on the other hand, affirmed that no such mechanism existed in Toledo, as he deduced from an undated document from the Simancas Archives, thought to be from 1585, which deals with damage to the *artificios*. Among the causes it states: "…[damages] due to the river carrying less water, or sometimes too much water …"[208] The same document also indicated what was done in periods of a weak river flow: "When the river goes low, it is necessary to close some of the channels of the waterwheels and one of the *artificios* …"[209]

205 A qualitative analysis of the amount of rain which fell annually during the sixteenth and seventeenth centuries can be found in: Ramón Gonzalvez, "El clima Toledano en los siglos XVI y XVII," *Boletin de la Real Academia de la Historia*. Tomo CLXXIV. Número II. Año 1977, pp. 305–333. Gonzalvez based his findings on ecclesiastic records of the different prayers: imploring for rain or for a damaging storm to cease and the churches income from tithes, related to the farmers' success.

206 Gibbs, "Federico Zuccaro," p. 51.

207 José Antonio García-Diego, "Una muerte y un Artificio," *Anales Toledanos*, IX, Toledo, (1974), pp. 289–304, p. 297.

208 AGS, *C & S. R.*, leg. 321, fol. 59, Relacion de Juan Ibarra, extracted from Nicolas García Tapia, *Ingenieria y arquitectura*, p. 289. Lhermite, *Le Passetemps*, Vol II, notes the same effect concerning floods (see p. 278).

209 Lhermite, *Le Passetemps*, Vol II, p. 278.

Technical Conclusions

The controversy surrounding the principle of operation of the *Artificio* has been heated, sometimes reaching personal animosities among the participants. Obviously, any model that strives to reproduce the device as it was must respect Ambrosio de Morales' description, take into consideration the official dossiers—such as Juanelo's contracts with the King and the city of Toledo, appraisals, bills of materials for repairs, bills of material concerning the dismantling of the artifice, as well as bearing in mind that the device was modified at the beginning of the seventeenth century. In order to analyze the feasibility and logic of each theory a list of the parameters which have to be fulfilled in order to prove its validity must be compiled, taking into consideration all available sources. Subsequently an assessment of the degree of compliance of each model against the list has to be performed. In any case, among the key elements which have to be taken into consideration we may record the similarity to a Valturio scale; inherent complexity requiring a balanced design obtained after numerous calculations and a large number of parts; smooth and continuous water transfer without any spills; friction between parts necessitating frequent lubrication; and the capability to be adaptable to different gradients of slope of the terrain. Not surprisingly, while submitting almost identical lists, all competing theorists attribute to their own theory full compliance with the detailed parameters.

The options presented by Escosura, and one century later perfected by Juan Luis Peces Ventas, are closely related and are based on the concept of an inclined plane. Movement is transmitted in this case by a mechanism resembling a modified Valturio scale, where different positions of the scale raise or lower the buckets in such a way that water is poured each time into the bucket attached to the following arm. The primary source that supports this theory is Ambrosio de Morales. Although it is certainly viable, and slopes can be overcome by this system, it is not in accordance with other contemporary sources related to the *Artificio*. Its configuration does not resemble a man who steps alternatively with his feet, as graphically described by Digby, and does not correspond to the sketch included in Severim's description.

Ladislao Reti's interpretation seems to be much closer to historical reality: its design is innovative and rather unique, and it closely follows both Morales and the Chantre of Évora. But even this theory does not accord with some statements found in other documents: it neither allows for two water alleys nor can it be compared with a man stepping on grapes, as in Digby's description. These reasons would be sufficient to discard the theory, but when García Tapia's modifications are taken into account, apparently most criteria are met.

It complies with all the details of the different descriptions, much more so than Reti's explanation. It also agrees with Ambrosio de Morales, who does not say that it is a Valturio mechanism, but rather that it has cross logs hinged as in a Valturio scale. The addition of the second water alley improves the symmetry and the balance of the machine. Finally, it allows a more convincing explanation for the elevations in each stage—a greater difference in level only required an increase in the number of sequentially connected towers. García Tapia's less convincing element in his explanation of the mechanism to stop the device while water is transferred from adjacent cups by means of a clock mechanism, of which no real evidence exists.

The devices in all theories were conceived to overcome any height difference since, as the water ran in contact with the air, no excesses of pressure occurred due to pumping, as would happen in a usual pipe system. It was necessary to raise the water by 100 meters, but also to displace it horizontally by 300 meters. Obviously the inclined plane system is more efficient to achieve this dual goal, as water moves all the time in the same direction, whereas in the vertical system it has to be carried from one tower to another by connecting channels, which require additional level differences. The accumulation of these differences would probably amount to 5–10 meters, which anyhow does not preclude the effectiveness of the vertical concept.

The external appearance of the building, better appreciated in the figure 3.26, shows that in the lower part of the *Artificio*—from the building housing the waterwheels up to about a third of the total height, the height of the buildings containing the towers would have been about 4.5 meters needed to raise the water, which probably resulted in about 7 meters total height of each stage. According to this same drawing, the last two stages, just before reaching the *Alcázar* are in a smoother slope and do not have the height required for a tower.

In the inventory taken in 1639 there is no mention of the channels which would have connected basins of the towers, which on average would have been at least ten meters long. In the inclined plane system, few shorter channels would have been needed. Moreover, it is important to note what is not said in the contemporary descriptions: no mention of towers is present in any of them, and no description of the mechanism giving motion to the whole machine is given. In a vertical system, the driving mechanism would be prominent, and would have been referred to. In an inclined plane, it would be probably hidden under the moving arms, and it is more plausible to accept the lack of reference to it.

Xavier Jofre's model has not yet been thoroughly studied by specialists, who hopefully could cast further light on its feasibility. My belief is that the practical

FIGURE 3.26
Enlarged detail of fig. 3.12.[210]

hurdles such a complex mechanism would encounter are insurmountable, as the level of accuracy needed to have such an operating device would have been unattainable in those days, especially when the main available building material was wood. Nevertheless, a final opinion would depend on producing a working model, as both previous competing theories have reached recognition when their models were shown to perform their intended task. In spite of some facts weighing in favor of the inclined plane concept, in my opinion, even if the working model built by Peces Ventas demonstrates its feasibility, scaling it up to the real dimensions required by the topography of Toledo would make it impossible to operate, much for the same practical constructive reasons related to Jofre's concept. Thus, though there is no consensus among historians of technology, the leading opinion is that Reti/García Tapia's vertical model represents the real *Artificio modus operandi*, a view to which I personally concur. Nevertheless, since on a purely theoretical level, all conflicting systems could have performed the feat of raising water to the high city of Toledo, the controversy as to which was the one Juanelo actually used will probably live on, unless new and conclusive information would be brought to light.

The *Artificio*, as Seen by Contemporary Writers

A document in the Archivo de Simancas, dated about 1590, summarizes the popular contemporaneous assessment of Spaniards towards Juanelo's achievement. It asserts, "... the *Artificio* is a royal undertaking among the most distinguished in Europe, and the most important of this city of Toledo for the

210 Colmenar, *Les Délices de l'Espagne et du Portugal*, p. 321.

embellishment and glory of its *Alcázar* …"[211] In all cases contemporaneous Spaniards referred to Juanelo as one of their own, warmly praising him. A testimony is found in a book by Antonio Campo, written in 1585, in which extraordinary claims are made about Juanelo's capabilities. Some of the facts told [such as Juanelo's having become very wealthy] are evidently wrong.[212]

The most prominent Spanish writers of that period—Cervantes, Tirso de Molina, Quevedo, Gracián, Lope de Vega, and Góngora—lavishly extolled Juanelo's Ingenio, immortalizing the waterworks in their poems.[213] They placed the *Artificio* in the gallery of the most amazing monuments in Spain, also attributing to its wonderful mechanism aesthetic values. A typical literary testimony is Agustín de Rojas' *El Viaje Entretenido* (1604), one of whose dialogues refers to the *Ingenio*: "Solano: 'This work is the most illustrious and the most ingenious of its kind in the whole world, whose inventor was Juanelo Turriano, native of Cremona, who only for this work deserves the same glory as Archimedes of Syracuse …'"[214] Other writers also took pride in comparing their supposedly better technology with France's, as in Ambrosio de Zalazar's writings, where he states that: … this water in Toledo reaches twice the height [of Paris' device] and raises water as the body of an ox.[215]

211 AGS, *C. y S. R*, leg. 272–1, fol. 89.

212 Antonio Campo, *Cremona fedelissima città et nobilissima colonia de Romani rappresentata in disegno col suo contato, et illustrata d'una breve historia delle cose più notabili appartenenti ad essa, et dei ritratti naturali de duchi et duchesse di Milano, e compendio delle lor vite da Antonio Campo*, Milano: Bidelli, 1645, originally written in 1585, pp. 198–199.

213 Jose Cristobal Sanchez Mayendia, "El Artificio de Juanelo en la Literatura Espanola," *Cuadernos Hispanoamericanos* 103, (1948), pp. 73–92. This comprehensive work presents a detailed record of the admiring mentions of Juanelo in Spanish sixteenth and seventeenth centuries literature.

214 Agustín de Rojas Villandrando, *El viaje entretenido*, original written in 1603, (2 volumes), Vol. 1, Madrid: B. Rodríguez Serra, 1901, p. 296.

215 The original paragraph reads: "Aquí veras el grande fuerte y nombrado Alcaçar morada de los Reyes Moros, veras también subir el agua por cierta muencion del Rio Tajo arriba de dich alcaçar (como ha hecho Henrrique quarto Rey de Francia llamado el grande de buena memoria) en el puente nuevo su ciudad de Paris en dos figuras que hay de bronce muy bien hechas, la una de nuestro salvador, y la otra de la Samaritana, mas esta agua de Toledo sube dos vezes mas alta y sube tanta agua como el cuerpo de un buey." Ambrosio de Salazar, *Almoneda general de las más curiosas recopilaciones de los reynos de España*, Paris: Antonio du Brueil, 1612, p. 210. Zalazar uses the expression "como el cuerpo de un buey" [as the body of an ox] to imply figuratively a huge quantity [of water].

The *Ingenio* of Juan Fernández del Castillo

If the "*Artificio* de Juanelo" became famous while relegating its creator to extreme poverty, the inventor of the *ingenio* that succeeded it, notwithstanding the general opinion that it worked admirably, was simply forgotten. As already stated, Juanelo Turriano's grandson, named Juanelo as well, succeeded his grandfather in the responsibility to maintain the device. Young Juanelo died in 1597 and the following year Juan Fernández del Castillo, a servant to the King, was appointed as responsible for the maintenance of the *Ingenio*, with the same daily salary that his predecessor had received: four *reales* daily.[216] But in view of the costly repairs that the device constantly required,[217] he decided to build at his own expense a model of another type of machine, based on the principle of the Ctesibius pump.[218] The model was completed in 1602, but offered to the king only in 1605 after rigorous trials, as its designer explained that it was to be easier to build and maintain, supplying four times as much water, and had been tested at his expense for three years.[219] Moreover, Juanelo's first *Ingenio* was already ruined, its beams and timber work rotten and the foundation building sunk, so the *Alcázar* officials commissioned the royal architect, Francisco de Mora, to look into it and take a decision regarding its reconstruction or replacement.[220] In a memorandum, dated 5 July, 1605, the officers of the *Alcázar*, Lorenzo Oliveiro and Juan Bautista Monegro recommended replacing Juanelo's first *Ingenio* with Castillo's device.[221] Francisco de Mora visited the *Artificio* in August 1605 and proposed to keep the second Juanelo's *Ingenio* forever, and to dismantle the first one, using what remained of it for the construction of Castillo's device.[222] Although there is no expressed mention in Mora's report of the mechanism that would impart motion to Castillo's pumps, it was implicit in it that the same *Stangenkünst* mechanism that drove the numerous

216 Cervera Vera, *Documentos biográficos*, p. 262.

217 AGS, *C & S. R.*, leg. 304, fols. 72, 488 y 495.

218 Though the first references to this matter appear in Llaguno y Amirola, *Noticias de los arquitectos*, pp. 258–259, this discussion and the reference footnotes which allowed reviewing the original documents in the archives, are almost in their entirety based on Nicolás García Tapia, *Ingenieria y arquitectura*, pp. 292–299. This pioneer Spanish historian of technology, has almost singlehandedly researched Castillo's technical achievements, as he later did with Pedro de Zubiaurre's *Ingenio* of Valladolid, discussed further on.

219 AGS, *C & S. R.*, leg. 304, fol. 491.

220 AGS, *C & S. R.*, leg. 304, fol. 492.

221 AGS, *C & S. R.*, leg. 304, fol. 493.

222 AGS, *C & S. R.*, leg. 304, fol. 494.

stages of Juanelo's first *Artificio* would be used to drive Castillo's multiple pumps, which would reduce the total investment in the device. A later document confirms that that was the case.[223]

According to the usual procedure, a questionnaire was submitted to the experts to choose the most convenient solution, considering the initial cost and expected maintenance expenses.[224] In his report, Juan Bautista Monegro concurs with Francisco de Mora's idea of keeping the second Juanelo device and installing Castillo's pumps in-lieu of Juanelo's first *Artificio*, justifying this opinion in an ironic tone: "… for grandeur without utility and so expensive, one suffices …". He also praises Castillo and the proven operation principle of his pumps and the fact that it possessed five stages, a unique feature in those days.[225] In similar terms Nicolás de Vergara, surveyor of the works of *Alcázar*, described the first device made by Juanelo, writing that "… it is very deteriorated, … its very thin wood is rotting and would be too costly to repair … The second one is better built and kept, with a better foundation and materials …" He recommended to keep it "… more for its reputation than its utility." As for the *tisibica*, he says that since Antiquity, it was first implemented to raise water in Augsburg, and later in London and Turin,[226] where they worked very well. He expressed his belief that the five planned stages up to the *Alcázar* represented the solution, since the concept had already been tested in two stages up to the Monastery of el Carmen (the Carmelites), halfway up the hill in the direction to the old city center.[227]

Opposing these opinions was another expert—Francisco de Cuevas, who believed that in no case Juanelo's *Ingenio* should be dismantled, as "… the water rises naturally and not violently and … no machine can be found to get the water so high …," whereas "… the *tisibica* forces the water, therefore a short life can be expected …, the five stages concept planned by Castillo is unproven, … the water in such a device tastes bad …". He even accused Castillo and those who supported him of seeking unfair economic benefit, and Castillo

223 AGS, *C & S. R.*, leg. 303, sin foliar, 8 December, 1613.

224 AGS, *C & S. R.*, leg. 303, s/f, 15 May, 1606.

225 AGS, *C & S. R.*, leg. 303, s/f, 2 June, 1606. Spanish historian García Tapia attributes great importance to this statement, saying that apparently it was the first piston pumps in several stages built in the world. This is arguable, as Agricola's *de re Metallica* shows several multi steps applications already in 1550, although driven by a single axle (Agricola, *de re Metallica*, book VI, pp. 149–218).

226 For the evolution of water related enterprises in Turin, see: Stefano A. Benedetto, "Macchine idrauliche e attività artigianali a Torino nel XV secolo," in Giuseppe Bracco, (ed.), *Acque, ruote e mulini a Torino*, (2 volumes), Vol. 1, Torino: Archivio storico della città di Torino, 1988, pp. 174–194.

227 AGS, *C & S. R.*, leg. 303, s/f, 6 June, 1606,

specifically of neglecting his responsibility to maintain Juanelo's *Ingenio* in order to promote his own project.[228] De Cuevas' diatribe against the *tisibicas* is virulent. In a letter to the king he forcefully debates with Castillo's supporters, stressing that Juanelo's device had been abundantly proven.[229] In De Cueva's mind, the increase in water pressure which takes place inside the cylinders was an expression of 'violence vs. nature', compared with the human-like smooth transfer of water from bucket to bucket in Juanelo's *Ingenio*. Nevertheless, the technicians understood the principle of Ctesibius' pumps and convinced King Felipe III to order the dismantling of Juanelo's old *Ingenio* and its substitution by Fernández del Castillo's.[230] Felipe III decision probably resulted from his own experience, as he had resided at Valladolid, where his gardens enjoyed the water raised from the River Pisuerga by a Ctesibius pump built in 1603 by Pedro de Zubiaurre, to which we shall refer in chapter IV.

Construction began, taking advantage of the materials of Juanelo's first device while repairs of Juanelo's second machine, which was relatively in good shape, progressed under a team comprising a carpenter, an ironmonger, a welder and a helper. Castillo's device did not require additional construction, as it was less cumbersome, and fitted within the existing building.[231] The budget for Castillo's device was 37,900 *reales*, including all the materials which could not be utilized from Juanelo's first *Ingenio*.[232] By 1617 much of the materials of Juanelo's old *Ingenio* had already been sold and Castillo's device was near completion.[233] However, by the following year, the allocation for Juanelo's second *Ingenio* repairs ran out.[234] About this time two German engineers, Slun and Giraldo Palman, appeared in Toledo. They claimed to have a solution to raise water to the *Alcázar* using a new kind of invention (which they kept secret), presenting to the king a proposal in this regard. It was examined by the same persons responsible for the works of the *Alcázar*—Lorenzo Oliver, Francisco Gutiérrez de Luxan and Juan Bautista Monegro, who rejected it in June 1619, since Castillo's device appeared in their eyes to be a much better system.[235]

Despite continuing economic difficulties, the construction of Castillo's device progressed and by 1620, only one pumping stage was missing for the water

---

228 AGS, *C & S. R.*, leg. 303, s/f, 7 May, 1606.

229 AGS, *C & S. R.*, leg. 303, s/f, 8 June, 1606 and 20 December, 1606.

230 Cervera Vera, *Documentos biográficos*, p. 274.

231 AGS, *C & S. R.*, leg. 303, sin foliar, 8 December, 1613.

232 AGS, *C & S. R.*, leg. 303, sin foliar, undated (1613–1614?).

233 AGS, *C & S. R.*, leg. 303, sin foliar, 12 February, 1617.

234 AGS, *C & S. R.*, leg. 327, fol. 307 (27 April, 1618); fols. 322 y 323 (20 September 1618).

235 AGS, *C & S. R.*, leg. 328, fols. 262–265.

to reach the *Alcázar*.[236] Less fortunate was the quest to repair Juanelo's second *Ingenio*, plagued by numerous failures of the complex mechanisms. Wheels were worn out and loose; in winter, floods caused the water to be murky, and made odor and taste much worse. Thus, in January 5, 1624, officers from the *Alcázar* decided to stop the operation of the second Juanelo *Ingenio*.[237] This would be the beginning of the end of Juanelo's second *Ingenio*, which in a few years would cease to function altogether. Notwithstanding the fact that Castillo's device worked flawlessly, some people set themselves up as defenders of the *Ingenio* built by the Cremonese genius. There are plenty of testimonies about Juanelo's talents, and none about Castillo's less spectacular achievement. Thus, in October 1625, the *Corregidor* of Toledo called for the conservation of Juanelo's *Ingenio* arguing that both Juanelo's and Castillo's *Ingenios* necessitated caretakers. The capacities were similar, he observed, but the uniqueness of Juanelo's made it worthier.[238] The fact is that in 1622 Juan Fernández del Castillo had successfully supplied water to the *Alcázar* with his multi-staged pumps. However, fate seemed to be cruel with those who succeeded to raise water from the Tajo, and as the unhappy Juanelo ended up in poverty, so was to be Juan Fernández del Castillo's destiny. Repeating Juanelo's saga, he invested all his means in financing the construction without any reward. He submitted memorandum after memorandum to the King, demanding long overdue monies, and in one case, the possibility of being partially compensated by exploiting an unused water channel belonging to the first *Artificio*, all that to no avail.[239]

In 1625, already old, after 27 years of service, in a letter to the king delivered through the king's secretary, Castillo requested that his son, Juan Fernández del Castillo Rivadeneira, be nominated responsible for the upkeep of the *Ingenio*, justifying his request by his son's experience in dealing with the device.[240] A few months later, on January 26, 1626 Juan Fernández del Castillo died, and was indeed succeeded by his son, Juan [Jr.], as master of the *Ingenio*.[241] As was the case with the heirs of Juanelo Turriano, the Castillos lived in poverty, as can be seen from letters sent to the king by Inés Rivadeneira, widow of Juan del Castillo, who begged to receive the same salary that the wife of Juanelo '*el mozo*' (the younger) had received after his death.[242] Meanwhile, the only

236 AGS, *C & S. R.*, leg. 328, (18 September, 1620).

237 AGS, *C & S. R.*, leg. 331, fol. 332.

238 AGS, *C & S. R.*, leg. 331, fol. 332.

239 AGS, *C & S. R.*, leg. 303, sin foliar, undated (1621?).

240 AGS, *C & S. R.*, leg. 331, fol. 253 (12 July, 1625).

241 AGS, *C & S. R.*, leg. 332, fol. 426.

242 AGS, *C & S. R.*, leg. 332, sin foliar (9 July, 1626).

Juanelo device that was left decayed and ended up as a mere attraction for visitors—Juanelo's heirs desired to collect their money no longer primarily by water supply, but by charging fees to visitors to the Ingenio.[243] It was even considered to use its driving mechanism for Castillo's device, and put it in motion only when Castillo's device would be under repairs.[244]

The definite end of Juanelo's *Ingenios* came about in February 1639, when almost all the expensive brass parts were stolen, causing the king to order a thorough investigation.[245] As a result, Juan Fernández del Castillo Rivadeneira's wife, Felipa Almaguer, who was then responsible for the maintenance of the extant Juanelo's *Artificio*, as well as her children and several brass dealers, were indicted as accomplices in the theft. They all went to jail as a result of the ensuing trial.[246] On March 18, 1639, Felipe IV granted the title of master of the *Ingenios* of Toledo to Luis Maestre, a resident of the city, 'silversmith in gold' and watchmaker of the Cathedral.[247] Juanelo's second *Ingenio* had ceased to operate by then, whereas Castillo's mechanism did work, but was constantly affected by breakdowns. In the eighteenth century new solutions to the water supply problem were considered, with no results.[248] A British engineer, Richard Jones, attempted to install an effective pump and piping system to supply water from the Tajo to the city in 1723, but this initiative was first delayed due to bureaucratic inefficiency and finally abandoned due to Jones' premature death.[249] Another entrepreneur, the French Pierre Curton, tried to achieve the same objective in 1765, but desisted for lack of adequate funding.[250] Finally, in the same century, the remaining pipes were removed to the gardens of Aranjuez, and only the ruins of the *Artificio* buildings remained.[251]

---

243 Memorandum sent by Juanelo's heirs AGS, *C. y S. R*, leg. 271, fol. 248.

244 AGP, *C. R.*, tomo 13, fol. 270 (15 February, 1639).

245 AGP, *C. R.*, tomo 13, fol. 270 (15 February, 1639).

246 García Tapia, *Ingeniería y arquitectura*, p. 299.

247 Julio Porres Martín-Cleto, "El artificio de Juanelo en 1639," *Anales Toledanos*, XIV, Toledo, (1982), pp. 175–186, p. 176.

248 These attempts were first described in a doctoral thesis, an abstract of which was later published: Gabriel Mora del Pozo, "Un Ingenio del Agua en Toledo en el siglo XVIII," *Anales Toledanos*, t. XIII, (1980), pp. 113–124.

249 Richard Jones, *Articles of agreement between the imperial city of Toledo and Mr. Richard Jones: for supplying the inhabitants with water, by an engine to be erected on the River Tagus: confirmed by His Catholick Majesty, with the approbation of his royal and supreme council of Castile*, London: [s.n.], 1723.

250 Jesús Carrobles Santos, *Historia de Toledo*, Toledo: Editorial Azacanes, 1997, p. 335.

251 Julio Porres Martín-Cleto, "El final del artificio de Juanelo," *Toletum*, No. 10, Toledo, 1980, pp. 171–176, p. 172.

Land Reclamation and Sewerage in Spain

In Spain, due to orographical, geological and climatic reasons, wetlands are concentrated in the coastal areas in the south-east corner of the Iberian Peninsula. In the 1500s–1650s only a small number of individual, municipal and seigniorial initiatives took place, mainly the digging of ditches to discharge water from flooded areas. The peasants usually delimited their submerged fields with small dams and filled them with earth transported in boats or extracted from neighboring channels. Very modest progress was achieved, most probably due to demographic disasters (expulsion of the Moors from 1609 to 1614, the plague of 1647–1651, huge floods in 1651 and 1653) and, above all, the economic crisis. These land reclamation initiatives encountered opposition, sometimes violent, from the owners of oxen, which grazed in the grassy banks of the lagoons and from fishermen whose working area was reduced. Therefore, to ensure the success of such operations, as in England and France, it was necessary to recruit powerful sponsors to overcome the resistance of interest groups.[252] Following the demographic contraction suffered in the mid-seventeenth century, by 1700 the number of Spanish inhabitants reached once again the level of 1600. During the course of the eighteenth century the population increased drastically, and with it the demand for arable land. The need to dig large ditches demanded the provision of major capitals and in consequence the involvement of local oligarchies, though municipal authorities refrained from any investments. Political factors are also to be considered. The State had not played an active role, except in respect to the concession of plots for cultivation, though the Borbonic monarchy intervened more forcefully in economic life in order to boost production of grain.

Almost no clues can be found about any type of sewerage system in Toledo in the medieval and early modern period, and what exists is very imprecise. There are extant testimonies for sewerage channels and archeological finds of the Roman era—in two streets in the city center parts of a Roman sewer network have been discovered.[253] We find among a long series of ordinances, a

252 Guy Lemeunier, "Drenaje y crecimiento agrícola en la España mediterránea (1500–1800)," *Areas: revista de Ciencias Sociales*, Vol. 17 (1998), pp. 31 -41, p. 34.

253 References extracted from the Website of the Consorcio de Toledo, https://consorciotoledo.wordpress.com/2013/06/23/la-cloaca-romana-de-la-calle-alfonso-x-el-sabio/ and https://consorciotoledo.wordpress.com/2012/11/05/alcantarilla-romana-en-la-calle-de-la-sal/ accessed November 15, 2015.

specific one signed by Queen Isabel [Elizabeth] la Católica in 1501[254] instructing the *corregidor* of Toledo to enhance the city's beauty by building *alcantarillas* [it is not clear if the term refers to a conduit, a sewer, or a gutter] and broadening the city central square. We learn from contemporaneous writings that by the end of the seventeenth century the only reference to waste disposal in Toledo had the Romans as protagonists.[255] As happened with water supply, the city would ultimately be one of the last ones in Spain to deal with this problem. Between the late medieval period and the beginning of the 19th century, we do not find any data about sanitation in the city, which could mean that no progress had been achieved in that period of time. Barcelona too, like other Spanish cities, had a Roman sewerage infrastructure, which decayed and was abandoned during the Middle Ages.[256] We can learn from the case of Madrid, on the contrary, how Spanish authorities struggled to deal with the maintenance of streets and conservation of water. Already in 1202, el *Fuero*, a set of rules and legal bylaws governing life in the city,[257] had provisions devoted to the maintenance of streets and conservation of water. These legal provisions are indicative of the interest that the city council had for hygienic and sanitary issues. This interest was maintained during the thirteenth and fourteenth centuries and increased during the fifteenth century, when the city reached about 60 hectares, in conjunction with a remarkable demographic growth. The endless string of regulations, declarations and bylaws continued during the fifteenth and sixteenth centuries, even after the arrival of the court to the city in 1561.

Madrid, which reached 140,000 inhabitants in the middle decades of the seventeenth century, generated such a volume of garbage, feces and human, animal and waste from economic activities, that taking care of its disposal was a daunting task, but not impossible to achieve with the means then available.

---

254 Diego Clemencín, *Memorias de la Real Academia de la Historia: leido en la Junta Pública que celebró la Real Academia de la Historia el día 31 de julio de 1807. Elógio de la Réina Católica, doña Isabel,* (14 Tomos), Tomo 6, Madrid: Imprenta de I. Sancha, 1821, p. 261.

255 Pedro de Rojas (Conde de Mora), *Historia de la imperial, nobilísima, inclita y esclarecida ciudad de Toledo: Fundación, antigüedades, grandezas y principio de la Religión Católica en ella y de su Santa Iglesia ... Vidas de sus arçobispos y Santos y Cosas memorables*, Madrid: Diego Diaz de la Carrera, 1654, pp. 98–99.

256 Montse Martinez, Pere Malgrat, Cristina Vila, Roman Llagostera and Miguel Salgot, "History of the Sewerage System in Barcelona, Spain: From its Origenes to Garcia Faria Plan," in Andreas N. Angelakis and Joan B. Rose, *Evolution of Sanitation and Wastewater Technologies Through the Centuries*, 1st. Edition, London, IWA, 2014, pp. 383–400.

257 Antonio José Cavanilles, *Memoria sobre el fuero de Madrid, del año de 1202*, Madrid: Real Academia de la Historia, 1852, p. 13.

Nevertheless, one of the most important factors that affected the effectiveness of street cleaning was undoubtedly the lack of abundant water resources, as the few available were used almost entirely to cover basic needs of the population—to quench the thirst of people and animals, domestic and craft work, and minimum personal hygiene. By the beginning of the seventeenth century the solution adopted was to provide the city with a series of culvert or pipes that followed the layout of the streets to facilitate the evacuation of waste. These were not part of an integrated network but a series of specific infrastructures built to evacuate large concentrations of water and sludge in specific areas of the city.[258]

The Spanish Attitude Towards Science and Technology during the Sixteenth and Seventeenth Centuries

During the reigns of Carlos I and Felipe II, the growing international power of Spain and the annexation of new territories caused a propensity to innovative activities, attracting engineers, architects and teachers of great renown, who in many cases came to settle temporarily or permanently in the Iberian Peninsula. As we have seen, Juanelo Turriano achieved popularity on a European scale and other, less well-known foreign figures such as Jorge Ulrich and Louis de Foix were likewise in high demand by the Spanish monarchy.[259]

Among modern historians the almost consensual view purports the 'ignorance' of the Spanish inventors, especially in the most brilliant period of the *Siglo de Oro*, so named for literature or art, which some Spanish historians maintain could also have been applicable to technology. Their verdict is clear: Spain was technologically backward in the sixteenth century, chronically afflicted by economic crises. Its people showed no interest in matters of scientific and technological concern ...[260] This position is supported by the lack of

258 This short discussion is based on Virgilio Pinto Crespo, Rafael Gili Ruiz and Fernando Velasco Medina, *Historia del Proyecto de investigación Saneamiento de Madrid*, Madrid: Centro de Documentación para la Historia de Madrid. Colaboración entre la Fundación Canal de Isabel II y la Universidad Autónoma de Madrid, marzo 2012–diciembre 2014, pp. 70–73.

259 About the named engineers, see: García Tapia, *Ingeniería y arquitectura*, pp. 41–45. For a comprehensive list of inventors active in Spain during the sixteenth and seventeenth centuries, which includes a large proportion of foreigners, see: García Tapia, *Patentes de Invención Españolas*, pp. 13–37.

260 J. H. Elliott, *Imperial Spain, 1469–1716*, New York: Mentor books, 1966, especially chapter VIII, pp. 281–316 and Alberto Marcos Martín, *España en los siglos XVI, XVII y XVIII:*

qualified technicians and scientists in Spain at that time, as is quite evident in a letter of Francés de Álava, ambassador of Felipe II in the Court of France, dated March 9, 1581, in which he remarks that all the engineers in the service of the King were foreigners, which he considered much better than the Spaniard ones.[261] Furthermore, no technological heritage was left by the foreign engineers, or if it was, was not really absorbed by native technicians. As an early nineteenth century historian put it, "The works of Juanelo Turriano ... aroused the imagination, but not the will to know, and Juanelo did not leave disciples in Spain."[262]

Political, and sometimes religious factors induced decisions that greatly influenced technology diffusion. Even for the manufacturing of guns, though Felipe II preferred to recruit German founders, "[their] religious beliefs had to be carefully investigated; Felipe did not want to pollute Spain with Protestants of questionable loyalty, however great their technical skill."[263] This attitude undermined the knowledge base of imperial mining activities in America. The cessation of relations with Protestant Europe prevented technical renovation and caused a sharp decline in productivity. Mining yields and economic results could only be maintained with the exploitation of human beings, cheaper than advanced technology, and consequently, bringing natives in the New World close to slavery.[264]

---

economía y sociedad, Barcelona: Editorial Critica, 2000, p. 293. The assertion that Spain was untouched by the Scientific Revolution is challenged by, among others, William Eamon and Victor Navarro Brotons, in "Separata," *Más allá de la Leyenda Negra: España y la Revolución Científica / Beyond the Black Legend: Spain and the Scientific Revolution*, Valencia: Instituto de Historia de la Ciencia y Documentación Lopez Piñero, Universitat de València—C.S.L.C., 2007, pp. 8–14.

261 García Tapia, *Ingenieria y arquitectura*, p. 41. Translation extracted from Emilio Bautista, José Luis Muñoz and Javier Echávarri, "Juanelo (1501–1585)," in Marco Ceccarelli, (ed.), *Distinguished Figures in Mechanism and Machine Science: Their Contributions and Legacies*, (2 Parts), Part 2, pp. 95–110, Heidelberg: Springer, 2009, p. 98.

262 "Las obras ejecutadas por Juanelo Turriano que vino á la península reinando Carlos V, y escribió una descripción de sus ingenios y máquinas que se conserva entre los manuscritos de la biblioteca real, excitaron la admiración, pero no el deseo de saber, y Juanelo no dexó discípulos en España." Extracted from Diego Clemencín, *Memorias de la Real Academia*, p. 418.

263 David C. Goodman, *Power and Penury: Government, Technology and Science in Philip II's Spain*, Cambridge: Cambridge University Press, 2002, p. 110.

264 Julio Sánchez Gómez, "La minería," in Manuel Silva Suárez, (ed.), *Técnica e Ingeniería en España I: El Renacimiento*, (6 Volumes), Vol. 1, Zaragoza: Real Academia de Ingeniería, Institución Fernando El Católico y Prensas Universitarias de Zaragoza, 2004, pp. 439–476.

Opposing this view, Nicolás García Tapia raises the argument that Spain of the Golden Age was not a backward country in science. He refutes the claim that the prohibition, imposed in 1559 by Felipe II on all Spaniards to study or teach in universities in countries that were at war with Spain, reflects the 'obscurantism' of the king and the Spanish Church. In his view, the ban was not cultural or religious, but military, imposed in order to prevent the enemy from acquiring information on Spanish nautical, cosmography or weaponry knowledge.[265] In any case, the fact is that this royal act resulted in an isolation of Spain from scientific activities that were carried out in other parts of Europe.[266] Other scholars support Garcia Tapia's views, providing a long list of instances in which Spanish scientific and technological inventiveness were proven, mainly in the areas of geographical surveys, shipbuilding and cosmography, driven by the need to strengthen their grip in the New World.[267]

Furthermore, García Tapia, in a controversial article, convincingly argues that the author of "*Los veinte y un libros de los yngenios y máquinas de Juanelo, los cuales le mando escribir y demostrar el catholico rei d. Felipe Segundo Rey de Hespanas y nuebo mundo*,"[268] was Juan Pedro de Lastanosa, a Spanish architect and fortifications expert (c.1500–1576), and not a foreigner, as commonly believed by other scholars.[269]

---

265 For a comprehensive analysis of the rationale behing this policy on what concerns cosmography know-how in general and geographical data on the New World, see Marià Portuondo, *Secret science: Spanish cosmography and the New World*, Chicago and London, The University of Chicago Press, 2009, especially pp. 5–8.

266 For the complete text of the ban, enacted as a "Pragmática [an executive order by a competent authority] de Aranjuéz del 22 de noviembre de 1559," see Bernardo García Olmedo, *En torno a la Universidad española. Crónica no autorizada de un conflicto, Appendix "C,"* pp. 174–175. Website of Yumpu Digital Magazine, https://www.yumpu.com/es/document/view/37126013/bernardo-garcia-olmedo-en-torno-a-la-universidad-espanola, accessed 2.7.2016.

267 See: Raquel Álvarez Peláez, "Felipe II, La Ciencia y el Nuevo Mundo," *Re* J. H. Elliott, *Imperial Spain, 1469–1716*, New York: Mentor books, 1966, especially chapter VIII, pp. 281–316 and Alberto Marcos Martín, *España en los siglos XVI, XVII y XVIII: economía y sociedad.*

268 Pedro Juan de Lastanosa [or Pseudo Juanelo Turriano], *Los veinte y un libros de los yngenios y máquinas de Juanelo, los cuales le mando escribir y demostrar el catholico rei d. Felipe Segundo Rey de Hespanas y nuebo mundo*, BNM, MSS 3372/3376.

269 Nicolás García Tapia, "Pedro Juan de Lastanosa y Pseudo Juanelo Turriano," *Llull*, vol. 10, (1987), pp. 51–74. The Spaniard José Antonio García-Diego—the original editor of the *Los veinte y un libros*, was the first historian to demonstrate that Juanelo could not have been its author (See: José Antonio García-Diego, "The Chapter on Weirs in the Codex of Juanelo Turriano: A Question of Authorship," *Technology and Culture*, Vol. 17, No. 2

García Tapia takes this example as further proof that the explanation for the supposed Spanish 'backwardness' in the early modern era is a result of the poor technological comprehension of later historians, who failed to appreciate the importance of the data kept in the Spanish archives.[270] Notwithstanding this passionate defense, other historians contend that from 1580 and, especially, since the defeat of *The Armada Invencible* in 1588 and the death of Felipe II in 1598, a socio-cultural involution, an exacerbation of religious intolerance and economic decline presage one of the most critical periods of Spanish history, characterized by the impairment of science, technology and knowledge.[271]

(1976), pp. 217–234 and José Antonio García-Diego, "El manuscrito atribuido a Juanelo Turriano de la Biblioteca Nacional de Madrid," in *Actas del I Congreso de Historia de las Ciencias, Madrid: Sociedad Española de Historia de las Ciencias*, 1981, pp. 449–461, p. 460. Nevertheless, Ladislao Reti maintained until his death in 1973 that the manuscript was written by Juanelo Turriano—see: Ladislao Reti, "The Codex of Juanelo Turriano (1500–1585)," *Technology and Culture*, Vol. 8, No. 1 (1967), pp. 53–66. For an analized summary of Lastanosa's monumental work, see Angel Moreno Santiago, "La tecnica en el siglo XVI según los manuscritos de Juanelo Turriano," *Provincia—Revista de la Diputación Provincial de Toledo*, No. 74, Toledo, (June 1971), separata. For a list of the historians who have studied this issue, see Javier Goicochea Zala, "Azudes, molinos y otros aspectos de los Veintiún Libros de los Ingenios y Máquinas," *Revista de Obras Públicas*, No. 3396, (March 2000), pp. 59–68, p. 61. In this study, following the nowadays commonly accepted authorship, the book is attributed to Lastanosa.

The hipothesis that this work is based in a former manuscript by Giovanni Francesco Sitoni originally published in 1599 in Milan—José Antonio García-Diego and Alexander G. Keller, (eds), *Giovanni Francesco Sitoni, ingeniero renacentista al servicio de la Corona de España : con su códice inédito "Trattato delle virtù et proprietà delléacque," en su idioma original y traducido al castellano*, Madrid: Fundación Januelo Turriano, Ed. Castalia, 1990—is analized in an in depth comparison in María Isabel Ostolaza Elizondo, "Los veintiún libros de los ingenios y de las máquinas—composición, reconstrucción arqueológica, proceso de copia, copistas, datación," *Archivo de filología aragonesa*, Vol. 48–49, 1992–1993, págs. 225–262.

270 Nicolás García Tapia, "Ciencia y técnica en la España de los Austrias. Una visión desde la perspectiva de las investigaciones actuales," *Cuadernos Historia Moderna*, No. 15, pp. 199–209, Editorial Complutense, Madrid, 1994.

271 J. Patricio Sáiz, "El peluquero de la Reina," *Serie teoría económica e historia económica*, Departamento de análisis económico, Universidad Autónoma de Madrid, (January 2005), p. 15.

LONDINVM FERACISSIMI ANGLIAE REGNI METROPOLIS

CHAPTER 4

London

Tudor London in the 1580s

Unlike many European cities, in London the Reformation did not bring about mass bloodshed. The shift from Catholicism to Protestantism was relatively gradual, though the abolition of the monasteries and chantries caused much of the property within the City and its vicinity beyond its walls to change hands. The Crown redistributed land and buildings to its supporters, who converted them into houses for themselves, or developed them for industrial, residential or commercial purposes. The result was a property boom, a consequence of rapid population growth, with housing of all sorts becoming jammed in the old city and into the suburbs. The number of London's inhabitants exploded, from an estimated 50–75,000 in 1500, 150–200,000 in 1600 and up to 750,000 in 1700.[1] Though most of the people came in from the country, many migrants arrived from abroad, amongst them religious refugees. The new inhabitants worked in London's workshops, in the port, or in domestic service. London became one of Europe's great commercial centers, its trade spreading to the East and West.

Although the organs of power at the state and municipal level seem to have been as complex as, or more than, the administrative structure of Toledo, they proved themselves remarkably effective. A substantial divergence existed between the theoretical and real authorities in London in the sixteenth and seventeenth centuries.[2] The Common Council and Common Hall, which were

---

Opposite page—"London in the late 1560s." Georg Braun and Frans Hogenberg, *Civitates Orbis Terrarum*, (Ausgabe *Beschreibung vnd Contrafactur der vornembster Stät der Welt*), (6 volumes), Vol. I, Köln: Kempen, 1582, plate IV. 7.

1 For estimates of London's population in the sixteenth and seventeenth centuries, see: Norman G. Brett-James, *The Growth of Stuart London*, London: G. Allen & Unwin, 1935, chapter XX. The lower figures are given by Jan De Vries, *European Urbanization 1500–1800*, Cambridge (MA): Cambridge University Press, 1984, p. 270. According to Richard Holt and Gervase Rosser, "Introduction: The English Town in the Middle Ages," in Richard Holt and Greased Rosser, (eds), *The Medieval Town in England 1200–1540*, London and New York: Routledge, 2014, pp. 1–18, p. 6, the population by the late thirteenth century had peaked at 80,000.

2 This discussion is largely based on Roy Porter, *London, a Social History*, Cambridge (MA): Harvard University Press, 1995 and John Noorthouck, *A New History of London: Including Westminster and Southwark*, (Two books), Book 2, chapter II, London: R. Baldwin, 1773. We

 | DOI 10.1163/9789004312425_006

formally the representative elements in the constitution of the Corporation of London, officially and legally named the Mayor and Commonalty and Citizens of the City of London—the municipal governing body of the City of London—had very limited powers.[3] The Common Council was the City's legislative body, generally numbering 212 members, 6 or 8 elected by male freemen in each of the 26 wards,[4] usually five times a year, and convened and dismissed whenever the Lord Mayor decided to do so. Though in theory the Common Council had wide powers, in practice its approval of City laws, expenditure and taxes was a mere formality. Its elections and meetings were dominated by the Court of Aldermen, the elected body constituted by twenty-five aldermen and the Lord Mayor—an alderman as well. This body had evolved from the ancient Court of Husting—the supreme court of the medieval City, which possessed administrative and judicial functions.

The Congregation, or Common Hall, was an electoral body of all the liverymen[5] of the City (London's largest assembly, numbering about 4,000 men in the beginning of the sixteenth century), who met at the Guildhall, though rather infrequently. Its right to elect the Lord Mayor, other City officials and two of the City's four Parliament members was limited by various traditions and rules, which meant that in effect men who belonged to the elite, approved or nominated by the Court of Aldermen (described below), were always chosen. Every year Common Hall would propose the names of two candidates for the office of Lord Mayor—usually hand-picked senior aldermen; the aldermen themselves then made the final decision. The election was a complex procedure, which took place on Michaelmas-day (9 September), and the new supreme magistrate entered into his high office the following 9 November. Lord Mayors were elected for one-year terms, according to custom they did not serve more than one consecutive term.[6] In reality, the highest members of the hierarchy:

---

will focus only on the authorities relevant to the engagement and approval of projects that are the subject of this study.

3 The official name—'Mayor and Commonalty and Citizens of the City of London," remains until today the formal name, though it is generally replaced by 'The City of London.'

4 A self-governing unit, covering a small area within the wider city. For a detailed account of the wards, see: John Smart (of Guildhall), *A Short Account of the Several Wards, Precincts, Parishes, &c. in London*, London: N.A., 1741.

5 A liveryman was a freeman of the City of London, entitled to wear the livery of the ancient guild or city district to which he belonged and to vote in the election of Lord Mayor and other municipal officers.

6 For details on how the election was conducted, see: John Wilkes, (compiler), *Encyclopaedia Londinensis, or, Universal dictionary of arts, sciences, and literature*, (24 volumes), Vol. 13, London: J. Adlard, 1815, p. 595.

the Lord Mayor and the sheriffs were chosen among the ranks of the aldermen. These officials, together with the rest of the aldermen, de-facto formed a self-perpetuating oligarchy which operated as the undisputed executive governing body of the city, answerable only to the House of Lords and the king.

Aldermen held office for life. When vacancies occurred, candidates, though formally put forward by the ward, were selected by the Court of Aldermen. Promotion for aldermen came only after long service in the pyramid of office, and depended upon reputation and on meeting wealth qualifications that became harder to achieve over the years. In many cases their careers included terms of service as members of Parliament. The responsibility for order in their respective wards rested on the aldermen's shoulders, while senior aldermen, in their capacity as justices of the peace, presided over London's numerous courts and jails. A wide range of responsibilities were borne by the Court of Lord-Mayor and aldermen—they executed all leases, issued licenses and heard grievances. Aldermen also controlled the City's treasure and charities, including properties and bequests. In spite of this concentration of power in the hands of the Lord Mayor and the Court of Aldermen, a profusion of lesser bodies at the ward and parish levels caused a high degree of popular participation in running the day to day affairs of the population, a practice that was encouraged by the higher authorities. Attendance at the ward assemblies (ward motes), at which elections took place, was compulsory. Prominent residents were appointed to offices such as aldermen's deputies, instrumental in the settling of disputes within the affected neighborhoods.[7]

Among the most important officers that took care of the smooth running of the city was the office of Chamberlain, who acted as the city treasurer, receiving and paying all monies belonging to the corporation. Another influential functionary was the Common Serjeant, who attended the Lord Mayor and Court of Aldermen on court-days, and served as counselor to them. He took care of orphans' estates, and signed their legal contracts, before their submission to the Lord Mayor and Court of Aldermen.

Overall, notwithstanding the strains caused by the drastic upsurge in its population and the disappearance of sacramental rituals erased by the Reformation, which had played an important role in everyday life, Tudor London seems to have been politically and socially remarkably stable.[8] Its local government system, which constituted the oldest continuous municipal democracy in the world, had predated Parliament. Its constitution was rooted in the ancient rights and privileges enjoyed by citizens before the Norman

---

7 Porter, *London, a Social History*, pp. 51–52.

8 Holt, "Introduction: The English Town in the Middle Ages," p. 2.

Conquest in 1066. London's importance as a centre of trade, population and wealth secured it rights and liberties conceded by the Crown earlier than other towns and cities.[9] Guilds regulated trade and employment conditions, integrating into their ranks outsiders and giving some semblance of social mobility, exemplified by the myth of Dick Whittington, the apprentice boy who rose to become Lord Mayor.[10]

Early Attempts to Provide the City with a Water Supply System[11]

The earliest engineered water supply systems in Britain were built during the Roman period—mainly aqueducts which supplied civil and military centers from wells and springs. The London city proper was situated within the old Roman walls of Londinium, where the underlying geology consisted of terrace deposits of loams and gravel, lying above impermeable clay. These terrace gravels were water-bearing strata, and supplied water to the city's wells. They also fed a number of natural springs outside the city. In view of such an abundance of streams and wells with pure water, which satisfied all their needs, the Romans were not required to demonstrate in London their famous technical prowess in building aqueducts. Adam Gladstone, in an amusing book (1884) dealing with the history of Water Supply to London, observed: "... neither in the legendary nor the authentic history of our Metropolis does there appear ever to have been a time before the arrival of Julius Caesar when the inhabitants found a scarcity of water.[12]

Medieval water systems were constructed for monasteries as early as the twelfth century. As monks strove to locate abbeys where nearby sources of water at a sufficient altitude would enable its supply and drain, the development of

9 "History of the government of the City of London," Website of the City of London Corporation: http://www.cityoflondon.gov.uk/about-the-city/history-and-heritage/mansion-house/Pages/History-of-the-Government-of-the-City-of-London.aspx, accessed 24 August 2013.

10 Walter Besant, *A History of London*, Bremen: Europäischer Hochschulverlag, 2010 (originally published in London, 1894), p. 101.

11 For the most complete narration on this subject see: Henry Winram Dickinson, *Water supply of Greater London*, London: Leamington Spa, printed for the Newcomen Society, 1954. Part of this summary is based on this work.

12 Adam Gladstone, *History of the London Water Supply, from the creation of Man to CE 1884*, London: [s.n.], 1884, p. 1. Immediately following these remarks Gladstone relishes in sarcastically attacking his contemporaneous Metropolitan Water Board for its ineffective management of the sewage problems of London, pp. 2–3.

water-raising technologies was not a priority.[13] Similar conduit systems were built for some medieval towns. These early systems depended on gravitational flow from a spring to a conduit head, structures that often had considerable architectural elaboration. Most medieval towns in England and Wales were supplied by at least one well or spring. Some conduits, as at Exeter (fourteenth and fifteenth centuries),[14] were laid underground; some (as at Wells) ran in the street.[15]

During the Middle Ages it also became common practice to use the Thames as a source of culinary water: "The Thames was a silver river abounding in salmon, and providing good drinking water for those who could fetch it."[16] Houses distant from the river were usually built over beds of gravel, which enabled water flow to shallow wells. Water was also collected from the streams of Fleet, Holborn and Walbrook, which flowed down to the Thames within the City area.[17] Thus medieval Londoners had easy access to water from the river and its tributaries, and an abundant supply of groundwater. Until the thirteenth century water was provided to individual households by carriers, commonly called *cobs*, whose business was to deliver water from the river or primitive conduits to customers. Equipped with pails and pitchers, they took water from the shores of the Thames, the Walbrook, the Oldbourne, the Longbourne streams and from the Holywell, Clementswell, Clerkenwell, and other wells for their supplies.[18]

---

13 For an account of the importance given by monasteries in England to water supply, see: J. Patrick Greene, *Medieval Monasteries*, London and New York: Continuum, 2005, chapter 5, "Water management," pp. 109–132. Lucas, in *Ecclesiastical lordship*, p. 29, notes that frequently monasteries moved to new premises, due to the lack of water in their original location.

14 For the development of water supply in Exeter, see Mark Stoyle, *Water in the City: The Aqueducts and Underground Passages of Exeter*, Exeter: University of Exeter Press, 2014.

15 This subject is extensively discussed in Roberta Magnusson, *Water Technology in the Middle Ages: Cities, Monasteries, and Waterworks after the Roman Empire*, Baltimore: John Hopkins University Press, 2001.

16 William Garnett, *A Little Book on Water Supply*, Cambridge: Cambridge University Press, 1922, p. 107. For a comprehensive paper on water related matters in the Middle Ages, see Derek Keene, "Issues of water in medieval London to c.1300," *Urban History*, 28 (2001), pp. 161–179. The Roman built embankments, pile and quays influenced the natural flow of the river and its waterfront—see Roberta Magnusson, "Public and Private Urban Hydrology: Water Management in Medieval London," in Steven A. in Walton, (ed.), *Wind & Water in the Middle Ages*, pp. 171–187, p. 172.

17 For a description of the ancient London streams, see: Garnett, *A Little Book*, pp. 105–110.

18 A comprehensive description of the ancient wells of London is given in Septimus Sunderland, *Old London's spas, baths, and wells*, London: J. Bale, sons & Danielsson, 1915, pp. 11–37.

As the city grew the increase in population and the consequent pollution of streams and wells, it became necessary to seek farther sources of supply. To the north and west of the city other streams and springs were available, offering clear drinking water, which led to the construction of several conduits delivering it to the city. William FitzStephen, in his *Description of London*, 1180, praised the spring waters: "Round the city again, and towards the north, arise certain excellent springs at a small distance, whose waters are sweet, salubrious, clear, and, '... whose runnels murmur o'er the mining stones.'"[19] Usually the head of the conduit was placed near a natural spring and its water was used to fill a nearby cistern or tank. From the cistern, the water flowed by gravity through pipes running for about a mile or more, to be stored in a larger cistern in the city, equipped with dispensing taps in its lower part. Water drawn from the storage tanks was subsequently provided to individual households by *cobs*.[20]

In 1235 the Corporation of the City of London acquired land at the Manor of Tyburn from Gilbert de Sandford, with the right to build a reservoir in site, and lay pipes to convey the water of the Tyburn stream to fixed places in the City for the use of the citizens. Authorized in 1236, construction of the first building of what was to be called the Great Conduit was begun in 1245.[21] The water was supplied through six-inch lead pipes to a conduit head at West Cheap.[22] In his monumental description of London's urban development, John Strype (1643–1737) explains that the notable increase in population was the motivation for the construction of the conduit, which he dates to 1285.[23] The stone building housing the Great Conduit was long and low, containing a large leaden

19 William Fitzstephen, *FitzStephen's Description of the City of London, Newly Translated from the Latin Original; With a Necessary Commentary. A dissertation on the Author, Ascertaining the Exact Year of the Production, is prefixed: and to the Whole is Subjoined, a Correct Edition of the Original, with the Various Readings, and Some Useful Annotations*, by an Antiquary, London: B. White, 1772, p. 27.

20 Nicola Tynan, "London's Private Water Supply, 1582–1902," in Paul Seidenstat, David Haarmeyer and Simon Hakim, (eds), *Reinventing Water and Wastewater Systems: Global Lessons for Improving Water Management*, Hoboken (NJ): J. Wiley, 2002, pp. 341–360, p. 343.

21 The discussion about the Great and Little Conduits is based on Alfred Stanley Foord, *Springs, Streams and Spas of London; History and Associations*, New York: Frederick A. Stokes Company, 1910 specially pp. 252–263.

22 Part of the pipe can be seen today at the *Kew Water Museum*.

23 John Strype, *A Survey of the Cities of London and Westminster, Containing the Original, Antiquity, Increase, Modern Estate and Government of Those Cities*, (2 volumes), Vol. 1, London: A. Churchill, 1720, Book 1, chapter 5, p. 24.

cistern, mostly subterranean, from which water was dispensed from a spigot into a stone basin at the eastern end.[24] The historian Gwyn Williams has suggested that already at that early stage: "The council's major achievement was the establishment of a civic water supply ... The management of this public service became a model for all civic enterprise."[25] On practical terms, the city put in place a dual pricing scheme. All citizens, both poor and rich, could collect water from the conduits for free, while the commercial users were required to pay an annual fee, amounting to ten shillings in 1327, equivalent to about £350 in 2010.[26]

What was to be called the Little Conduit is also mentioned in a deed for a parcel of land, dated February 20, 1355, granted to the Mayor and the Commonality of the City, to serve as a fountain-head for the 'conduit of London.'[27] In 1390 permission was granted to build an extension to the existing conduit "... provided that the same work should not be injurious or harmful to the Great Conduit aforesaid."[28] However, the actual construction seems to have begun only over five decades later, as reported by Strype: "The Conduit in West Cheap, by Pauls Gate, (commonly called, The Little Conduit) was Builded about the Year 1442."[29]

The conduits were very popular features during major festivals. *The French Chronicle of London, Edward I*, mentions the Cheapside Conduit and the special occasion when King Edward and his wife were crowned at Westminster in 1273, and the Conduit in Chepe ran all the day with red wine and white wine to drink.[30] Gifts or posthumous benefactions were not uncommon, sometimes given in charity, for the purpose of building, maintaining or repairing a

---

24 Foord, *Springs, Streams*, p. 255.

25 Gwyn Williams, *Medieval London: from Commune to Capital*, London: University of London, Athlone Press, 1970, p. 84.

26 Debora Spar and Krzysztof Bebenek, "To the Tap: Public versus Private Water Provision at the Turn of the Twentieth Century," *Business History Review* 83 (Winter 2009), pp. 675–702, p. 680. The present value was calculated using professors Lawrence H. Officer and Samuel H. Williamson's purchasing-power calculator at their website: http://www.measuringworth.com, accessed 20 June, 2012.

27 Foord, *Springs, Streams*, p. 253.

28 Henry Thoman Riley, (editor and translator), *Memorials of London and London life, in the XIIIth, XIVth and XVth Centuries*, London: Longmans, Green and Co., 1868, p. 521.

29 Strype, *A Survey of the Cities*, Vol. 1, Book 1, chapter 5, p. 24.

30 Henry Thomas Riley, (ed.), "The French Chronicle of London: Edward I," in *Chronicles of the Mayors and Sheriffs of London: 1188–1274*, pp. 237–248, London: [s.n.], 1863, p. 237.

conduit.[31] The authorities appointed 'keepers' to ensure water was not stolen and brewers, cooks and fishmongers would have to pay for water use, at the discretion of the keepers. Many regulations were drafted to prevent a reckless use of water. Already in 1310 the Keeper of the Conduit had to take an oath that: "... he will well and trustily cause the conduit to be kept so that neither brewers nor fishmongers shall waste the water thereof, nor will he sell the water to any one by night or by day on peril of losing his freedom."[32] Those who took more than their share were subject to heavy punishments and several instances in which offenders were punished are recorded.[33] As an example, in 1345 an 'Ordinance that Brewers shall not waste the water of the Conduit in Chepe' was issued by the City. As was customarily the justification for the act preceded the regulation and subsequently the text details the prohibition and the harsh penalties for its infringement.[34] These measures were not effective enough. Wealthy Londoners living near the route of a conduit pipe sometimes managed to obtain authorization to bring a connection or 'quill' into their homes. Though water was supposed to be piped only into the dwellings of those having official permission, illegal tapping of the conduits was rampant.[35]

Water-carriers, besides filling their tankards from the conduits, used the river water to supply the houses of citizens for a small remuneration. Carts also conveyed water in still greater quantities from the Thames. However, the river tides caused frequent distress, causing the water to become salty and direct discharges into the river caused pollution, as did the accumulation of dung on the river banks. In the City, ordinances were made to regulate the rates at which the population would be charged by the water-carriers, per tankard or per cart. The *cobs* were organized under a water carriers' union or guild, mentioned by John Strype in his description of the Augustine Friers Church: "Statutes of the Brotherhood of St. Christopher of the Water Bearers: founded in the Friers Augustine: Confirmed by Thomas Brent, Dr. of Laws, Commissary

---

31 Foord, *Springs, Streams*, p. 261, quoting some examples from the *Calendar of Wills*.

32 Reginald R. Sharpe, (ed.), *Calendar of letter-books preserved among the archives of the Corporation of the City of London at the Guildhall*, London: J. E. Francis, 1912, pp. 236–237.

33 Claud Mullins, *London's Story*, London: G. Bell and Sons, 1920, p. 130.

34 Riley, *Memorials of London*, p. 225, quoting: 19 Edward III. CE 1345. Letter-Book F. fol. cvii. (Latin).

35 Charles Henry Richards, William Henry Christopher Payne and John Philpott Henry Soper, *London Water Supply: Being a Compendium of the History, Law, & Transactions Relating to the Metropolitan Water Companies from Earliest Times to the Present Day*, London: P. S. King & Son, 1899, p. 5.

of London, Oct. 20, 1496."[36] The inhabitants did not buy water itself but its transport, as the carrier collected it freely from the Thames or from public fountains. Indeed, the mayor of London, as were municipal authorities of other British cities, was concerned to maintain free access to water to poor inhabitants for drinking and cooking.[37] The conduit buildings were not only used for storing and dispensing water but also served for centuries a secondary function as billboards for moral instruction, such as when James I passed through the city on accession to the throne in 1603, on the walls of the Little Conduit were these verses: "Life is a dross, a sparkle, a span, / A bubble: yet how proud is man![38]

A parliamentary act in 1544—The London Conduit Act[39]—empowered the Mayor and Corporation to take steps for the improvement of the situation, allowing the city to bring water to London from Hampstead and other surrounding areas.[40] The Act describes as the background of its enactment that the water coming from old springs was "sore decayed, diminished and abated…"[41] Thus, a legal mechanism was put in place, but a practical implementation of the objectives of the act was yet decades away.

By the end of the sixteenth century, apart from water from the Thames, nine conduit systems existed in London, all on the western side of the Walbrook stream, which flowed down to the Thames through the center of the city.[42]

---

36 Strype, *A Survey of the Cities*, Vol. 1, Book 2, chapter 7, p. 115.

37 Mark. S. R Jenner, "L'eau changée en argent? La vente de l'eau dans les villes anglaises au temps de l'eau rare," *XVVe siècle* 2003/4, No. 221, pp. 637–651, p. 643.

38 Foord, *Springs, Streams*, p. 261.

39 The act was 35 Henry VIII, cap. 10. Extracted from Helen Miller, "London and Parliament in the Reign of Henry VIII," *Historical Reseach*, Volume 35, Issue 92, (November 1962), pp. 128–149, p. 146.

40 Henry Berry, "London's Water Supply—The Coming of the Companies," a lecture given on 19 February 1943 to the Metropolitan Water Board, London, 1943, pp. 1–17, p. 3; William Matthews, *Hydraulia, An Historical and Descriptive Account of the Water Works of London*, London: Simpkin, Marshal, 1835, p. 10.

41 John Raithby, (ed.), *The statutes at large, of England and of Great Britain: from Magna Carta to the union of the kingdoms of Great Britain and Ireland*, (20 volumes), Vol. 3, London: G. Eyre and A. Strahan, 1811, p. 436.

42 The list is detailed in John Stow, *A suruay of London: Conteyning the originall, antiquity, increase, moderne estate, and description of that city, written in the yeare 1598. by Iohn Stow citizen of London. Since by the same author increased, with diuers rare notes of antiquity, and published in the yeare, 1603. Also an apologie (or defence) against the opinion of some men, concerning that citie, the greatnesse thereof. VVith an appendix, contayning in Latine Libellum de situ & nobilitate Londini: written by William Fitzstephen, in the raigne of Henry*

The eastern side was supplied by wells.[43] The two main conduit houses were located at Cheapside,[44] which was in those days a large square. The Great Conduit, which had been rebuilt and enlarged in 1479[45] by Thomas Ham (one of the sheriffs), is clearly seen in the Agas Map (fig. 4.1), dated around 1560, in the east end of Cheapside, with tankards lying in the street in front of it.

In a view delineated in 1585 by the Ralph Treswell, leaden pipes can be seen along Cheapside, conveying water to various sites (fig. 4.2). Large tankards, holding from two to three gallons, in which water was fetched from the reservoir for domestic use, are seen around the Little Conduit.

Two dispensing taps are fixed to the sides of the building (fig. 4.3). It is obvious that Treswell copied from Agas, a phenomenon common among cartographers of that period.[47]

FIGURE 4.1 *The Great Conduit, detail of Agas map (c.1560).*[46]

---

the second, London: Iohn Windet, 1603, pp. 17–18. The capacities of the cisterns in the conduit-houses are not given by Stow or any other historian.

43 Foord, *Springs, Streams*, p. 253.

44 'Cheap' is Anglo-Saxon for 'market'.

45 Foord, *Springs, Streams*, p. 255.

46 Attributed to Ralph Agas, "Civitas Londinum: A Survey of The Cities of London and Westminster, The Borough of Southwark and Parts Adjacent in the Reign of Queen Elizabeth," paper on linen (75×185 cm), c.1560. For the sake of clarity, shown is a copy attributed George Vertue, 1737, which bears an additional tilte: "Civitas Londinum Ano Dni circiter MDLX," 1737.

47 Felix Barker, *The History of London in Maps*, London: Random House, 1990, p. 19.

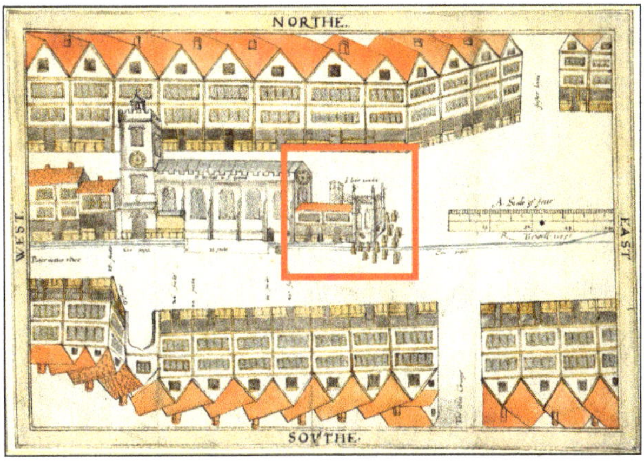

FIGURE 4.2 *The Little Conduit, Map of Cheapside (1585).*[48]

FIGURE 4.3
Detail, Cheapside map.[49]

48 Ralph Treswell, "West-cheap" (i.e. Cheapside), London, Bird's-eye view with church and houses shown in elevation, paper, pen and ink, watercolour, (27.1×39.3 cm), 1585, BM, DPD, Registration number: 1880,1113.3516, Crace Collection. (The collection is catalogued in Frederick Crace, *A Catalogue of Maps, Plans, and Views of London, Westminster & Southwark*, London: Spottiswoode & Co., printers, 1878).

49 Robert Chambers, *The Book of Days: A Miscellany of Popular Antiquities in Connection with the Calendar, Including Anecdote, Biography, & History, Curiosities of Literature and Oddities of Human Life and Character*, (2 volumes), Vol. 2, Volume London & Edinburgh: W. & R. Chambers Limited, 1832, p. 392.

Since the thirteenth century, for more than three centuries, the Lord Mayor and Commonalty, assisted by the private gifts of many citizens, continued to supply London with water. "Thus the corporation of London, down to the reign of Elizabeth, regarded it as one of their principal duties to supply the town with water, and see to the erection and preservation of conduits, to which the poor could freely resort."[50] But conduits were neglected and decayed with time, causing supplies to be neither sufficient nor reliable for the rapidly increasing city population, especially under the Tudors, a situation that undoubtedly contributed to the epidemics that often devastated the city for lack of hygiene. According to Gladstone, "The cost of water transport made personal hygiene unaffordable to the poor, and even to the wealthiest thorough washing was an occasional and expensive indulgence."[51]

By the mid sixteenth century London's drastic growth had once again made its water supply completely insufficient. Commercial and industrial users, such as brewers, took advantage of the free water, impairing supplies to the population, having no incentive to reduce their own consumption during periods of high demand.[52] Water scarcity affected the cobs as well: a petition was presented to the House of Commons by the water-tankard bearers of London in which it is stated that they and their families numbered 4,000 souls. The petition is undated but by its writing and spelling it is thought to have been drawn about 1600. It describes problems of operation and maintenance affecting London's conduit system. They complained that most of the water was taken and kept from the conduits by pipes laid into private dwellings.[53] Foreign visitors to the city, such as the Duke of Württemberg, Friedrich I, who visited London in 1592, showed sympathy for the cobs hard work, who carried water in heavy wooden vessels bound with iron hoops.[54] Nevertheless, the cobs had a tainted fame—"The water-carriers constituted [about 1600] a large class, and seem to have formed a rather unruly part of the population."[55] When water in the conduits

50 Joseph Fletcher, "Historical and Statistical Account of the present System of Supplying the Metropolis with Water," *Journal of the Statistical Society of London*, Vol. 8, No. 2 (June 1845), pp. 148–181, p. 149.

51 Gladstone, *History of the London*, p. 5.

52 Tynan, "London's Private Water Supply," p. 345.

53 Clifford, *A History of Private Bill Legislation*, Vol. 1, p. 61.

54 Friedrich I (Duke of Württemberg), "Travel Journal" (1592), in William Benchley Rye (editor and translator from the German), *England as seen by foreigners in the days of Elizabeth and James the First*, London: John Russell Smith, 1865, pp. 1–53, p. 8.

55 William Benchley Rye (editor and translator from the German), *England as seen by foreigners in the days of Elizabeth and James the First*, London: John Russell Smith, 1865, note 18 to page 8, p. 189.

FIGURE 4.4
A tankard [tanker] bearer (c.1650).[56]

became scarcer, disputes among cobs led to violence, so much so that special orders had to be issued against the use of weapons in order to secure the first place at the dispensing cocks. The city had its own supply of tankards for rent, stamped to provide official guarantee of their measure. When smaller quantities of water were required, servant-girls and apprentices were sent to the conduits, and stories abound about them as gossiping-places, where quarrels frequently arose.

Tankard-bearers were well-known London characters, and appear in a pictorial series of the professions of London preserved in the British Museum, drawn in the reign of James I (fig. 4.4). The City leaders were finally compelled to solve the water supply problem, and much alike many modern authorities, they were keen to transfer the responsibility for these projects to private entrepreneurs who were profit motivated. This attitude would result in the first steps taken by private initiative in the water supply business.[57]

56 John Thomas Smith, *The Cries of London: Exhibiting Several of the Itinerant Traders of Antient and Modern Times*, London: John Bowyer Nichols And Son, 1839, p. 17.

57 Felipe II of Spain adopted a similar strategy, when commissioning the construction of the Toledo waterworks to Juanelo Turriano. For more about the development of this concept in London, see: Spar, "To the Tap: Public versus, pp. 679–683.

The Elizabethan Patent System

Compared to other contemporary monarchs, such as Henri IV in France or Felipe II in Spain, Queen Elizabeth had a different attitude towards the hundreds of persons requesting support for their initiatives. A tight circle of courtiers played the role of intermediaries between the entrepreneurs and the queen. She neither met nor corresponded with engineers, medical experts or figures dealing with natural sciences.[58] First and foremost among these 'brokers' was William Cecil, First Baron Burghley, twice Secretary of State and Lord High Treasurer from 1572, who acted as the chief advisor of Queen Elizabeth I for most of her reign. Cecil's task included the identification of individuals who really possessed the know-how and skills to succeed, and then take care of supervising their work. He had on one hand to 'import' techniques unknown or unused in England, while on the other hand be careful not to threaten guild privileges.[59] Cecil monitored personally the public works projects, though when the number of proposals drastically grew, he delegated authority to a network of assistants, constantly requiring clearer plans and profit projections from entrepreneurs. This network allowed him to vet proposals on a relatively reliable basis. Visual proofs, such as the demonstration, carried out by Peter Morris, of the jet of water over the St. Magnus steeple (to be discussed presently), constituted specially convincing arguments.[60]

An almost unanimous view is that the Patent System was introduced into England as a structured system in the second year of Elizabeth's reign (1559).[61] Letters-patent granted by the queen, usually in accordance with Cecil's advice, constituted legal documents which were issued for different objectives, such as the establishment of corporate bodies, etc.[62] Basing themselves on the Venetian precedent, they became a tool for incorporating immigrant technicians, many of them protestant Dutch and French fleeing in the 1560s and 1570s from religious persecution, thus promoting innovation by encouraging

---

58 Deborah F. Harkness, *The Jewel House, Elizabethan London and the Scientific Revolution*, New Haven and London: Yale University Press, 2007, p. 147.

59 Ibid., p. 149.

60 Ibid., p. 157.

61 David Richard Seaborne Davies. "Further Light on the Case of Monopolies," *Law Quarterly Review*, Vol. 48, Issue 3 (July 1932), pp. 394–414, p. 396.

62 For a history of patents development, which includes a vast source list, see: Sa Yu, "Political Privilege, Legal Right, or Public Policy Tool? A History of the Patent System," International Association for the Advancement of Teaching and Research in Intellectual Property (ATRIP) Essay Competition 2009, 31 August 2009. Website of ATRIP: http://www.atrip.org/Content/Essays/Sa%20Yu.pdf, accessed 9 May 2012.

new projects.[63] The process was not transparent, as reflected in the sharp criticism expressed by the legal historian, Thomas Terell: "In the days of the Tudor Sovereigns this prerogative was abused and monopolies were granted to Court favorites without any pretence of merit."[64] The Stuarts don't fare better: "In the corrupt ages of the Stuarts, it is not to be wondered at, that the prerogative of the crown to grant monopoly rights to first and true inventors, should have been made a lever for assuming a prerogative to grant monopoly rights in trade generally."[65]

The Queen was not interested in ostentatious shows of technical prowess, as her fellow monarchs in the continent were, but on practical developments, in strengthening the currency and recovering land. Cecil translated this philosophy into three parameters when analyzing schemes for new projects: utility, economy and novelty—utility being the most important. During the 1560s and up to the 1580s the flow of proposals was intense, though as many of them failed, royal support diminished, while all the time speculators invested in waves, according to their ever changing financial possibilities.[66] For technically outstanding immigrants letters-patent represented a means of smoothing the road to denization, the equivalent of resident-alien status.[67] The combination of skilled workers and wealthy investors caused the City of London not only to become a center for the development of large-scale projects but also a model for financing, administering and managing them.[68]

Peter Morris

According to historian Charles Welch, the earliest surviving references to an attempt to use pumps at the London Bridge are held in the Bridge House

63 Joan Thirsk, *Economic policy and projects: the development of a consumer society in early modern England*, Oxford: Clarendon Press, 1978, p. 43. For an in-depth discussion, see: Laura Hunt Yungblut, *Strangers settled here amongst us: policies, perceptions, and the presence of aliens in Elizabethan England*, London: Routledge, 1996, especially pp. 95–113 and Harkness, *The Jewel House*, p. 150.

64 Thomas Terrell, *The law and practice relating to letters patent for inventions*, 6th ed., revised and rewritten by Courtney Terrell and Arthur Jaffe. London: Sweet and Maxwell, 1921, p. 1.

65 Thomas Terrell, *The law and practice relating to letters patent for inventions*. 3rd. ed., edited by William Peter Rylands, London: Sweet and Maxwell, 1896, p. 3.

66 Harkness, *The Jewel House*, p. 151.

67 Ibid., p. 154.

68 Ibid., p. 158.

records of the fifteenth and sixteenth centuries,[69] such as the item 'expenses of laborers at the waterworks', which appears in the accounts for 1479–80. A novel hydraulic machine of Flemish invention was described [rather incomprehensibly] by the bridge accountants in 1497.[70] There are mentions of repairs to the waterworks in the accounts for 1510, and in 1558 and 1559 the cost of various tools employed at the works are accounted for. The master carpenter, the warden and the mason of the works complained in 1523 that the woodwork of the bridge was being damaged as consequence of the obstacle to the current of the river created by "the mill now lying in one of the gullies which breaketh the right course of the stream."[71] Welch indicates that this mill probably formed part of the waterworks. Pumping had been tried before in England. Already in 1486–7 the Benedictine monks of the Finchale priory spent almost £10 on the construction of a horse driven pump to drain water from a mine at a pit in Moorhouseclose.[72] But it was during the second half of the sixteenth century that pumps and 'gins' (bailing engines) became common.[73]

Until the early 1580s, with the exception described by Charles Welch, none of the attempts to solve the problem of water supply to the city dwellers had been based on pumping water from the river. In 1582 a complex machine, located under one of the arches of the London Bridge and worked by the Thames tidal flow, began operating. Its objective was to lift water from the river up to the city for general consumption. It was designed and built by the Dutch engineer Peter Morris,[74] and later became to be known as The London Bridge

69 Charles Welch, *History of the Tower Bridge and of other bridges over the Thames built by the Corporation of London. Including an account of the Bridge House Trust from the Twelfth century, based on the records of the Bridge House Estates Committee*, London: Smith, Elder and Co., 1894, p. 86–87. Bridge House Estates was established by Royal Charter in 1282 with responsibility for the maintenance of London Bridge. The records occupy 47.5 linear meters in LMA shelves.

70 Welch, *History of the Tower Bridge*, p. 86.

71 Ibid., p. 57.

72 John Hatcher, *The history of the British coal industry*, (5 volumes), Vol. 2, Oxford: Clarendon Press, 1993, p. 213.

73 Louis Francis Salzmann, *English industries of the Middle Ages being an introduction to the Industrial history of mediaeval England*, London: Constable, 1913, p. 9.

74 Henry Thomas, in *The wards of London: comprising a historical and topographical description of every object of importance within the boundaries of the city. With an account of all the companies, institutions, buildings, ancient remains ... and biographical sketches of all eminent persons connected therewith*, London: J. Gifford, 1828, (2 volumes), Vol 1, p. 232 refers to Morris as German. So does Paul Hentzner, in *Paul Hentzner's Travels in England: During the Reign of Queen Elizabeth*, translated by Horace, late earl of Oxford, London: E. Jeffery, 1797, p. 29. Different authors spelled Morris name differently, either as Maurice, Morrice,

Waterworks. The project embodied a revolutionary concept: supplying houses on a large scale through wooden pipes laid in the streets.

The extant information about Morris is scant. He has been referred to as Fleming, Dutch or German, almost indistinctly, though his technical expertise in draining flooded lands reinforces the assumption he originated in the Netherlands. Morris had the status of 'free denizen' and as such he could hold letter-patents, which allowed him to purchase land and enjoy other rights.[75]

Morris had already tried his hand at water supply, as the Records of the City of Chester bear witness. He signed an agreement in 1574 with the Mayor and the Aldermen and Citizens of Chester to build a leaden conduit from a nearby spring to the city center. He was allocated the necessary quantities of lead and was paid £70.[76] He resurfaces in a petition to the Crown dated 1575: "Peter Morrice ... solicits a Patent for him for the sole right of making and employing certain hydraulic engines for the raising of water, draining marshes, &c."[77] The Patent granted on January 28th, 1578, gives the justification for its granting:

> ... Whereas oue welbeloved subjecte Peter Moris hath by his great labor and charges founde and learned the skill and coning to make some newe kynde and manner of engines to drawe and raise vp waters higher then nature of yt selfe only serveth out of any manner of fenne groundes or other places not nowe or heretofore ...[78]

It extended Morris a "... license for 21 years to make and put into use such engines..." though "... the license to be void if the licensee do not within three years from the present date put it in use".[79] Being Dutch, it would have been

---

Morice, Morrys, Moris, Moritz or Morys. Hereunder quotations will use the spelling as it appears in the respective originals. In the text Morris shall be used.

75 An alien admitted to rights of citizenship. See Steve Rappaport, *Worlds within Worlds: Structures of Life in sixteenth Century London*, Cambridge: Cambridge University Press, 1989, p. 42.

76 Rupert Hugh Morris, *Chester in the Plantagenet and Tudor Reigns*, New York: Read Books, 2008, p. 284.

77 Robert Lemon, (ed.), *Cal. SPD, of the Reigns of Edward VI, Mary, Elizabeth, 1547–1580*, London: Longman, Brown, Green, Longmans, and Roberts, 1856, CVI, No. 62, p. 512.

78 Extracted from Rhys Jenkins (in collaboration with E. Wyndham Hulme), "Notes on the London Bridge Waterworks," in *Links in the History of Engineering and Technology from Tudor Time*, London: Ayer Publishing, 1971, pp. 131–140, p. 132. The original patent is in the National Archives, Patent Roll, 20 Elizabeth, p. 10, membrane 34.

79 Margaret Post, (ed.), *Calendar of the Patent Rolls, Elizabeth I, Volume VII, 1575–1578*, London: HMSO: 1982, No. 3672, p. 540.

natural if Morris' expertise would have been dedicated to marshes draining, but instead, all his efforts were directed to water supply technology, while surrendering (or maybe selling), his invention for the draining operations, as we learn from a letter of the Privy Council to the Bishop of Ely, George Carlton and Humfrey Michell, written in 1580, requesting the allocation of workmen.[80]

From a letter delivered by Sir Christopher Hatton[81] to the Lord Mayor and Court of Aldermen, dated May 26th, 1580, in which he requests on behalf of "his servant, Peter Morice," a postponement of the deadline for the completion of the works, we can learn that Morris had previously applied to city officials for permission to convey water from the Thames to Leadenhall and that an agreement had been reached in this respect.[82] Some historians state that 1574 was the year in which "the Corporation arranged for a supply of water with a Dutchman named Morice"[83] This is curious, as it even predates Morice's above-mentioned request for his 1575 patent. That a former agreement existed is quite obvious and is also supported by an ensuing letter, dated July 5th, 1580, in which the Lords of the Common Council write to the Lord Mayor, Aldermen, and Commonalty, following a complaint by Morris about unpaid monies to him.[84]

The City had also agreed to provide the land for the erection of the Morris' works, but not having done so in time, they granted him license to attend his own business until such ground was provided. By 1580 Morris desired to proceed with the work according to his agreement, and had already spent £200 in preparing piles and stones for the foundation. As the City declined to complete the agreement, the Common Council demanded on 5 July 1580 to be informed about their reasons for refusing to do so.[85] As will be discussed later,[86]

80 Jean Roche Dasent, (ed.), *Acts of the Privy Council of England, new series, Volume XII, CE 1580–1581*, London: HMSO, 1896, pp. 68–69. For an account of the use of the patents by Carlton and Michell, see: Clive Holmes, "Drainage Projects in Elizabethan England: The European Dimension," in *Eau et développement dans l'Europe moderne*, edited by Salvatore Ciriacono, Paris: Ed. de la Maison des sciences de l'homme, 2004, pp. 87–102, p. 89.

81 Sir Christopher Hatton was an English politician, the Lord Chancellor of England and, according to Mary, Queen of Scotts—Queen Elizabeth's lover.

82 W. H. and H. C. Overall, (eds), *Analytical index to the series of records known as the Remembrancia, Preserved among the Archives of the City of London, CE 1579–1664*, London, E. J. Francis & Co., 1878, "Water I-28," p. 550.

83 Cecil T. Carr, "Introduction," in Cecil T. Carr, (ed.), *Select Charters of Trading Companies CE 1530–1707*, London: Bernard Quaritch, 1913, pp. xi–cxxxvi, p. cxxii, and Clifford, *History of Private Bill Legislation*, Vol. 2, p. 52.

84 Dasent, *Acts of the Privy Council*, pp. 85–86.

85 Overall, *Remembrancia*, "Water I-102," p. 551.

86 Infra, p. 194, footnote 124.

John Strype was the first historian to state that Morris orchestrated, probably in 1580, a public demonstration of the power of his pump by throwing a jet of water over the steeple of St. Magnus church, located at the bank of the Thames, at the northern approach of the Old London Bridge. This display impressed city officials, and according to Strype, greatly contributed to their positive attitude in allowing Morris to build his machine. Although no primary sources have been found to substantiate this story, it is generally accepted by researchers of the period.[87]

The definite reasons that finally caused the Lords of the Council to express clear support and come together to convince the city leaders to act can be inferred from a much later letter, dated December 1582, which records the developments that brought about the final approval of the project.[88] The Lord Mayor wrote to the Lord Chancellor:

> Bernard Randolph, Common Serjeant of the City, had lately charitably agreed to bestow a large sum of money for bringing water out of the River Thames, by an engine to be constructed by Peter Morice, from London Bridge to Old Fish Street, in like manner as he had already brought the water to Leadenhall, and by the way to supply the private houses of the Citizens, which offer had been approved by the Court of Aldermen. Thereupon Mr. Randolph had entered into an agreement with the Company of Fishmongers for such charitable deed. The matter being afterwards brought before the Common Council, they had granted the necessary license to the said Peter Morice for the carrying out of the work, which would profit the whole City, and be no hindrance to the poor water-bearers, who would still have as much work as they were able to perform so far as the water of the conduits would satisfy.[89]

As can be inferred from the letter above, there was at first opposition from the city's water carriers, but this hurdle was overcome. The document shows that Morris received payments from one of London's largest guilds—the Fishmongers' Company, an interest group that obviously understood the benefits of such an investment. The phrase in the letter transcribed in Remembrancia: "... in like manner as he had already brought the water to Leadenhall..." is intriguing, as there is neither a record whatsoever of such a

87 Overall, *Remembrancia*, "Water I-28," p. 551, footnote No. 1, states this feat was performed in 1580.

88 Dickinson, *Water supply*, p. 22.

89 Overall, *Remembrancia*, "Water I-449," p. 553.

feat being accomplished before 1582 nor any graphical sign in contemporaneous views of the city. In our opinion, for the sake of consistency, the correct reading should probably be: "... in like manner as had been proposed to bring water to Leadenhall..."

It seems that the commitment of Bernard Randolph to bestow a substantial amount for the completion of the project was crucial to the grant of a 500-year lease on one arch of the bridge by an indenture between Queen Elizabeth, John Branche,[90] Lord Mayor of the City of London, and the Commonalty and Citizens of the City, of the one part and Peter Morrys, of the other, dated 30 May, 23 Elizabeth (1581).[91] The conditions were very favorable: "... at a yearly Rent of ten Shillings, for the use of the Thames Water, and one Arch and a Place for fixing his Mill upon..."[92] The indenture was recorded in the Repertories of the London Court of Aldermen (fig. 4.5). Thus established, the road was open to provide the local population with a copious supply of water.[93]

When the lease was granted, construction seems to have already been at an advanced stage, as the first water pumping system began operating on Christmas Eve, 1582. It included two forcing pumps, built under the northernmost arch of the London Bridge, and operated by an undershot waterwheel turned by the river stream. Being moored in the arches of the bridge, they could take advantage of the stronger current there. The openings in medieval bridges were usually small and inadequate, making navigation difficult but favoring this application. The wheels could turn in either direction so that it

90 John Branche, a draper by profession, was first elected as alderman by the Cripplegate Ward in November 6, 1571. He became sheriff in the years 1571–1572, and Lord Mayor in 1580. He was the grandson of an alderman—another case showing the nepotism pervading the aldermen elections. Extracted from: Alfred B. Beaven, *The aldermen of the city of London: with notes on the parliamentary representation of the city, the aldermen and the livery companies, the aldermanic veto, aldermanic baronets and knights, etc*, (2 volumes), Vol. 2, London: Eden Fisher & Co, 1913, p. 173.

91 Carr, *Select Charters*, p. cxxiii, footnote No. 3. Mark S. R. Jenner, "From Conduit Community to Commercial Network?" in Mark S. R. Jenner and Paul Griffiths, (editors), *Londinopolis: Essays in the Cultural and Social History of Early Modern London*, Manchester: Manchester University Press, 2000, pp. 250–265, p. 269, gives the present location of the indenture as: Thames Conservancy Papers, 181/4/30. Despite efforts by the LME staff during the author two visits in 2008 and 2010 and following intense correspondence up to 2016, they were unable to locate the original document.

92 William Maitland, *The History and Survey of London from its Foundation to the Present Time*, (2 volumes), Vol. 1, London: T, Osborne and J. Shipton, 1756, p. 51.

93 Tynan, "London's Private Water Supply," p. 343.

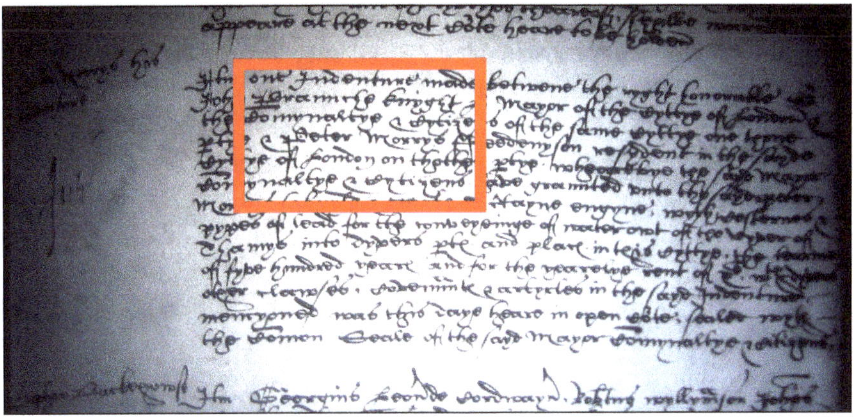

FIGURE 4.5 *Record of the indenture between Queen Elizabeth, the City and Peter Morris.*[94]

could be driven by the flowing and ebbing tide; the pumps ceased working for a short time when the tide was turning. The project was crowned with success: "And the Citizens, soon experiencing the Benefit of this Invention, granted him [Morris] a like Lease two Years after for another Arch; By which Means he grew very wealthy; and it continued in his Family, under various Improvements 'till the year 1701…"[95] The second lease for an additional waterwheel in an adjacent arch was granted on 24 December, 25 Elizabeth (1583).[96] It is not clear when exactly were the additional waterwheels and pumps installed, as they are not yet shown in drawings dated around 1600, such as Norden's view of 1597 (fig. IV-11) and the anonymous' view of c.1600 (fig. IV-12). This second waterwheel and the leases of additional arches evidently multiplied the waterworks capacity.

The London Bridge Waterworks in Contemporary Sources

The reference closest in time to the installation of the waterworks is in the second edition of Holinshed's Chronicles, continued after his death in 1580 by

94 CLRO, *The Repertories of the Court of Aldermen 1495–1835*, 58 microfilm reels, (3 Parts), Part 1, Repertories 1–24 in reels 1–18 (1495–1599), Bristol (England): Research Publications Inc., 1986, Repertory 20, fol. 201.

95 Maitland, *The History and Survey*, p. 51.

96 John Raithby, (ed.), *The Statutes of the United Kingdom of Great Britain and Ireland, From CE 1819; 60 George III–to CE 1822; 3 George IV*, London: John Eyre and Andrew Strahan, 1822, p. 1049.

Abraham Fleming and others, and published in 1587. The text represents a key element in understanding the operation and effectiveness of the system, and is therefore worth quoting in some detail:

> This yeare Peter Moris, free denizen, having made an engine for that purpose, conueied Thames water in pipes of lead over the steeple of St. Magnus Church, at the North end of London Bridge, and so into diverse men's houses in Thames Street, new Fish Street, and Grasse-street, up unto the North-west comer of Leaden hall (the highest ground of the citie of London),[97] where the waste of the first maine pipe ran first this yeare, one thousand five hundred eightie and two, on Christmasse eeven: which maine pipe being since at the charges of the Citie brought up into a standard there made for that purpose, and divided there into foure severall spouts, ran foure waies, plentifullie serving to the use of the inhabitants neere adioining, that will fetch the same into their houses, and also clensed the chanels of the streets, north towards Bishopsgate, east towards Aldgate, south towards the Bridge, and west towards the Stocks market. No doubt a great commoditie to that part of the Citie, and would be farre greater, if the said water were mainteined to run continuallie, or at the least at everie tide some reasonable quantitie, as at the first it did; but since is much aslaked, thorough whose default I know not, sith the engine is sufficient to conveie water plentifullie …[98]

It is not clear what was the reason for the 'aslaked'[99] performance of the device and for how long its water delivery was reduced. Peter Cunningham, when describing the Standard in Cornhill, wrote: "The water ceased to run between 1598 and 1603 …," but gave no details about his sources.[100]

97 The site was called Cornhill. The approximate height in that point is Ordnance Datum (the mean tide level at Liverpool) 60.5, or 48 feet over the Trinity High Marks (the Datum from which all the levels and contours are taken in Ordnance maps), according to Cyril Edward Nowill Bromehead, "The Influence of Its Geography on the Growth of London," *The Geographical Journal*, Vol. 60, No. 2, (1922), pp. 125–135, p. 129.

98 Raphael Holinshed, *The Chronicles of England, Scotland & Ireland*, (6 volumes), Vol. 4, (originally published 1587), London: J. Johnson, F. C. and J. Rivington and others, 1808, p. 496.

99 'Appeased,' in Middle English. See Geoffrey Chaucer, *The Canterbury Tales*, (originally published 1475), edited by D. Laing Purves, Auckland (New Zealand): The Floating Press, 2012, p. 86.

100 Henry B. Wheatley, *London past and present, its history, associations, and traditions*, (3 volumes), Vol. 3, London: John Murray, 1891, p. 301. Wheatley's book is based upon the handbook of London by Peter Cunningham, *Handbook of London, past and present*,

John Stow, who had been one of the contributors to the second edition of the *Chronicles*, adds in the first edition (1598) of his *Survey of London* an important detail to Holinshed's narration, by calling the 'engine' an 'artificial forcier', a term generally understood to mean a force pump.[101] Subsequent Stow's editions (1603 and 1633) do not include changes to the 1598 text in this respect. John Strype greatly expanded Stow's Survey, and published in 1720 *A Survey of the Cities of London and Westminster* in which new information is added: "Of later times Thames water was conveyed into mens houses by pipes of lead … It was observed, that the water conveyed into houses by this mill, did sooner become fine and clear than the New River water, and was ever a clearer water…."[102] Strype also offered a plausible explanation to Holinshed's obscure statement about water being conveyed over the steeple of St. Magnus Church: "For the Maior and Aldermen … saw him throw the water over S. Magnus steeple. Before which time, no such thing was known in England, as this raising of water.[103]

During the construction phase, Morris, as a private entrepreneur, had initially covered all costs by himself. Reginald Sharpe writes that "the civic authorities were so pleased with the results of his first efforts that they assisted him with a loan of £1,000 to perfect his work."[104] The immediate success of Morris' device brought about open criticism of the authorities for the impediments they had raised in front of the entrepreneur, and their slow reaction time. A typical article published by Hugh Platt in 1593 mocked this turnaround in the magistrates' attitude: "What graveman or magistrate of our Mertropolitan Citi, when that excellent water-force at the bridge was first attempted by that worthie Enginer Peter Morris, would either lend his eare or open his purse to the accomplishment of so memorable a worke."[105]

London: John Murray, 1850, p. 467. Jenner in "L'eau changée en argent?" p. 640, states that water was supplied only three times a week.

101 Stow, *A survey of London* (written in 1598, reprinted from the 1603 edition), p. 71.

102 Strype, *A Survey of the Cities*, Vol. 1, Book 1, chapter 5, p. 25. Although the style in which Strype's book is written makes it difficult to understand if this last paragraph refers to Stow's time, the comparison with the New River, inaugurated in 1613, shows it refers to Strype's time.

103 Strype, *A Survey of the Cities*, Vol. 1, Book 1, chapter 5, p. 27.

104 Reginald Sharpe, *London and the Kingdom*, (3 volumes), Vol. 2, London and New York: Longmans, Green and Co., 1894, p. 19. He gives as his sources CLRO, *Jour. 21*, fol. 251; *Jour.* 22, fols. 47 and 53b.

105 Hugh Platt, *A briefe apologie of certaine nevv inventions, whereof there hath bene a publicke viewe taken in London by some of her Maiesties priuie Counsell, and diuerse other gentlemen and citizens of good worship and account, with the authors excuse for not publishing the same, to the generall good of his country*, 1 of 3 sheets, London: Richard Field, 1593.

The establishment of the water companies in London, and later in the whole of England, put in motion a process of profound social change. The concept of levying money for water was being resisted by the population. As a consequence of this public pressure, ordinances regulating the use of public wells and fountains were established—for example, the supply connection to Essex House, where water was wasted in stables and laundry, was cut;[106] horses were forbidden to drink from public fountains;[107] several requests for connections by prominent citizens were denied.[108] These measures were part of a sort of 'moral economy', as Edward P. Thompson has called it, whereby the basic necessities needed to be preserved for the benefit of the community.[109] However, such acts were not considered sufficient by the majority of the population. Popular sentiment was strongly against the new companies: John Stow, a close friend and admirer of Hugh Myddelton, when referring to the opponents of the New River project, characterizes them in blunt words: "… those enemies to all good endeavors, danger, difficulty, impossibility, detraction, contempt, scorne, derision…"[110]

Some cases are recorded in which servants or water drawers planned to disrupt private taps grafted to pipes feeding fountains, therefore drying them up several months per year and leaving residents without water.[111] In addition, water carriers led vigorous campaigns for the maintenance of public fountains and against competition, denouncing as unfair the practices of the distribution companies. Although these men had a reputation as troublemakers, their protests were heard at the beginning of the seventeenth century because they could count on the loyalty of customers they interacted with daily. Some wealthy citizens spoke of 'their' water-carrier and they were taken care of in their wills, just like any other servant.[112] Thus, subscribing to a distribution company meant that the household broke the bonds of charity, loyalty and community. It was a decision that affected all social relationships much more than comfort at home.[113]

In general the opposition to pumping water from rivers was widespread, led by farmers and millers: "The first claimed their fields would be flooded, while

106 Overall, *Remembrancia*, "Water II-321," 8 June, 1608, p. 554.

107 Jenner, "L'eau changée en argent?" p. 646, footnote 36: CLRO, *Jour. 13*, fol. 206 v.

108 Overall, *Remembrancia*, "Water," pp. 554–561.

109 Edward P. Thompson, "The Moral Economy of the English Crowd in the Eighteenth Century," *Past and Present*, No. 50, (1971), p. 79.

110 Stovv (Stow), *The survey of London*, (1633), p. 12.

111 Jenner, "L'eau changée en argent?" p. 643.

112 Ibid., p. 645.

113 Ibid., p. 645, footnote 47.

the latter claimed that their mills would no longer have enough water left to rotate."[114] To counteract these trends, contractors and utility companies when submitting projects often resorted to rhetoric of public good to neutralize criticism. Somewhat later, when Hugh Myddelton first proposed the project of the New River to the City of London, he spoke little of the sale of water, but stressed the possibility of that the water would help in cleaning the large stinking pits in the city.[115] The company also emphasized—and rightly so—the asset represented by the water supply in the fight against fire.[116] Nevertheless, many people doubted that the companies were primarily concerned with the general interest. As proof, when the City made a financial contribution to the company, voices were raised to denounce the fact that the municipality had allowed Myddelton and his associates to hijack a public asset for private gain.[117] Without the rhetoric of public well being and the municipal support that resulted, it is unlikely that companies like the New River Company would have overcome the formidable obstacles put in the way of their projects. Ben Jonson understood the deep change brought about by Morris and others, when he wrote in 1598: "We have water-companies now instead of water-carriers..."[118]

The authorities were rightly concerned that, unlike those who could afford to pay, the poor would not be able to afford services from the private sector and some poor areas did not have supplies at all.[119] As a consequence, token gestures on the side of the private sector were made: though it was reluctant to supply water to the poor, some water was provided free through charities in the form of public fountains. It was only much later, in the early 1800s, that the London-bridge Waterworks company instituted a sort of Roman-style cross subsidies for the supply of water: extra charges were levied from brewers and tradesmen, while keeping the lower fees for commoners.[120] By the seventeenth

114 Jenner, "L'eau changée en argent?" p. 646.

115 Bodleian Library, Oxford, ms. Tanner 98, fol. 47 r., extracted from Jenner, "L'eau changée en argent?," p. 646, footnote 50.

116 J. V. Lyle (editor), *Acts of the Privy Council of England 1616–1617*, volume 35, London: HMSO, 1927, p. 99.

117 British Library, London, microfilm 282, Duke of Northumberland, ms. 8, fol. 171 r., extracted from Jenner, "L'eau changée en argent?" p. 646, footnote 52.

118 Ben Jonson, *Everyman in his humour*, written in 1598, first acted in 1601, first published London: William Stanley, 1616, an unidentifiable edition quoted in: John Timbs, *Curiosities of London: exhibiting the most rare and remarkable objects of interest in the metropolis*, London: D. Bogue, 1855, p. 551.

119 Fletcher, "Historical and Statistical Account," p. 175.

120 Naren Prasad, "Privatisation of Water: A Historical Perspective," *Law, Environment and Development Journal*, Volume 3/2 (2007), pp. 217–233, p. 221.

century, Morris' and later established companies had transformed the status and cultural significance of water.

Technical Operation

Although no original drawings have survived, scholars generally agree as to the operational principle of Peter Morris' system. The most important primary source available is a clearly understandable conceptual sketch, though with no dimensions indicated and not scaled, drafted by John Bate, a technically-gifted writer who examined the waterworks in 1633, after a fire occasioned by the carelessness of a maid-servant in placing some hot coals under a pair of stairs had destroyed the upper floor of the building in which the device was housed:[121]

The appearance of the bridge after the 1633 fire, as Bate would have seen it, is reflected in a Wenceslaus Hollar view, shown in fig. 4.6. Bate describes both the circumstances that enabled him to record its main features and its operation.[122] We can rely on Bate's description if we suppose that between the

FIGURE 4.6 *London Bridge after the fire of 1633 and before the great fire of 1666.*[123]

121 George Lillie Craik, "London Bridge," in Charles Knight, (ed.), *London*, (6 volumes), Vol. 1, London: H. G. Bohn, 1851, pp. 73–96, p. 83. For a very detailed account of the fire development and results, see: Gordon Home, *Old London Bridge*, London: John Lane, 1931, pp. 210–215.

122 John Bate, *The Mysteries of Nature and Art*, 3rd. edition, London: R. Bishop for Andrew Crook, 1654, p. 51. The original book was written in 1635.

123 Wenceslaus Hollar, "London before and after the Fire: two long views extending from the Temple on the west to the Tower on the right, with buildings including Baynard's castle,

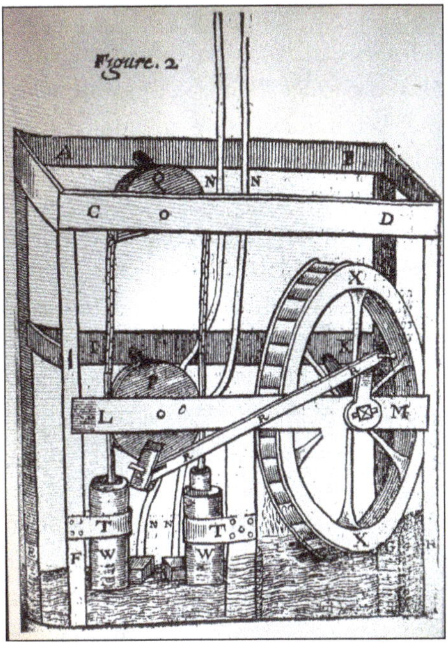

FIGURE 4.7 *John Bate's model of the waterworks (1633).*[124]
A large waterwheel (XX) is set in two brass sockets fixed onto middle beams, which are part of a timber frame. To these beams wheel P is attached, and above it is set the half-wheel Q. Two metal barrels (WW) are attached to the posts of the frame, each barrel fitted with a well leathered forcer. To the top of the forcers are attached two wooden rods, 2 feet long and 2 inches thick, prolonged by iron chains that, on their part, are linked up to an iron band around the half-wheel Q. The half-wheel Q is driven by wheel P through the iron chains linked unto it, which in turn imparts a reciprocating motion to the forcers, displacing the water inside the barrels to the output pipes (NN). A wooden bar (RRR), is attached, on one end, to the handle of the main wheel XX and on the other, to the spoke of wheel P, transforming the rotating motion of XX into an alternating motion to wheel P.

construction of the device in 1582 and the fire in February 1633 only minor repairs were performed and no drastic changes were introduced (fig. 4.7).

Stiliard, Guildhall, St Paul's Cathedral, and London Bridge; a later state with Overton's address," (25.4×70.8 cm), 1666, BM, DPD, Registration number: Q,6.58. The figure shows a detail of the upper part, before the Fire. Arthur Mayger Hind, *Wenceslaus Hollar and his views of London and Windsor in the seventeenth century*, London: John Lane, 1922, plates XXVI and XXVII, describes the maps in pp. 49–50.

124 Bate, *The Mysteries of Nature*, p. 53. Adaptation of description in Bate, *The Mysteries of Nature*, p. 52, in a summary in Abraham Wolf, *A History of Science, Technology and Philosophy in the 16th and 17th Centuries*, New York, Harper Torchbooks, Harper & Brothers, 1959, p. 533, with additions by the author of this study.

It is obvious that the pumps were simply force pumps, the wooden plungers covered with leather seals to prevent leakages. It is probable that the whole framework floated, so as to maintain the correct depth of the waterwheel as the river flowed and ebbed, but the sketch doesn't give any indication of it. Bate's description partly explains the fact, later noted by Strype, that the water was apparently clearer than the one supplied by the New River, since it is said to have been filtered through a fine mesh strainer before distribution.[125] Cleverly, as Bate indicates, continuous water supply to users was provided by a high reservoir, which acted as a buffer for the times of the day when the changing of the tide caused the current of the river and consequently the waterwheel to stop.[126] But not everyone thought that the water was of good quality. An Italian visitor in Stow's time, noted, for example, that: "it was hard, turbid and stinking."[127]

Chance contributed to support Bate's interpretation. The foundations of what was to be called the New London Bridge were excavated in 1828.[128] From a letter written in 1849 by a certain Mr. W. Baddeley—who apparently took part in the excavations—to the editor of the *Mechanic's Magazine*,[129] we learn that remains of pumps were found in the northern end of the bridge. Mr. Baddeley, whose writing reveals his technical prowess, attributed the remains to Morris' waterworks.[130] In figures 4.8 and 4.9 it can be seen that the New Bridge traversed the river passing over a point coinciding exactly with the location of Morris' waterworks. This fact reinforces the assumption that the remains found belonged to Morris' engine. According to Baddeley, the relics consisted of two cast iron pump barrels about 120 cm. long, with a bore 14.5 cm. in diameter, with protruding trunnions to hold them in place. The valves were fitted in two separate chambers, and as the suction pipes were only 10 cm. long, they had obviously been immersed in the river water. With the barrel was found a large square iron shaft which appeared to have carried a cross arm, to each

125 Bate, *The Mysteries of Nature*, p. 52.

126 Lynn White places the appearance of the first tidal mills in Europe in the eleventh century: White, *Medieval Technology*, p. 84.

127 Rosamond Joscelyne Mitchell and Mary Dorothy Rose Leys, *A history of London life*, London: Longmans, Green, 1958, p. 234.

128 By the end of the 18th century the old London Bridge needed to be replaced. The new bridge was built 30 m west (upstream) of the original site. The old bridge continued in use while the later called New Bridge was being built, and was demolished after the latter opened in 1831. The current London Bridge dates from 1972.

129 Joséph Clinton Robertson, (ed.), *The Mechanics' Magazine, Museum, Register, Journal and Gazette*, Vol. IX, London, 1828, page 155.

130 Joséph Clinton Robertson, (ed.), *The Mechanics' Magazine, Museum, Register, Journal and Gazette*, Vol. LI, London, 1849, p. 581.

FIGURE 4.8 *Top view of the northern half of the old London Bridge (c.1634), enlarged detail.*[131]

FIGURE 4.9 *St. Magnus Church, Old and New London Bridges (c.1830).*[132]

131 Anonymous, "Plan of the northern half of the old London Bridge, incorporating a view of the water works, on the west side of the bridge opposite St Magnus the Martyr to enable Thames water to be piped to citizens' homes," watercolor/pen, paper, (111×41 cm.), c.1634, LMA, *Main Print Collection*. Catalogue No.: q6886938. The plan post-dates the fire of 1633.

132 George Scharf (The Elder), "View in the late 1820s, south across the River Thames towards Southwark," watercolor, (49×67 cm.), 1831, in Celina Fox, (ed.), *London—World City, 1800–1840*, New Haven (CT) and London, Yale University Press, 1992. The original is kept in the London Metropolitan Archives. The old London Bridge is on the left and the new London Bridge is on the right.

end of which one of the pump rods was attached. All the remains bore marks of fire, as was clearly distinguishable in one of the oak piston rods. The axles of the shaft were "much worn on the underside, as if from having performed a reciprocating movement in an arc of about 90 degrees."

Though unfortunately, the scarred findings were not preserved, thus preventing the use of modern techniques to precisely date them, the information on their form and dimensions allow us to roughly estimate the quantity of water lifted daily by the waterworks in 1582.

Supporting Evidence for Bate's Interpretation of Morris' Operational Principle

A second completely unrelated source which indirectly corroborates Bate's graphical description was found in the framework of the present research, and is described below. This supporting evidence refers to a unique water supply undertaking in Valladolid, Spain.[133]

In 1601, an event completely transformed the signs of decay that the city had began to show towards the late sixteenth century, when King Felipe III and his *válido* the Duke of Lerma decided to move the court from Madrid to Valladolid.[134] In preparation for the King's and his entourage arrival and the installation of the Court in the city, a number of houses and palaces were conditioned, constituting a compound named 'La Ribera'.[135] A considerable amount of water was needed to feed the fountains and the gardens surrounding the huge residence, and given the remoteness of high springs, this could only be achieved by raising water taken from the adjacent river Pisuerga by some device. The problem would be solved by the Basque General Pedro de Zubiaurre, a daring sailor, who had been imprisoned in 1584 in London when the English discovered he had been plotting to capture for Spain the Dutch city of Vlissingen. He was kept in prison for two years in London and later in Holland, from where he was

---

133 This section is largely based on Nicolás García Tapia, "El ingenio de Zubiaurre para elevar el agua del río Pisuerga a la huerta y palacio del Duque de Lerma," *Boletín del Seminario de Estudios de Arte y Arqueología*, Tomo 50, 1984, pp. 299–324.

134 Juan Antolínez de Burgos, *Historia de Valladolid, publicada, corregida, anotada y adicionada con una advertencia por Juan Ortega y Rubio*, Valladolid: Libreria Hijos de Rodríguez, 1887, pp. 180–182.

135 Antonio Feros, *El Duque de Lerma. Realeza y privanza en la España de Felipe III*, Madrid: Marcial Pons, 2002, p. 170.

released in 1588.[136] At the time London was already supplied water by Morris' waterworks. Before his imprisonment Zubiaurre had had the opportunity to carefully study the waterworks mechanism, to memorize their different components and to produce small scale models. A plausible explanation for his interest in Morris' device, according to the Spanish historian Nicolás García Tapia, is that he intended to sabotage them.[137]

By 1603 Zubiaurre resided in Valladolid, where the Spanish Court was now established. He proposed to the authorities to build a device to supply water to the city by raising it from the Pisuerga by an engine constructed according to what he had learned in London. He submitted his models (no records exist about these being accompanied by drawings or calculations) to the consideration of Felipe III. The reception was very positive and the project approved.[138] The works were completed in 8 months, and in March of 1604 waters raised by Zubiaurre's *Ingenio* irrigated for the first time the gardens of La Ribera.[139]

Valladolid attracted visitors, and among them the Portuguese Tomé Pinheiro da Veiga, who was invited to visit the *Ingenio* and was amazed by its performance:

> As no springs were available for the fountains, water was taken from the river with an invention, easily lifted to a height of 150 palms[140] or more, with metal pumps, equipped with plungers driven by wheels turned by the river flow, which after seeing it, is simple and costless ...[141]

---

136 J. M. Lacunza, "Pedro Zubiaur y su relación con Rentería," *Revista Oarso*, Ayuntamiento de Errenteria, No. 49, (2004), pp. 89–91, p. 90.

137 Nicolás García Tapia, *Patentes de Invención Españolas en El Siglo de Oro*, Madrid: Oficina Española de Patentes, 1994, p. 11.

138 Juan Agapito y Revilla, *Los abastecimientos de aguas de Valladolid: apuntes históricos*, Valladolid: Imprenta La Nueva Pincia, 1907, p. 34.

139 Manuel Gracia Rivas, "En el IV Centenario del fallecimiento de Pedro Zubiaur, un marino vasco del siglo XVI," *Itsas Memoria. Revista de Estudios Marítimos del País Vasco*, 5, Untzi Museoa-Museo Naval, Donostia-San Sebastián, (2006), pp. 157–171, p.169.

140 Equivalent to about 30 meters, probably 25% less than the height achieved by Morris' pumps.

141 "Y, como no había manantiales para la fuente, se hizo una invención, con que muy fácilmente la llevaron del río, y está corriendo sin intermitencia y elévase del río ciento cincuenta palmos o más, con mucha facilidad, con unas bombas de metal, con bombardas y unas ruedas que se mueven con la corriente del río, cosa, después de vista, muy fácil y de ningún coste". Tomé Pinheiro da Veiga, *Fastiginia, Vida Cotidiana en la Corte de Valladolid*, (original written in Portuguese in 1605), translated into Spanish by Narciso Alonso Cortés, Madrid: Ambito—Fundacion municipal de cultura, 1989, p. 70.

Zubiaurre openly acknowledged to Pinheiro that he had copied from and improved on Morris' bigger-scale technology: "He said it was not his invention; he had seen it in London during his imprisonment and had introduced some improvements..."[142] A clue about the operation principle of the device is found in the Archivo General de Simancas, where a document dated June 21, 1607 details the inventory of paintings and art objects in the *Finca* [Manor house] de la Ribera, including among the different items:

> An Artifice for supplying water to the Ribera, consisting of two big wheels and two smaller ones, with their chains and four ctesibic [pumps], made of brass with four connecting rods and two other wheels, with pipes [the pipes through which the water was supplied] partially made ...[143]

The list corresponds exactly to Bate's description of the device installed in the London Bridge, but with two identical waterwheels instead of one, and gives credibility to Zubiaurre's replication claim. Based on the description of Pinheiro, the above mentioned parts inventory, Bate's conceptual drawing and the scant data about the building, Nicolás García Tapia virtually reconstructed the operation principle of Zubiaurre's device (fig. 4.10).

Conceptually, two large waterwheels wheels drive two small ones and their chains through a mechanism consisting of long sliding rods transmitting a rocking motion to the wheels. The two upper half-wheels are driven by the lower ones by a chain, which in turn alternately lift the pistons of the four pumps that serve to force water through pipes on to the fountains and gardens of La Ribera.[144] It may be argued that there is an element of circular argumentation in performing such a reconstruction, as one description reinforces the other, but yet in our opinion it is a valid exercise.

Surviving graphic testimonies show that the device was built next to a bridge, taking advantage of the water drop caused by an existing weir built for some mills operating on the other margin of the river.[145] This was a very

142 "Y dijo que él no la inventó, sino que la vió hecha en Londres, estando allí cautivo, y el enmendó algunas cosas. Y ya ha hecho otro individuo cosa semejante, pero con una sola rueda, cosa facilísima...". Pinheiro da Veiga, *Fastiginia* ..., p. 70.

143 "Un Ingenio para dar agua a la rrivera, con dos ruedas grandes y quatro pequeñas con sus cadenas y quatro tisívicas, de bronce con quatro boquetones de hierro y otras dos ruedas, con parte de caños que se empezaban a hacer..." AGP, Caja 10.977/7. There is a copy in the AGS, *C. & S. R*, leg. 303, sin foliar (see footnote No. 276 in García Tapia, *Técnica y poder*, p. 187).

144 García Tapia, "El ingenio de Zubiaurre," p. 309.

145 García Tapia, "El ingenio de Zubiaurre," Appendix A 2, p. 313.

FIGURE 4.10 *García Tapia's reconstruction of Zubiaurre's Ingenio.*[146]

successful location, as the weir was close to the arches of the bridge. The obstruction caused by the bridge pillars accelerated the water flow, as in London, improving the efficiency of the *Ingenio*.[147]

The External Appearance of the London Waterworks

What is generally considered to be the earliest genuine and full view of the London bridge was drafted by an anonymous artist about the year 1600 (fig. 4.11).[148] As we shall see presently, a careful examination with special

146 García Tapia, "El ingenio de Zubiaurre," p. 323.

147 See: Ventura Seco, "Plano de Valladolid," 1738, redrawn by Juan Agapito y Revilla in 1901, Archivo Municipal de Valladolid and a sketch of La Huerta del Rey (the later name of La Ribera), drawn in or about 1629 by Juan Antolínez de Burgos, *Historia de Valladolid que dexó manuscrita Juan Antolinez de Burgos, vezino y natural de la misma ciudad*, extracted from García Tapia, "El ingenio de Zubiaurre," Appendix A 3, p. 326. No indication of the folio number in the original is given.

148 John Richard Green, *A Short History of the English People, Illustrated Edition*, edited by Mrs. J. R. Green and Mrs. Kate Norgate, (4 volumes), Vol. 3, London: Macmillan and Co., 1903, caption in the frontispiece, c.1600. The original image is in the Pepys's Collection, Magdalene College, Cambridge. Richard Thomson (An Antiquary-Librarian of the British

attention to the chronology of the construction of one particular building on the bridge shows that John Norden's "The view of London Bridge from East to West," drawn in 1597 (fig. 4.12), predates the aforementioned one.[149] In any case both drawings are the first visual testimonies of the appearance of the waterworks.

John Bate, a reliable witness, tells in his 1635 waterworks description about a water tower, as we have seen and quoted above: "… which engine by the ebbing and flowing of the Thames, doth mount the sayd water unto the top of a turret…." On the basis of Bate's testimony and the graphic evidence from Norden's view from the East and the anonymous views of the bridge, it can be safely assumed that a water tower—the *Water Worke*, was in use continuously at least from the last years of Elizabeth's time onwards, and perhaps since the construction of the waterworks.

FIGURE 4.11 *Detail of London Bridge seen from the West* (*c.1600*).[150]

Institution), *Chronicles of London Bridge*, London: Thomas Tegg, 1839, p. 259, gives as its dimensions 62×12 cm.

149 On the other side of the bridge [not shown here] new corn mills are shown in the 1600 anonymous painting but do not yet appear in Norden's 1597 drawing (fig. 4.12), thus indirectly establishing their approximate date of construction.

150 Green, *A Short History*, detail from frontispiece, c.1600.

FIGURE 4.12
Detail of John Norden's view of London Bridge, East to West, 1597.[151]

The *Water Worke* is not mentioned either in the original Stow edition of 1598, nor in the last one of 1633; but appears in Strype's description of 1720.[152] Strype's description perhaps refers to the wooden building erected after the Great Fire of 1666, which is clearly shown in a later view dated 1749. The cistern at its top was at a height of 128 feet, fed by a main pipe that had a diameter of 12 inches.[153]

Richard Thomson, in his *Chronicles of the London Bridge* (1839), describes the northern side of the bridge: "At the north end, also, of this most interesting prospect, against the first sterling, is a high square building, like a tower, having a low wooden gallery in front of it and a single water-wheel turning beneath it; which are, most probably, intended for the waterworks and tower

151 John Norden, "A long view of old London bridge with houses built along it, with three columns of explanatory letterpress printed beneath." This is an antiquarian copy, made c.1804, of the original plate by Norden of 1597, engraving, (29.9×52.5 cm), right side, BM, DPD, Registration number: 1859,1210.1004.

152 Strype, *A Survey of the Cities*, Vol. 1, Book 2, chapter 11, p. 180.

153 Thomson, *Chronicles* p. 340. For the view, see: Samuel and Nathaniel Buck, "Bird's eye view of the north end of London Bridge from the Thames, showing the waterworks" fraction of a larger print, etching, (12.6×15.7 cm), 1749, BM, DPD.

at London Bridge."[154] According to Thomson, who strangely doesn't mention John Bate at all in his book, the tower may have been as old as the waterworks themselves, though he stresses that this suggestion is not based on documented information.[155]

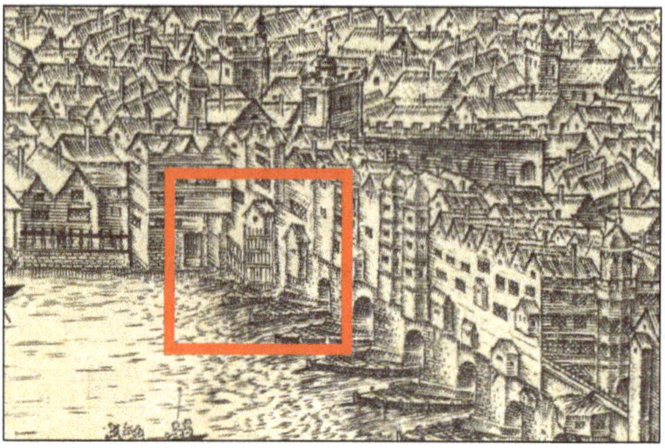

FIGURE 4.13
Detail of Civitas Londini, bird's eye view of the City of London, after John Norden, 1600, showing the Waterworks building exterior.[156]

FIGURE 4.14
Detail of "View of London from South Bank" by Claes J. Visscher, 1616, showing the protective barrier.[157]

154 Thomson, *Chronicles* p. 261.
155 Ibid., p. 262.
156 John Norden, "Civitas Londini," Bird's eye view of the City of London from Hampstead to St. Dunstan in the East, paper, engraving, (36.5×69.3 cm), London, 1600, BM, DPD, Registration number: 1880,1113.1120, photograph taken from the original by the author.
157 Claes Jansz Visscher, "View of London from South Bank," (43×216.9 cm), published by Jodocus Hondius II, 1616, BM, DPD, Registration number: 1880,1113.1124.1–4. For the sake

A few early graphic sources provide additional details about different periods in the life of the waterworks, two of which are shown in figs. 4.13 and 4.14. In 1600 John Norden delineated *Civitas Londini*, a bird's view of the city from Hampstead to St. Dunstan in the East and in 1616 Claes Jansz Visscher produced *View of London from South bank*, which shows old St Paul's in the mid-distance at centre left and London Bridge on centre right. In both representations we can see the external appearance of the building which housed the 'first generation' of the device—from its construction until the first fire that destroyed it in 1633.

The Consequences of the Installation of Waterwheels in the London Bridge

The London Bridge had 19 small arches, with wide starlings built around the piers, which caused strong river currents and almost totally impeded navigation.[158] Those who attempted to 'shoot' the bridge (to sail upstream) took a great risk, and many drowned in the effort. The effective narrowing of the river had begun very early in the life of the bridge. Strype tells, referring to the late twelfth century, that waterwheel-driven mills installed in the bridge arches were "... taken away for the less annoying of the River."[159] At the beginning of the sixteenth century, corn mills were erected upon the Thames, near the southern end of the bridge, where corn was stored in ten warehouses. No information is available about how long they operated, but they were not in place by the eight decade of the century.[160] When the waterworks in the first arch of the bridge were built

---

of graphic clarity, this image is a photograph taken by the author from a copy of the original Visscher's view made in 1848 by J. Pullam. Etching printed from four plates joined as continuous strip, (44.5×222.7 cm), also kept in the BM, DPD, Registration number: 1880,1113.1125.

158 This effect was well known to Arab engineers in the Middle Ages. Ervan G. Garrison, *A History of Engineering and Technology: Artful Methods*, Boca Raton (Florida, USA): CRC Press, 1999, p. 102. For the impact of the bridge construction on the hydrology of the Thames, see Roberta Magnusson, "Public and Private Urban Hydrology: Water Management in Medieval London," in Steven A. Walton, (ed.), *Wind & Water in the Middle Ages*, pp. 171–187, p. 179.

159 Strype, *A Survey of the Cities*, Vol. 1, Book 1, chapter 9, p. 41.

160 Robert and James Dodsley, *London and Its Environs Described: Containing an Account of Whatever is Most Remarkable for Grandeur, Elegance, Curiosity, Or Use, in the City and in the Country Twenty Miles Round It. Comprehending Also Whatever is Most Material in the History and Antiquities of this Great Metropolis*, (6 volumes), Vol. 4, London: Printed for R. & J. Dodsley, 1761, p. 146.

in 1582, effectively further narrowing the passage of water, to the roar caused by the rushing water[161] was added the strong sound of the rotating waterwheel.[162] The wooden waterwheel creaked as the wood was distorted by the weight of the water, "… and [the wheel] endless grinding and reverberation added materially to the overwhelming noise of the city."[163] For these reasons the leasing of one additional arch to expand the waterworks was vigorously opposed.

The situation was aggravated when the need to help the poor to grind their grain encouraged the authorities to reinstall corn mills under the bridge in 1588.[164] The commission appointed to study the potential inconveniences resultant from such mills ruled that: "… the erecting of the said mills would not be hurtful or prejudicial to the Thames any way…."[165] Nevertheless Strype relates that problems arose after their construction: "… to which as soon as they were set up some exception was taken, and complaint made, as it seems to the [Fire] Court; as that they might prove injurious to the bridge or to the river."[166] The corn mills are shown in the 1600's anonymous painting (fig. 4.15), but do appear yet in Norden's 1597 drawing (fig. 4.12), thus indirectly establishing their construction approximate date. The corn mills were eventually dismantled and the mills converted to pump water to Southwark.[167]

FIGURE 4.15
Corn mills at the southern side of the London Bridge, c.1600.[168]

161 Bruce R. Smith, *The Acoustic World of Early Modern England: Attending to the O-Factor*, Chicago and London: University Of Chicago Press, 1999, p. 57.
162 Mitchell, *A history of London life*, p. 133.
163 Peter Ackroyd, *London: The Biography*, London: Chatto and Windus, 2000, p. 74.
164 Ibid.
165 Ibid., p. 41.
166 Ibid., chapter 13, p. 57.
167 Patricia Pierce, *Old London Bridge: The Story of the Longest Inhabited Bridge in Europe*, London: Headline Book Publishing, 2002, p. 150.
168 Green, *A Short History*, detail from frontispiece, right corner.

The end result of the obstruction caused by the big starlings of the bridge, the waterworks and the corn mills was that the river flowed at no more than half of its width, which was about 550 meters, causing the head[169] at high tide to be about 30 cm. and at low tide about 130 cm. Navigation both upstream and downstream became practically impossible.[170] Among the arguments raised in support of the state of things was a curious one—that the bridge was originally built to restrain the ebbing of the tide, facilitating in such a way the navigation of the river upstream.[171]

The Later Development of the London Bridge Waterworks

Morris' pumps and other machinery were destroyed in the Great Fire of 1666: "Not only was the Thames out of reach, but the fire destroyed the waterworks at London Bridge, where equipment there could have raised a supply that might have been effective."[172] After the Fire, and as would be the case during their long lifetime, the waterworks received preferential treatment from the authorities. Charles II enacted 'An Act for the Rebuilding of the City of London,' which, among other regulations, forbade any type of timber to be used in new construction within the city, but exempted the Waterhouse: "... it shall and may be lawfull for the water house, called Mr. Thomas Morris his water house, formerly adjoyning to London Bridge, to be rebuilt upon the place it formerly stood with Timber..."[173]

The reconstruction by Morris' grandson was swift, owing to the wisdom of the Fire Court,[174] which brought about a compromise in a financial dispute between members of the Morris family who had inherited the waterworks. In order to finance the reconstruction of the waterworks the Morrises succeeded

169 The difference in elevation between two points in a body of fluid.

170 Kenneth J. Major, "Waterworks of London Bridge," *International Molinology*, No. 65, (2002), pp. 23–25, p. 23.

171 Francis Bolton, *London Water Supply*, London: W. Clowes & Sons, 1888, p. 54.

172 John Richardson, *The Annals of London*, Berkeley (CA), University of California Press, 2000, p. 142.

173 Danby Pickering, *The Statutes at Large from the Twelfth Year of King Charles II to the Last Year of James II*, (46 volumes), Vol. 8, London: Joséph Bentham, 1763, p. 248.

174 A special Court of Fire Judges (Fire Court) was set up to deal with disputes between tenants and landlords and promptly decide who should rebuild, without the usual legal formalities. See David A. Weiss, *The Great Fire of London*, London: Trafford Publishing, 2012, pp. 89–90.

in borrowing 2,000 pounds, giving as collateral a portion of future profits.[175] As the Broken Wharf Waterworks had also been destroyed and the New River infrastructure did not cover the area served by the London Bridge Waterworks, the authorities understood that rebuilding them was the only solution to provide the water necessary for the reconstruction of houses.[176] Morris' engine was quickly rebuilt. The positive attitude of the City and Parliament continued even when an act for the relief of the orphans and other creditors of the city of London[177] was enacted by King William II in 1694, declaring that "... [the act] shall not extend to the water works of Master Thomas Morris, at or near London Bridge, which he holdeth and enjoyeth by virtue of a grant made to Peter Morris his late grandfather..."[178]

But not surprisingly, two hundred years after their construction, the cash-thirsty tax authorities successfully obtained payments from the Waterworks' owners, as we learn from a case brought against them in July 1783. The arguments put forward to support taxation sound very similar to present practice: "This is not a property arising from the ingenuity merely of a man's head, or the work of his hands: the above therefore are to be conclusive arguments of rateability."[179] The Waterworks were renovated by the famous engineer George Sorocold[180] in 1701,[181] and included at the time four waterwheels installed in the four northern arches of the bridge. In 1729 a first technical description of

175 William Robert Scott, *The constitution and finance of English, Scottish and Irish joint-stock companies to 1720*, (3 volumes), Vol. 3, Cambridge: Cambridge University Press, 1911, p. 12.

176 Thomas Fiddian Reddaway, *The Rebuilding of London After the Great Fire*, London: Jonathan Cape, 1940, p. 99.

177 In essence—additional emergency taxation to cover monies already spent, which had been taken from funds belonging to orphans but appropriated by government, plus an interest of four percent to be paid upon those monies).

178 Danby Pickering, *The Statutes at Large from the Fifth Year of K. William and Queen Mary to the Eighth Year of King William III*, (46 volumes), Vol. 9, London: Joséph Bentam, 1764, p. 274.

179 Thomas Caldecott, *Report of Cases Relative to the Duty and Office of a Justice of the Peace from Michaelmas Term 1776, inclusive to Trinity Term 1785, inclusive*, London: His Majesty's Law Printers, 1786, p. 334.

180 For a detailed account of Sorocold's engineering achievements, see: Rhys Jenkins, "George Sorocold: A Chapter in the History of Public Water Supply," *The Engineer*, Vol. 126 (1918), pp. 333–4 and Frederick Williamson, "George Sorocold of Derby: a pioneer of water supply," *Journal of the Derbyshire Arceological & Natural History Society*, Vol. 57, 1936, pp. 43–93.

181 Arnold Pacey, *The Maze of Ingenuity: Ideas and Idealism in the Development of Technology*, Cambridge (Mass.): MIT Press, 1992, p. 137.

the revamped waterworks was given by Stephen Switzer in his *Introduction to a General System of Hydrostaticks and Hydraulicks*.[182]

In 1731 an English engineer, Henry Beighton, published a much more detailed description (fig. 4.16), which included a detailed calculation of the quantity of water raised by the Waterworks, based upon a careful check of all its dimensions and functional parameters.[183] He calculated a maximum theoretical potential of approximately 500 cubic meters per hour for all the engines. Due to the change of the flow and ebb of the tide the waterwheels turned only 12 hours per day.[184] In addition, the inevitable losses due to friction and the imperfect valves and seals, resulted in a real volume of water supplied that may have been not more than one third of the theoretical one.

FIGURE 4.16 *The Waterworks in Beighton's time.*[185]

182 Stephen Switzer, *An Introduction to a General System of Hydrostaticks and Hydraulicks, Philosophical and Practical*, London: T. Astley, 1729, pp. 317–320.

183 Henry Beighton, "A Description of the Water-Works at London-Bridge, Explaining the Draught of T a b. I." *Philosophical Transactions*, Vol. 37, (1731–1732), pp. 5–15. Two English historians maintain that Switzer described the waterworks before Sorocold's overhaul and Beighton after its completion, though in our opinion there is not enough data to substantiate this claim. See Albert Edward Musson and Eric Robinson, *Science and Technology in the Industrial Revolution*, Manchester: Manchester University Press, 1969, p. 43.

184 John Williams, *An historical account of sub-ways in the British metropolis, for the flow of pure water and gas into the houses of the inhabitants*, London: Carpenter and Son, 1828, p. 413.

185 Beighton, "A Description of the Water-Works," p. 13.

It seems reasonable to assume that the waterwheels diameter in Beighton's time would have been similar to the ones in Morris' time, as the height of the arches under which the wheels were placed had not been altered. The speed at which they turned should also have been similar, as it depended on the flow speed of the river, which we shall deem unchanged for the sake of our approximated calculation,[186] thus rendering an equal number of strokes per minute for the pumps. Allowing for the different number of 'forciers', or forces pumps, numbering 52 in Beighton's time, compared to only 2 in the first Morris device, and for the different volume of water in each 'forcier,' due to the dimensions of the barrels in the renovated waterworks vis-a-vis the dimensions in the barrels recovered during the excavations of 1828, a relationship between the quantities can be obtained. The end result of this calculation is that Morris' waterworks lifted not more than 200 cubic meters of water daily.[187] This quantity increased proportionally to the number of arches later leased to Morris and his successors, reaching about 400 cubic meters daily when the second waterwheel was installed by the end of the sixteenth or the first years of the seventeenth century, and about 4800 cubic meters daily when four arches were used, at the time Beighton analyzed the waterworks in 1731.

Foreign travelers and historians have lavishly praised Morris's undertaking, though frequently confusing the improved performance of the device as it was during their time with the output of the first, much more primitive machine, as designed and built by Morris in 1581–1582. The list of admirers is long—Luigi Ferdinando Marsigli in his "Voyage to London and Holland, 1721–1722," calls it: "… a most splendid machine, built at a huge cost …;"[188] There are even references in literature dealing with totally different issues, such as in William Paley's *Natural Theology* (1802): "The aorta of a whale is larger in the bore than the main pipe of the water-works at London Bridge, and the water roaring in its passage through that pipe is inferior in impetus and velocity to

186 This supposition can be contested, as the flow speed depended on factors that were subject to change, such as the quantity of water at the sources and the season, the use of water between the sources and London and the sediments deposited on the water bed.

187 For a different (but in our opinion wrong) estimation of merely 150 cubic meters daily, see: Edward Cresy, in *An Encyclopaedia of Civil Engineering, Historical, Theoretical and Practical*, (2 volumes), Vol. 1, London: Longman, Brown, Green, and Longmans, 1847, p. 545.

188 McConnell, Anita, "L. F. Marsigli's Voyage to London and Holland, 1721–1722," *Notes and Records of the Royal Society of London*, Vol. 41, No. 1. (1986), pp. 39–76, p. 61.

the blood gushing from the whale's heart.[189] Bernard Lambert in *The History and Survey of London and its Environs* (1806) refers to "An ingenious German, named Maurice, submitted a scheme to the lord major and the aldermen…"[190] John William Abbot, in his *History of London from the Earliest Period to the Present Time* (1821), characterizes the year 1582 as "… memorable for a scheme proposed by Peter Maurice, a German, to the lord major and aldermen, for supplying the city with Thames water by a mill…"[191] and William Clay wrote in 1849, in his book about the water supply to London "The first water works, properly so called—for previously to their establishment the only supply had been from wells and conduits, were those at London Bridge…"[192] In a London anthology published in 1851 it is written that: "It was not until 1582 that any mechanical power or skill was applied in providing London with water; but in that year Peter Morris, a Dutchman, made 'a most artificial forcier,' by which water was conveyed into the houses … These were the waterworks famous for so long a period as one of the sights of London."[193] The popular journal *Notes and Queries*, in 1861, refers to the waterworks as "… famous …," and "… one of the most curious and powerful systems of hydraulic mechanism ever constructed…"[194] In a similar vein no other than Charles Dickens, editor of the journal *All the Year Around*, wrote in 1862: "The art of supplying water to towns was in a very rude state until the appearance of Peter Morice, a Dutchman, in 1582…"[195]

189 William Paley, James Paxton and John Ware, *Natural Theology: Or, Evidences of the Existence and Attributes of the Deity, Collected from the Appearances of Nature*, London: Lincoln, Edwards and Co., 1833, p. 93.

190 Bernard Lambert, *The History and Survey of London and its Environs*, London: T. Hughes, 1806, (4 volumes), Vol. 2, p. 16.

191 John William Abbot, *History of London from the Earliest Period to the Present Time*, London: A. K. Newman, 1821, p. 108.

192 Willian Bart Clay, *Remarks on the water supply of London*, London: James Ridway, 1849, p. 5.

193 J. C. Platt and J. Saunders, "Underground," in Charles Knight, (ed.), *London*, (6 volumes), Vol. 1, London: Henry G. Bohn, 1851, pp. 225–240, p. 237.

194 "Waterworks at Old London Bridge," *Notes and Queries*, Vol. 11, 2nd S (266), (Feb 2, 1861), p. 90.

195 Charles Dickens, (ed.), "London Water," *All the Year Around*, Vol. 6, London (1862), pp. 137–140, p. 140.

The water-wheels and pumps remained operational under the bridge until the early nineteenth century. The Act for removing the Waterworks at London Bridge on the 26th of July, 1822 describes in detail the developments of its ownership through the 260 years in the course of which it had faithfully served the city.[196]

Other Initiatives

The success of this project caused other private Londoners to make proposals for similar water supplying devices and sophisticated business models for their operation. The establishment of a profit-oriented, private water supply enterprise indicated to other investors that individuals were willing to pay for piped delivery to their homes.[197] The London authorities actively encouraged other water supply entrepreneurs, often foreigners, to establish new pumps and piping networks, and were even ready to support such initiatives financially, because they received an annual income from each lease.[198] In 1591 the Italian engineer Frederico Gianibelli[199] obtained the consent of the Court of Aldermen to erect new waterworks at Tyburn. Strype details the extraordinary claims about the capabilities of the planned device, such as cleansing ditches, supplying wholesome water and quenching fires.[200] From the *Remembrancia* of the City of London we learn that it was an ongoing project: "The City were in treaty with Frederick Jenibella, skilled in waterworks, for the erection of a windmill at the fountain head to increase the supply, which, if successful, would induce that body to comply more readily with his request."[201] However, it was never put into effect, and no record exists explaining why.

196 John Raithby, (ed.), *The Statutes of the United Kingdom and Ireland, 3 George IV*, London: George Eyre and Andrew Strahan, 1822, pp. 839–846.

197 Tynan, "London's Private Water Supply," p. 348.

198 Ibid., p. 344.

199 Frederico Gianibelli was a 16th century Italian military engineer, famous for his attempt to destroy a pontoon bridge with exploding fire-ships during the siege of Antwerp in 1585. He is also referred to as Genibelli or Jenibella.

200 Strype, *A Survey of the Cities*, Vol. 1, Book 1, chapter 5, p. 27. See also Sharpe, *London*, p. 19, where the sources referred to are: CLRO, *Repertories of the London Court of Aldermen, Repertory 26*, fols. 270, 281, 376b.

201 Overall, *Remembrancia*, "Water I-656," p. 554, 23 April, 1592.

In 1593 Bevis Bulmer,[202] a predominant figure in mining engineering and other industrial projects,[203] obtained a lease from the Corporation of London allowing him to construct in Broken Wharf a pumping engine to supply Thames water to the Western part of the city. Stow reports succinctly: "One other new forcier was made neare to Broken Wharfe, to convey Thames water into mens houses of West Cheape, aboute Powles, Fleete street, &c., by an English gentleman, named Bevis Bulmer, in the yeare 1594."[204] Bulmer obtained a lease for 500 years and the Court of Aldermen granted him the use of the Green-Yard at Leadenhall for constructing his engine, whilst the court of Common Council advanced him the sum of £1,000 on easy terms.[205] The device, completed in 1594, consisted of 4 chain pumps worked by horses,[206] similar in its principle of operation to the one described in John Bate's *The Mysteries of Nature and Art*, but with horses turning the vertical axis, instead of the waterwheels shown in the figure (fig. 4.17).

Henry A. Harben, in his *Dictionary of London*, locates the engine within the gate of an old stone house belonging to the Duke of Norfolk.[207] The building is clearly noted in all the early views of the city: Norden's "Civitas Londini,"

202 It is probable that Bulmer was a foreigner as well, according to his name. His surname was sometimes spelled Bulmar, or Bulman. Hereunder quotations will use the spelling as it appears in the respective originals. In the text Bulmer shall be used.

203 For an extensive discussion about Bulmer's actions in the field of mining, see: John Calvert, *The gold rocks of Great Britain and Ireland*, London: Chapman and Hall, 1853; Hector Menteith Robertson, "Sir Bevis Bulmer: A Large-Scale Speculator of Elizabetan and Jacobean Times," *Journal of Economic and Business History*, volume 4, 1931, p. 99–120 and Rhys Jenkins, "Bevis Bulmer," in *Links in the History of Engineering and Technology from Tudor Time*, Freeport (NY): Books for Libraries Press, 1971, pp. 24–27.

204 Stow, *A Survey of London* (written in 1598, reprinted from the 1603 edition), p. 8. In his *Annales*, Stow describes Bulmer as: "… a most ingenious gentleman…," see: John Stow, *The Annales, or a General Chronicle of England, Begun First by Maister Iohn Stow, and After Him Continued and Augmented with Matters Forreyne, and Domestique, Auncient and Modern, Vnto the Ende of this Present Yeere 1624, by Edmond Howes, Gentleman*, Londini: Impensis Thomæ Adams, 1615, p. 768.

205 Sharpe, *London*, p. 20, gives as his sources: CLRO, *Repertories of the London Court of Aldermen, Repertory 23*, fol. 68; CLRO, *Jour. 23*, fol. 196; Everett Green, *Cal. SPD, (1591–1594)*, p. 576.

206 Dickinson, *Water Supply*, p. 16.

207 Henry Andrade Harben, "Water Supply of London," in *A Dictionary of London*, London: H. Jenkins, 1918, p. 615, under the entry: 'forcier'.

(fig. 4.18), Speed's "View of London" (1612),[208] Visscher's "View of London from South bank," (fig. 4.19), Merian's "A bird's eye view from Whitehall," (1638)[209] and Hollar's "London [the long view]" (1647).[210] In Visscher's and Hollar's representations the building is marked as 'The Water House'.

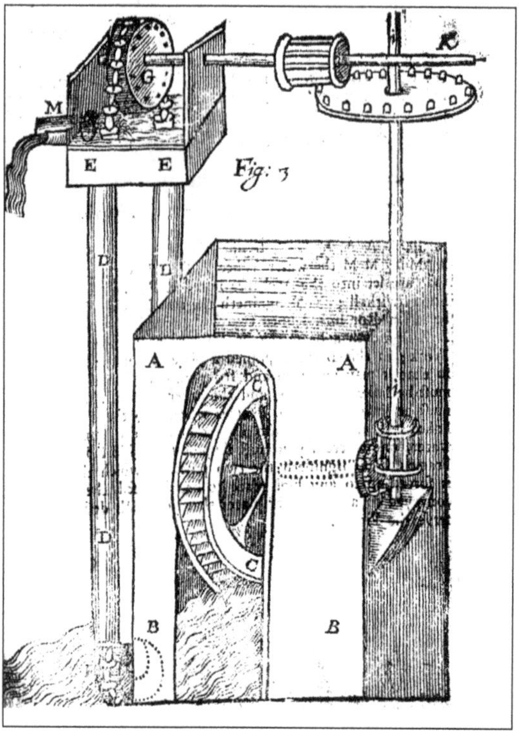

FIGURE 4.17 *Bate's chain pump.*[211]

208 John Speed, "View of London," in *The theatre of the empire of Great Britaine: presenting an exact geography of the kingdomes of England, Scotland, Ireland, and the iles adioyning: with the shires, hundreds, cities and shire-townes, within ye kingdome of England*, London: William Hall, 1611.

209 Matthäus Merian, "A bird's eye view from Whitehall to beyond the Tower, with Southwark in the foreground," paper, engraving, engraving from two plates, joined, (22.7×70.4 cm), 1638, BM, DPD, Registration number: 1880,1113.1132.

210 Wenceslaus Hollar, "London [the long view]," bird's eye view of the City of London from Hampstead to St. Dunstan in the East, paper, etching, Amsterdam, 1647, (46.6×39 cm), BM, *D.P.D.*, Registration number: 1864,0611.438. Hind, *Wenceslaus Hollar*, plates XV–XX, describes the map in pp. 44–47.

211 Bate, *The Mysteries of Nature*, p. 55.

FIGURE 4.18 *John Norden, "Civitas Londini," (1600).*[212]

FIGURE 4.19 *Visscher, "London from South bank," (1616), detail.*[213]

For this enterprise a blank form of a lease for a water connection of Thames water has been preserved, detailing also the number of dispensing cocks required by the proposed lessee and the obligations of both parties to the contract.[214] Discrepancies among historians about the useful life of Bulmer's device abound. A description of the city dated 1708 refers to the device as yet operating: "... Broken Wharf in Thames Str. near Old Fishstreet Hill, here is a Water-work for raising the Thames Water to an altitude whereby 'tis forced to supply several parts of the City".[215] Matthews wrote in 1835: "The expense ... being much greater in proportion to the supply than that of the other establishments, ... therefore the want of encouragement occasioned the discontinuance of the works."[216] Thomson concurs with Mathews in 1839 that: "... after working it [the pumping station] for a short time, it was laid aside,

212 John Norden, "Civitas Londini," photograph of the detail taken from the original by the author.

213 Claes Jansz Visscher, "View of London from South Bank," (43×216.9 cm), published by Jodocus Hondius II, 1616, BM, DPD, Registration number: 1880,1113.1124.1–4.

214 Mary Ann Everett Green, (ed.), *Cal.* SPD *of The Reign of Elizabeth, 1591–1594*, London: Longman, Green, Reader and Dyer, 1867, vol. CCL, No. 85 and 85.1, p. 576.

215 Edward Hatton, *A new view of London: or, an Ample Account of that City, In Eight Sections*, (2 volumes), Vol. II, London: J. Nicholson, 1708.

216 Matthews, *Hydraulia*, p. 30.

on account of its great charge both to the tenants and the proprietors..."[217] On the other hand, Dickinson noted that the device ceased functioning after the London Bridge Waterworks took control of it in about 1703,[218] most probably due to the high running costs of feeding the animals,[219] and Mark Jenner wrote that it supplied 600 houses by the 1650s.[220] In any case, the tower was reconstructed after the Great Fire of 1666, when it was completely destroyed, as Hollar's graphical testimony shows (fig. 4.20). When referring to this enterprise, Matthews describes what is an additional case in which the installation of new technology—a steam engine in the early 1800s—indirectly helped to discover a large cistern with a wooden trough communicating with the river and tons of large, heavy leaden pipes.[221]

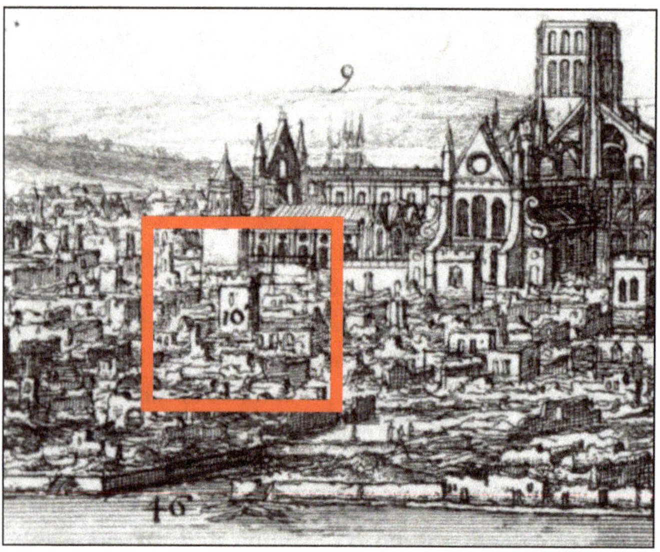

FIGURE 4.20 *Wenceslaus Hollar, "London after the Fire," detail.*[222]

217 Thomson, *Chronicles*, p. 255.

218 Dickinson, *Water supply*, p. 18.

219 Ibid., p. 15.

220 Mark S. R. Jenner, "From Conduit Community to Commercial Network?" in Mark S. R. Jenner and Paul Griffiths, (editors), *Londinopolis: Essays in the Cultural and Social History of Early Modern London*, Manchester: Manchester University Press, 2000, pp. 250–265. p. 257 and Dickinson, *Water supply* p. 17.

221 Matthews, *Hydraulia.*, p. 30.

222 Hollar, "London before and after the Fire," the figure shows a detail of the lower part, after the Fire.

Another prominent figure that took part in promoting water supply projects to the city was Edward Wright (1561–1615). He came from Garveston, Norfolk, and was a skilled mathematician, a Cambridge Master of Arts, an astronomer who formulated the essential principles of the Mercator projection and a friend of Lord Napier, the discoverer of logarithms.[223] About the year 1600 he endeavored to raise water from Botolph Wharf, situated to the east of the London Bridge, to supply houses at the eastern end of Leadenhall.[224] He was awarded a grant in April 25, 1598, conferring him the right for eight years "to make and utter mathematical instruments." In this case, the mathematical instrument was another water-raising device, obtained "by long and painful study of the mathematical sciences." The grant contains an enigmatic addition, according to which, 'a special work' for supplying water to London had already been undertaken by the patentee.[225] G. C. Berry describes a clash that took place with Morris's successor at London Bridge, resulting in a boundary agreement between both concessions. No record exists indicating what the fate of this venture was.[226] Some years later, it was Wright who made a detailed survey of the future course of the New River, personally directing the mathematical part of the work in the beginning of the project.[227] "He was the first undertaker of that difficult but useful work, by which a little river is brought from the town of Ware to supply the city of London with water…"[228]

The technologically-gifted Londoner, John Lanyon was reported by Samuel Hartlib in his *Ephemerides* in 1651 as "… making a Modell of a rare Water-mill

223 Geoffrey. C. Berry, "Sir Hugh Myddelton and the New River," in Denis Smith, (ed.), *Water Supply and Public Health Engineering*, in Joyce Brown, general editor of the series: *Studies in the History of Civil Engineering*, (12 volumes), Vol. 5, Aldershot (UK): Ashgate, 1999, pp. 46–78, p. 56.

224 CLRO, *Jour.* 25/180b, extracted from Berry, "Sir Hugh," p. 48.

225 E. Wyndham Hulme, "History of the Patent System Under the Prerogative and at Common Law—a Sequel" *Law quarterly review*, Vol. XVI, 1900, pp. 44–56, p. 51.

226 CLRO, *Repertories of the London Court of Aldermen, Repertory 26–1*, fol. 40, extracted from Berry, "Sir Hugh," p. 48.

227 Berry, "Sir Hugh," p. 63.

228 Alexander Chalmers, *The General biographical dictionary, containing an historical and critical account of the lives and writings of the most eminent persons in every nation; particularly the British and Irish; from the earliest accounts to the present time*, (32 volumes), Vol. 32, London, J. Nichols and Son, 1812, p. 233. For a detailed account of his contribution to the art of navigation, see Eric H. Ash, *Power, Knowledge, and Expertise in Elizabetan England*, Baltimore and London: The John Hopkins University Press, 2004, pp. 165–169.

in Brussels."[229] No implementation of this invention is recorded. During the Protectorate of Oliver Cromwell another pumping device was erected on a bank of the Thames by Sir Edward Ford.[230] In 1655 Cromwell issued letters-patent for the protection of Ford's invention of a machine that would produce, install and maintain engines, pipes, cisterns and auxiliary equipment for the cities of London and Westminster, since "… many parts in and about London and Westminster are in greate wante of water and thereby in the more danger of fire and other mischeiffs."[231] The device, which was apparently completed in 1656, was probably a real innovation, gaining the admiration of Samuel Hartlib, who was pleased when Ford obtained the patent, as is reflected in the *Ephemerides*.[232] Balthasar de Monconys, who drew the sketch shown in fig. 4.21 and described the operation of the mechanism in 1663, noted that from the top of the wooden water tower to which the water was raised the whole of the City could be seen, especially Somerset House and the Temple.[233] A year later, Samuel Joséph Sorbière, who visited the site in 1664, described it in this way "… It is similar to our Samaritane du Pont-Neuf, and has in addition to the suction pump an impulsion element … they have two horses which incessantly drive the machine, because the tide changes the course of the river twice a day."[234]

Although it was known to have been located close to Somerset House, only in 1920, when a map by Wenceslaus Hollar kept in the British Museum was reprinted, the exact point where it was situated has been clarified. It shows a pyramidal tower marked "Ye Waterhouse" just east of the Strand Bridge and in front of Arundel House (fig. 4.22). According to Rhys Jenkins, the presence of the building establishes the date of the map within a period of ten years:

---

229 Samuel Hartlib, "Ephemerides 1651," in *The Hartlib Papers (a Complete Text and Image Database of the Papers of Samuel Hartlib)*, Sheffield (UK): Humanities Research Institute, University of Sheffield, 2002, 28/2/14B, 15B, 18A, extracted from Hartlib, *Hartlib Papers 28/2/14B and 15B*, quoted in Jenner, "'Another epocha'?," p. 347.

230 The most comprehensive study on this undertaking is: Rhys Jenkins, "A Chapter in the History of the Water Supply of London: A Thames-side Pumping Installation and Sir Edward Ford's Patent from Cromwell," *Transactions of the Newcomen Society*, Volume 9, (1930), pp. 43–51. This discussion draws from Jenkins' article.

231 Jenkins, "A Chapter," p. 43. Jenkins states that this grant does not appear in the official publications of the Patent Office, and does not give his source.

232 Jenner, "'Another epocha'?," p. 352.

233 De Monconys, *Journal des Voyages*, p. 29.

234 Samuel Joséph Sorbière, *Relation d'un voyage en Angleterre, où sont touchées plusieurs choses, qui regardent l'estat des Sciences, et de la Religion, et autres matieres curieuses*, Cologne: Pierre Michel, 1666, p. 58.

1656–1666,[235] though another historian, W. R. Lethaby, places its depiction after the Great Fire of 1666.[236] The fact that the building overlooked Somerset House was probably the exclusive reason for the King's instruction for its demolition. The order, which was issued in November 16, 1664 by the Attorney General, allowed the transfer of the waterworks to a site between Temple Bar and Charing Cross, on the condition that they would be not more than 15 feet high.[237] They were ordered to remove it within three months, which they did, leaving no record attesting to a new location. What is clear is that the 15 feet limitation was absolutely non-practical, as not enough head would have been available to distribute the water. According to Mark Jenner, Edward Ford sold the water tower without mentioning to the buyers that he had already assigned part of the stock to others.[238]

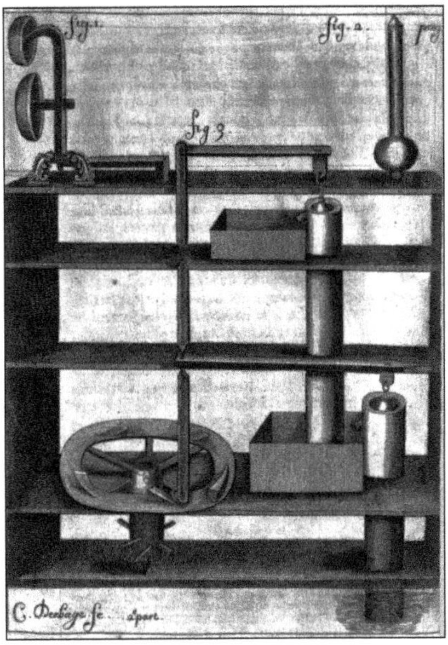

FIGURE 4.21
Ford's pump.[239]

235 Rhys Jenkins "Hollar's Map II," in T. Fairman Ordish, (ed.), *London Topographical Record*, illustrated, (5 volumes), Vol. 2, London: London Topographical Society, 1903, pp. 110–111, p. 110.

236 W. R. Lethaby, "Hollar's Map I," in Fairman Ordish, *London Topographical Record*, p. 109.

237 Mary Anne Everett Green, (ed.), *Cal. SPD of the reign of Charles II, 1664–1665*, (11 volumes), Vol. 4, London: Longman, Green, Longman, & Roberts, 1863, p. 72.

238 LMA, Acc. 2558/NR13/238, extracted from Jenner in "L'eau changée en argent?" p. 648.

239 Balthasar de Monconys, *Journal des Voyages de Monsieur de Monconys*, (3 volumes), Vol. 2, Lyon: Horace Boissat & George Remevs, 1666, plate between pages 28 and 29.

FIGURE 4.22 *Detail of Bird's eye plan of the west central district of London (c.1660).*[240]

All these undertakings reinforce the feeling that the Thames was still revered in the sixteenth and seventeenth centuries with unique fervor. The poet Daniel Lupton praises it in 1632: "... to speake truth of him, he is the priviledg'd place ...: he is a little sea, and a great river."[241]

The New River

By the end of the sixteenth century London had become a huge metropolis but despite its technological achievement, the London Bridge Waterworks could not provide sufficient water for the city population. Although in the last years of Elizabeth's reign the Corporation had already obtained permission from Parliament to dig a channel for conveying water to the City from Middlesex or Hertfordshire, they procrastinated for several years. According to Bernard

240 Wenceslaus Hollar, "Bird's-eye plan of the west central district of London, with Lincoln's Inn Fields at the top right the Covent Garden piazza left of centre, Holborn along the top, St Martin's Lane along the left, and the Thames from the Savoy to Essex Stairs at the bottom," (1660–1665), etching on paper, (34.4×45.5 cm.), BM, DPD, Registration number: Q,6.136. Hind, *Wenceslaus Hollar*, plate XIV, gives an explanation on the map in pp. 33–34.

241 Donald Lupton, London and the Countrey Carbonadoed and Quartred into Severall Characters, London: Nicholas Okes, 1632, in James Haliwell, (ed.), *Books of Characters, illustrating the Habits and Manners of Englishmen from the Reign of James I. to the Restoration*, London: J. E. Adlard, 1857, p. 274.

Rudden, it was the 1603 plague, which killed more than thirty thousand people out of a population of about a quarter of a million, that moved them to action.[242]

Captain Edmund Colthurst of Bath submitted a proposal in 1600 to bring water from springs in Hertfordshire. This proposal was approved by the Corporation in 1602 and since 1604 was also protected by letters-patent granted by the king for seven years "... in recompense for his longe service in Ireland."[243] Though Colthurst personally invested efforts and resources, the project was not implemented for lack of funding, though he would ultimately become one of the 'adventurers' [investors] in the New River project.[244] In January 1606, some other entrepreneurs, with the City's support, submitted a bill to the House of Commons, asking permission to tap the rivers Lea or Colne, which had been found suitable for consumption by surveyors sent by the City. The committee appointed to consider the bill included a Mr. Myddelton (presumably Hugh), his brother Robert Myddelton and three other distinguished members. Since all of them would eventually become investors in the New River scheme, one may wonder to what extent were they able to distinguish between the conflicting interests of the three constituencies: the City, Parliament and those striving to profit from the project. The springs of Amwell and Chadwell, in Hertfordshire, considered to be sufficiently pure and abundant, were the subject of a new Act in 1607, authorizing the conveyance of these waters to the city by an aqueduct, but subject to settlements with the owners and occupants of the land through which the ditch would pass. The Corporation had the opportunity to undertake the work itself, but declined to do so, on account of the financial risks.[245] Again, three years were wasted in negotiations with interested parties, but finally, on 28th March 1609, the Common Council committed Hugh Myddelton, a goldsmith and a friend of Sir Walter Raleigh, to undertake the project, keeping for himself any profit that he might accrue from the enterprise. As Bulmer before him, Myddelton was an expert in mining of precious metals, which gave him both insight into hydraulics and the necessary capital, and made him welcome at court. The statutory, legal and financial

242 Bernard Rudden, *The New River: A Legal History*, Oxford: Oxford University Press, 1985, p. 8.

243 Rudden, *The New River*, p. 9.

244 This short discussion is based on Gladstone, *History of the London...*, pp. 7–8 and Ceri Sullivan, "Thomas Middleton's View of Public Utility," *Review of English Studies*, Volume 58, Issue 234, April 2007, pp. 162–174.

245 Arthur Shadwell, *The London Water Supply*, London: Longmans, Green, and Co., 1899, p. 17.

arrangements took seven years, but the council ordered that work should be completed by the spring of 1613.

Peter Morris, Bevis Bulmer and others joined in ridiculing Myddelton's scheme, who was constantly harassed with obstructions by the owners of land across which he had to cut his trench.[246] The technical difficulty of overcoming the hilly grounds required a sinuous channel course, approximately double the length of a crow-flight line, increasing proportionally the cost of the work.[247] By the time Myddelton had reached just about half the distance to London, his project was stopped by lack of funds. Myddelton pleaded unsuccessfully on behalf of his scheme before the Lord Mayor and Corporation, appealing in vain to their duty to ensure sufficient water supply to the city. He applied later to the King, who had already taken an interest when the canal passed through the grounds of Theobalds.[248] The King agreed in August 1612 to pay half the capital costs of the work and reimburse half the expenditure incurred until that moment[249] in return for half of the ownership and half of the annual profits (though he was to take no part in management).[250] Myddelton did not have any choice but to accept, and subsequently James I granted the charter to the Company, under the title of the New River Company. The money provided by the King allowed Myddelton to resume work.

On April 10, 1613 the trench reached Islington, where the Round Pond (later to be called the New River head, fig. 4.23) was to be made. The inauguration of the new water supply system in Michaelmas Day (29 September), 1613, was attended by royal and civic personages of high dignity, who assembled at Clerkenwell to see the flood-gates open and the stream run into the reservoir.[251]

246 Samuel Smiles, "Life of Sir Hugh Myddelton," in *Lives of the Engineers, with an Account of their Principal Works: Comprising also a History of Inland Communication in Britain*, (3 volumes), Vol. 1, London: J. Murray, 1862, pp. 85–152, pp. 110–115.

247 The technical problem of waterways and canals building had already bothered English people in the Early Middle Ages, see: James Bond, "Canal Construction in the Early Middle Ages: An Introductory Review," in John Blair, *Waterways and Canal-Building in Medieval England*, Oxford: Oxford University Press, 2014, pp. 157–206.

248 Theobalds House was a prominent royal palace of the 16th and early 17th centuries in Hertfordshire.

249 Smiles, "Life of Sir Hugh Myddelton," p. 118.

250 Walter Harry Green Armytage, *A Social History of Engineering*, London: Faber and Faber, 1961, p. 72.

251 Alec W. Skempton, *A Biographical Dictionary of Civil Engineers in Great Britain and Ireland: 1500–1830*, London: Thomas Telford, 2002, p. 100.

Distribution to private houses and public conduits began shortly after the opening, but the domestic supply was provided to each area only two or three days a week. Supporters of ther project gave as the reason for this return to older but proven alternative technologies the quest for purer water.[252] Charles F. Partington sarcastically wrote, when referring to Thames water, "The intrepid drinkers do not seem to have given themselves much concern about the quality of the water, so long as the quantity was sufficient."[253] The fact that the New River water was cleaner can also be inferred from its sources: two-thirds from springs in Chadwell and one-third from the relatively unpolluted upstream of the River Lea.[254]

Myddelton had spent all his capital in the venture. The purchase of elm trunks, boring them and laying the pipe network had been grossly underestimated, and that meant nearly as much was spent on the local distribution infrastructure in the city as on bringing the stream to London.[255] The New River Company was incorporated in 1619, when Myddelton sold part of his shares to other 28 'adventurers' (investors). Although the records of the New River Company were destroyed by fire in 1769,[256] it is known that the company suffered financial difficulties until the early 1640s, paying no dividends for the first twenty years since its opening, but later became very profitable, with its shares increasing in value meteorically.[257]

The New River was a huge feat of engineering. The canal followed a winding course of almost forty miles into the terminal cistern, with some of the trenches sank to ten meters and some of the aqueducts raised seven meters. It had taken persistence, political acumen, technical skill, good connections and financial expertise to provide this water.

---

252 Kirby, *Engineering in History*, p. 148.

253 Charles F. Partington, *British cyclopedia of the arts and sciences*, (2 volumes), Vol. 1, London: Orr & Smith, 1835, p. 889.

254 Ibid.

255 Mathews, *Hydraulia*, p. 49.

256 Smiles, "Life of Sir Hugh Myddelton," p. 119.

257 Ibid., pp. 128–131. For an account of the New River Company in the second half of the seventeenth and eighteenth centuries, see Leslie Tomory, "London's water supply before 1800 and the roots of the networked city," *Technology and culture*. Vol. 56, No. 3 (2015), pp. 704–737.

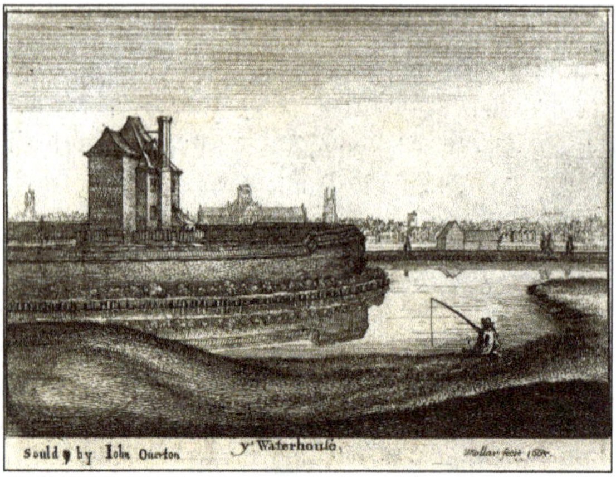

FIGURE 4.23 *Wenceslaus Hollar, The New River Head in Islington, 1665.*[258]

Competition between the Different Water Suppliers

In the first half of the seventeenth century the area served by the London Bridge Waterworks was South London (Thames Street, New Fish Street, and Grasse-street, up to North-west comer of Leaden hall),[259] Bulmer's Broken Wharf engine supplied the West Cheapside and St Paul's neighborhoods and later Fleet Street[260] and the New River operated in the City and northeastern London. The fact that there was almost no overlapping between the pipe networks of the different companies allowed for a relatively mild competition. This attitude would eventually change after the Great Fire, which destroyed the London Bridge Waterworks, enabling the New River Company to lay pipes and to take over some of their customers in parts of South London while the rebuilding was taking place. In 1667 the Morrises presented a petition to restrain the New River Company to continue doing so.[261]

---

258 Wenceslaus Hollar, "Ye Waterhouse" [The new River Head in Islington], print, (12×8.3 cm), London: John Overton, 1665, Registration number. Q,6.51. Hind, *Wenceslaus Hollar*, plate XLIV, describes the map in pp. 66–67.

259 Raphael Holinshed, *The Chronicles of England, Scotland & Ireland*, (6 volumes), Vol. 4, London: J. Johnson, F. C. and J. Rivington and others, 1808, p. 496.

260 Skempton, *A Biographical Dictionary*, p. 100.

261 Mary Anne Everett Green, (ed.), *Calendar of State Papers, Domestic, Charles II, 1667–8*, London: Public Record Office, 1893, ccxxix. 162, p. 132, extracted from Scott, *The constitution and finance*, p. 12.

The type of contract with, and prices to, customers were almost identical in the cases of Bulmer's undertaking and the New River.[262] Each customer signed a long-term agreement with the company, according to an extant lease, for a period of 21 years, at a rate of 20 shillings for connection and a 5 shillings quarterly rent.[263] This level of prices—a connection fee of what would have been a month's wages, followed by the rental of a week's wages every quarter (the rate Myddelton paid his laborers amounted to a shilling a day)—could only be attractive for rich people. There is no surviving data concerning the rates charged at this early period by the London Bridge Waterworks, but the City allowed Myddelton to charge individual customers any price agreed upon for pipe water, so it is logical to suppose that the three suppliers did not have restrictions concerning their rates. Market forces dictated that the alternative price charged for water carrier delivery represented the practical upper limit on the price in all three cases.[264] All sides—the companies lifting water from the Thames on one hand, and the New River on the other, claimed that 'their' water was cleaner and purer than their competitor's.

Water carriers, who continued to ply their trade for centuries, distributed New River and Thames water to those who had no private pipes connected, as can be seen in a figure delineated by Marcellus Laroon at the end of the seventeenth century, crying: 'Any New River water here!'[265] (fig. 4.24).

262 Suzanne Gossett, (ed.), *Thomas Middleton in Context*, Cambridge: Cambridge University Press, 2011, p. 84.

263 "The figures are from British Library MS Add. Ch. 56224, a printed form lease completed in favor of 'Hugh Merrick Esq' and 'Hugh Golde, Gent', who share a half-inch quill of water from the Seething Lane pipe from 17th March 1616. Other leases show even a higher rent, of 6s 8d per quarter." Extracted from Ceri Sullivan, "Thomas Middleton's View of Public Utility," *Review of English Studies*, Volume 58, Issue 234, April 2007, pp. 162–174, p. 167. For the text of an original grant by Sir Hugh Myddelton, dated 1616, see: David Hughson, *London; Being an Accurate History and Description of the British Metropolis and Its Neighborhood: To Thirty Miles Extent, from an Actual Perambulation*, (6 volumes), Vol. 6, London: W. Stratford, 1809, pp. 358–361.

264 Tynan, "London's Private Water Supply," p. 348. In John Trusler, *The London adviser and guide: containing every instruction and information useful to persons living in London*, London: printed by the author, 1786, pp. 24–26, comparative data for the second half of the eighteenth century shows that all ten companies existing then charged similar rates, from about twenty-four to thirty shillings a year, according to whether supplies were daily or three times a week. This price varied with the distance to the pumping device for Thames water, or if the water was to be conveyed to the second or third story.

265 William Hone, *The Table Book*, (2 volumes), Vol. 1, London: Hunt and Clarke, 1827, p. 733.

FIGURE 4.24 *A water carrier (1700s)*.[266]

In later years, when numerous water companies were chartered as purely commercial enterprises, the rivalry became fiercer. When different companies finally consolidated, the interests of consumers were forsaken. Supply became intermittent, flowing only a few hours each day, requiring storage-tanks in every house, being exposed to contamination, while the original source of supply was polluted with sewage and industrial refuse. Regulation was established by Parliament at a much later stage, about 1840,[267] by a statement that the principle of free competition should not be entirely applicable to water companies.[268]

266 Marcellus Laroon, "New River water," *The Cryes of the City of London Drawne after the Life*, etching, engraving, published in London, 1688–1750, plate 61, (24.7×16.3 cm), BM, DPD, Registration number: 1868,0808.13614.

267 Naren Prasad, "Privatisation of Water: A Historical Perspective," *Law, Environment and Development Journal*, Volume 3/2 (2007), pp. 217–233, p. 221.

268 Charles Frederick Wingate, "The Water supply of Cities," *North American Review 136*, New York, (1883), p. 365.

William Mathews, in his landmark *Hydraulia*, when referring to the London water supply undertakings, concludes: "… in the period referred to, when an important object gave the stimulus, there was no deficiency of projectors to propose plans for benefiting the inhabitants of the metropolis, though inferior to the number distinguishing later times, with purposes and schemes almost infinitely various, as well as enormously expensive."[269]

Land Reclamation and Sewerage in England

The confirmation of the *Charter of Romney Marsh* (1252)[270] by King Henry III established the basis for land drainage in various parts of England for several centuries. During the Elizabethan era the economic benefits of land reclamation and intensive mining exploitation through drainage was the motivation for several ambitious projects. Foreign experts, mainly Italian and Dutch, tried their luck in those enterprises, after being awarded exclusive patents to implement their inventions. On the last day of December 1562, a patent was granted for 20 years to John Medley for an instrument 'for the draynage of water' from flooded mines of tin, lead and coal. Another patent was granted for 20 years to Burchsard Cranick on 22 June 1563, to make engines for the draining of waters—the engine was stated to be unlike anything devised or used within the realm.[271] In 2 July 1565, a letter from William Humphrey to Sir William Cecil, concerning copper mines, "recommends an 'Almain' engineer, who can raise water one hundred fathoms[272] high, by a newly invented engine;"[273] a license was granted to Dan Houghsetter on 20 April 1569, for 21 years for setting up and using engines for mine drainage.[274] Another patent was granted for 20 years to Sir Thomas Goldinge on 5 July 1571, for an engine for land drainage and water supply. If the engine would not be erected in two years or if it failed to perform,

---

269 Matthews, *Hydraulia*, p. 30.

270 Henry de Bathe, *The charter of Romney-Marsh, or, The laws and customs of Romney Marsh, very useful for all professors of the law, and also for all lords of towns …/ framed and contrived by the Venerable Justice Henry de Bathe*, London: Printed by S. R. for Samuel Keble, 1686. Romney Marsh is the largest coastal wetland on the south coast of England.

271 E. Wyndham Hulme, "History of the Patent System Under the Prerogative and at Common Law" *Law quarterly review*, Vol. XII, (1896), pp. 141–154, p. 146.

272 Equivalent to two yards, or about 1.8 meters. *The Encyclopaedia Britannica, a dictionary of arts, sciences, literature and general information*, 11th ed. (32 volumes), Vol. 10, New York: Encyclopaedia Britannica, 1910, p. 201.

273 Lemon, *Cal. SPD, of the Reigns of Edward VI, Mary, Elizabeth*, XXXVI, 73, p. 254.

274 Hulme, "History of the Patent System," p. 149.

the grant would be void.[275] On 28 October 1573, Richard Candish received a twenty-year grant, extending over eight counties, for an engine for draining coal and iron mines.[276] The list goes on—a "description of … pumps for raising water from mines, by Peter Jordayne and his coadjutors" in 1575,[277] and in July three years later a suit of Gherard Honricke, a native of West Friesland, for a patent giving him and his assigns for thirty years the "sole right to erect certain engines invented by him for the draining of mines."[278] Honricke promised to 'put the same engines and instruments [to drain water from flooded mines] in practice' at 'his own charges' if only Elizabeth would grant him this monopoly.[279] Another was the case of Jacopo Aconcio,[280] a renowned military engineer, native of Trento, persecuted by the Catholic Church because of his Protestant beliefs, who applied for several patents on hydraulic machinery in the 1560s.[281] He was excommunicated by the Bishop of London, Edmund Grindal, accused of heretical tendencies. After he devised a successful plan for draining lands which had been flooded by the Thames, he was readmitted into the Church,[282] allocated a £60 annuity in February 1561[283] and granted the status of denizen on October 1561, under the anglicized name of James Acontius, though he had to continue paying customs as an alien.[284] This impressive string of patents did not bring substantial practical results, which were finally achieved with the arrival of experienced Dutch or Dutch trained engineers.

---

275 Ibid., p. 45.

276 Julius W. Pratt, "Machinery in Sixteenth-Century English Industry," *The Journal of Political Economy*, Vol. 22, No. 8. (Oct. 1914), pp. 775–790, p. 779.

277 Robert Lemon, *Cal. SPD, of the Reigns of Edward VI, Mary, Elizabeth*, CVI, 24, p. 509.

278 Ibid., CXXV, 50, p. 598.

279 Harkness, *The Jewel House*, p. 152.

280 For a detailed account of his technical achievements see: Lynn Townsend White, "Jacopo Aconcio as an Engineer," in *Medieval religion and technology: collected essays*, Berkeley: University of California Press, 1978, pp. 149–174. For the crucial role Aconcio seems to have played in establishing the theoretical basis for patent grants in England, see: Jeremy Phillips, "The English patent as a reward for invention: The importation of an idea," *The Journal of Legal History*, (1982), Volume 3, Issue 1, pp. 71–79.

281 Ciriacono, *Building on water*, p. 237.

282 William M. Jones, "Two Learned Italians in Elizabethan England," *Italica*, Vol. 32, No. 4 (Dec., 1955), pp. 242–247, p. 243, Published by the American Association of Teachers of Italian.

283 Mary Anne Everett Green, (ed.), *Cal. SPD of the Reign of Elizabeth, 1601–1603, with Addenda 1547–1565*, London: Longmans & Co., 1870, p. 495.

284 J. H. Collingridge and R. B. Wernham, (eds), *Calendar of the Patent Rolls, Elizabeth [I]: 1560–1563*, London: HMSO, 1948, p. 101.

Cornelius Vanderdelf was one of the first Dutchmen to undertake drainage work in England, followed in the beginning of the seventeenth century by other famous engineers like Joos Croppenburgh. The leading figure amongst them was Cornelius Vermuden, who was appointed Chief Engineer to the Fenland project after having established his reputation by draining Hatfield Chase, a low-lying area in South Yorkshire, in 1626. He was also instrumental in the drainage of nearly 1300 square kilometers of marshland known as the Bedford Levels in 1636.[285] The scheme was imposed despite huge opposition from locals who were afraid of losing their livelihoods, which were based on fishing and wildfowling. However, success was short-lived due to the lowering of field levels, caused by the shrinkage of the drained peat soil. The more effectively the marshes were drained, the worse the problem became, and soon the fields were lower than the surrounding rivers, so hundreds of windmill pumps had to be built to raise water into embanked rivers. Nature prevailed, and by the end of the 17th century, the land was under water once again. All the large scale projects dealing with draining of marshes and the reclamation of land from the sea were capital-intensive ventures, which led contractors, backed by a combined force of financiers and technical experts, to embark on grand schemes. The Dutch often provided not only the know-how, but the capital too, supplied by Dutch money-lenders or business houses, as in England (Jacob Cats, the van Valckenburghs, Constantijn Huygens and others) in France (Jan Hoeufft) and in northern Germany (Willem van den Hove).[286]

In London, as early as 1280, an edict was enacted making citizens who threw rubbish on the street or left timber abandoned liable to fine.[287] While the fines were initially small, they were gradually increased. The law was subsequently expanded to include watery wastes (kitchen and urinal) as well as solid.

285 "An Act for settling the Draining of the Great Level of the Fens called Bedford Level," 1663, reproduced in Samuel Wells, *The History of the Drainage of the Great Level of the Fens called Bedford Level*, London: R. Pheney, 1830, (2 volumes), Vol. 2, pp. 383ff. Margaret Albright Knittl, "The design for the initial drainage of the Great Level of the Fens: an historical whodunit in three parts," *Agricultural History Review* 55 (2007), pp. 23–50, challenges the accepted view that it was Vermuyden who designed and oversaw the draining work done in the Great Level of the fens when Francis, fourth Earl of Bedford, became its undertaker in January 1631.

286 E. E. Rich and C. H. Wilson, (eds), *The Cambridge Economic History of Europe: Volume V, The Economic Organization of Early Modern Europe*, Cambridge: CUP Archive, 1977, p. 68.

287 Reginald R. Sharpe, (ed.), 'Folios 110b - 135b,' in *Calendar of Letter-Books of the City of London: 1275–1298*, London: Her Majesty's Stationery Office, 1899, pp. 207–230, http://www.british-history.ac.uk/london-letter-books/vola/pp207-230, fol. 130, accessed November 15, 2015.

Industrial wastewater was the principal discharge, mainly from tanners and dyers. An official investigation into the state of the Fleet River in London in 1307 concluded that the main cause of pollution was tanning waste and butchers' offal from Smithfield market.[288] By the following decades refuse was removed weekly, supervised by the 'beadle' (petty parish officer of each ward), who employed 'rakers', to gather up the waste and remove it to refuse dumps, where householders could deposit their rubbish as well. The filth which accumulated in the streets facilitated the rapid reproduction of rats, which carried the plague bacillus.

Following a king's writ in 1347, a proclamation was issued forbidding anyone from throwing rubbish, earth, gravel, or dung into the Thames or the streams of the city. Filth was to be taken out of the city in carts or in dung boats, "without throwing anything into the Thames for the saving of the body of the river ... and also for avoiding the filthiness that is increasing in the water and upon the Banks of the Thames, to the great abomination and damage of the people."[289] In the mid-fourteenth century, the plague came into Europe from Asia breaking out frequently in London between 1349 and 1665. One-third of the city's inhabitants died or left town when the plague struck in 1349.[290]

The English Parliament enacted a comprehensive law against polluting in 1388.[291] In 1407 Londoners were instructed to keep their refuse indoors until the special collection day. Rewards for informants were included into the law in 1414. However, despite apparently extensive city efforts, ordinances were usually disobeyed. In 1427, the first English public Acts about the sewage issue was delivered.[292]

288 Adam Markham, *A Brief. History of Pollution*; London: Earthscan Publications, 1994, p. 4.

289 Henry Thomas Riley, (ed.), "Memorials: 1357," in *Memorials of London and London Life in the 13th, 14th and 15th Centuries*, London: Corporation of London, 1868, pp. 295–300, p. 299. The original document is: *31 Edward III, CE 1357. Letter-Book G. fol. lxxi*, (Latin and Norman French).

290 Henry Duff Traill and James Saumarez Mann, *Social England: a record of the progress of the people in religion, laws, learning, arts, industry, commerce, science, literature and manners, from the earliest times to the present day*, London: Casell and Company, Ltd., 1902, p. 944.

291 [No. IV.] 12 Richard II. c. 13, Extracted from William David Evans, Anthony Hammond and Thomas Colpitts Granger, *A Collection of Statutes Connected with the General Administration of the Law: Arranged According to the Order of Subjects, with Notes*, (10 volumes), Vol. 6, London: W. H. Bond, 1836, p. 218.

292 Burian, "Historical Perspectives of Urban Drainage," p. 7.

The main legislation dealing with drainage in Britain was the *Bill of Sewers (23 Henry 8 c 5), Statute of Sewers, A General Act concerning Commissions of Sewers to be directed in all Parts within this Realm*,[293] which had been passed by King Henry VIII in 1531, and created a Commission of Sewers to enforce these rules. Although amendments were made during the reigns of Edward VI and Elizabeth I,[294] it continued to be the legal basis for almost all sanitary sewerage works well into the nineteenth century.[295]

Several public latrines were constructed in Medieval London over streams and rivers.[296] Rich and influential Londoners had private or semi-private latrines, typically located over a running stream or a cesspool. But it was only in 1589 that the poet, courtier and translator Sir John Harrington devised Britain's first flushing toilet. The invention was not adopted on a wide scale in England, but became popular in France.[297] London's early sewers were basically open ditches sloped to convey any waste people would discharge into them to the Thames, and ultimately out to the sea. No one had second thoughts about discharging the contents of chamber pots and kitchen garbage into the gutters. The Thames was fouled from a combination of factors, all of them bad. The population growth boosted the production of urban wastes from many sources: domestic rubbish and ashes, refuse from ships, animal dung, inflow from polluted tributary streams, industrial wastes and human sewage. An obnoxious smell pervaded everywhere, even at a personal level—citizens washed their clothes in the river and its unpleasant odors were on their garments. Isabella Whitney's *Wyll and Testament*, published in 1573, is one of the literary

293 William John Broderip and Robert Callis, *The Reading of upon the Statute of Sewers, 23 Hen. VIII. C. 5*, J. Butterworth & Son, 1824, pp. 1–16.

294 Ibid., pp. 17–20.

295 Joshua Getzler, *A History of Water Rights at Common Law*, Oxford: Oxford University Press, 2004, p. 107.

296 The discussion on London is based on William Henry Wheeler and Leonard Charles Batty, *A History of the Fens of South Lincolnshire*, Boston: J. M. Newcomb, 1897; Ernest L. Sabine,"City Cleaning in Mediaeval London," *Speculum*, Volume 12, (1937), pp. 19–43 and Ernest L. Sabine, "Latrines and Cesspools of Medieval London," *Speculum*, Volume 9, (1934), pp. 303–321.

297 Giovanni De Feo, Georgios Pericles Antoniou, Larry Wesley Mays, Walter Dragoni, Hilal Franz Fardin, Fatma El-Gohary, Pietro Laureano, Eleni Ioannis Kanetaki, Xiao Yun Zheng, and Andreas N. Angelakis, "The Historical Development of Wastewater Management," *in* Saeid Eslamian, (ed.), *Handbook of Engineering Hydrology: Environmental Hydrology and Water Management*, Boca Raton (FL): CRC Press, 2014, pp. 163–218, p. 196.

testimonies that convey the altogether negative image of the city in the eyes of Londoners sensitive to the quality of life in their city.[298]

Unwisely, the government forbade in 1580 the erection of new buildings within three miles of any of the gates of the City,[299] driving many families to find shelter under one roof within the limited area of the City proper. A number of letters exchanged between the Lords of the Council and the Lord Mayor show the concern caused by the overcrowding of the City.[300] Even Queen Elizabeth declared in 1593 that "great mischief daily grow and increase by reason of pestering the houses with diverse families, harboring of inmates, and converting great houses into several tenements…"[301]

In the seventeenth century parish officers ('scavengers' or 'goungefermours'),[302] were appointed yearly to oversee the work of street cleansing and refuse removal. The collected waste was either immediately spread on common ground or taken to holding dumps, either at the urban periphery or on the Thames banks.[303] The most awkward proposal for waste disposal was John Lanyon's, who proposed in 1654 to take upon himself, as a private contractor, the huge

298 Albrecht Classen, "Introduction," in Albrecht Classen, (ed.), *Urban Space in the Middle Ages and the Early Modern Age*, Berlin: Walter de Gruyter, 2009, pp. 1–146, pp. 137–138. For discussion on the attitude in Early Modern England physicians towards the influence of the environment over human health, see Mary J. Dobson, *Contours of Death and Disease in Early Modern England*, Cambridge: Cambridge University Press, 2003, pp. 9–42.

299 Thomas Rymer, *Fœdera: conventiones, litteræ, et cujuscunque generis acta publica, inter reges Angliæ et alios quosuis imperatores, reges, pontifices, principes, vel communitates, ab ingressu Gulielmi I. in Angliam, CE 1066, ad nostra usque tempora habita aut tractata: ex autographis, infra secretiores archivorum regiorum thesaurarias, asservatis, aliisque summæ vetustatis instrumentis, ad historiam anglicanam spectantibus, fideliter exscripta*, (20 volumes), Vol. 16, Londini: A. and J. Churcill, 1704–35, p. 448.

300 See for example Robert Lemon, (ed.), *Cal. SPD, Elizabeth, 1581–90*, London: Longman, Green, Longman, Roberts, and Green, 1865, p. 29 and Overall, *Remembrancia*, "Buildings II, 17," p. 44.

301 "An Act for Restraint of new Buildings, converting of great Houses into several Tenements, and for Restraint of Inmates and Inclosures, in and near unto the Cities of London and Westminster," Anno 35, Elizabeth, c. 5, 6, CE 1593, extracted from John Raithby, (ed.), *The Statutes at Large, from Magna Carta to the Union of the Kingdoms of Great Britain and Ireland*, (20 volumes), Vol. 4, London: G. Eyre and A. Strahan, 1811, p. 484.

302 Emily Cockayne, *Hubbub: Filth, Noise & Stench in England 1600–1770*, New Haven and London: Yale University Press, 2007 p. 184.

303 Ibid.

problem of cleansing and maintaining the streets of London.[304] Lanyon had argued that the unpaid scavengers could not effectively supervise the work of public cleansing, and that the results were unsatisfactory; he offered to do the work for the same cost, provided he was offered a long term contract. This reform was purely administrative and involved no new technology. However, this arrangement seems to have broken down after the Plague and Great Fire in the following decade.[305]

By the early 1700s, nearly every home in London had a cesspit beneath it—and suffered from the unavoidable foul odors, which were especially bad during quiet nights. Cesspools and pits reaching the water-bearing gravels, contaminated the groundwater that supplied the wells.[306] The sanitation plight continued to afflict London, throughout the seventeenth and the beginning of the eighteenth centuries, as a vivid description by Jonathan Swift in 1710 exemplifies: "Sweepings from butchers' stalls, dung, guts, and blood, / Drowned puppies, stinking sprats, all drenched in mud, / Dead cats and turnip-tops come tumbling down the flood![307]

304 For details of this proposal see: CLRO, Misc. 11.25, extracted from Rosemary Weinstein, "New Urban Demands in Early Modern London," *Medical History*, (1991), pp. 29–40, pp. 31–32, footnote Nr. 11.

305 Peter Hounsell, *London's Rubbish*, Electronic edition, chapter 1, Stroud, Gloucestershire: Amberley Publishing Limited, 2014.

306 An extensive description of the urban waste problematic in medieval times is given by Roberta J. Magnusson, "Water and wastes in medieval London," in Terje Tvedt, (ed.), *A history of water*, (3 volumes), Vol. 1, London: I. B. Tauris, 2006, pp. 299–313. This discussion draws also from this article.

307 Gladstone, *History of the London*, p. 5, quoting verses from Jonathan Swift's "A Description of a City Shower" (1710). The complete poem can be found in: *The poetical works of Jonathan Swift*, compiled and commented by John Mitford (3 volumes), Volume 1, London: William Pickering, 1833, pp. 93–96. The quote refers to p. 96.

LE PLAN DE LA VILLE, CITE, VNIVERSITE ET FAVXBOVRGS DE PARIS AVEC LA DESCRIPTION DE SON ANTIQVITE
LA RIVIERE DE SEINE

CHAPTER 5

Paris

On March 15, 1528, king François I wrote to the municipality of Paris: "Our intention is to henceforth spend most of our time in our home and stay in the good town and city of Paris ... knowing our castle of the Louvre to be the most convenient place and time for us to live."[1] By the stroke of a pen, Paris was so proclaimed officially the capital of France, with all the consequences such an act carried: government employees and courtiers establishing permanent residence, a direct relationship between King and municipality, and the common need to increase the prestige of a city hosting the court.

Henri IV, who reigned in France between 1594 and 1610, proved to be a man of vision and courage. France, which had previously undergone a chaotic period of religious and civil wars, was transformed during his reign: foreign peace and internal order were reestablished, agriculture and commerce were benefited. Credit for this feat should be given both to Henri and to his faithful minister Maximilien de Béthune, Baron of Rosny, later to become Duc de Sully and known popularly as Sully. The king undertook projects to improve the lives of all subjects, which converted him into one of the country's most popular rulers ever.[2] A statement, popularly attributed to him, epitomizes the prosperity and relative peace Henri brought to France after decades of religious wars, and shows his deep understanding of the plight of the French workers and farmers: "If God gives me life, I will ensure that there is no working man in my kingdom who does not have the means to have a chicken in the pot every

---

Opposite: Paris in 1615—Matthäus Merian, l'ancien, "Le plan de la ville, cite, université et Fauxbourgs de Paris avec la description de son antiquité et singularités," (1615), engraved by Merian and published in Paris by Nicolas de Mathoniére. East-southeast is at the top, (50×101 cm), Bibliothéque historique de la Ville de Paris, catalogued under A 104.

1 "Pour ce que nostre intention est de doresnavant faire la pluspart de nostre demeure et sejour en nostre bonne ville et cité de Paris et alentour plus qu'en autres lieux du Royaulme, cognoissant nostre chastel du Louvre estre le lieu le plus commode et à propos pour nous loger." Alexandre Tuetey, (ed.), *Registres des Déliberations du Bureau de la Ville de Paris*, (15 volumes), Volume 2 (1527–1539), Paris: Imprimerie nationale, 1886, p. 17.

2 For a description of Henri IV's tactical actions to further his agenda, see: Vincent J. Pitts, *Henri IV of France, His Reign and Age*, Baltimore: John Hopkins University Press, 2009, chapters IX and X.

 | DOI 10.1163/9789004312425_007

Sunday."[3] After generations of domination by the Valois dynasty, which extorted from the French people everything they had in order to pay for the royal family's luxuries, Henri's attitude won wide support from the Frenchmen. Voltaire expressed his admiration eloquently: "France enjoyed since the peace of Vervins[4] happiness as she had almost never known. The Catholic and Protestant factions were calmed by the wisdom of this King, who would be regarded as a great statesman, whose courage and goodness didn't eclipse his other virtues.[5]

Henri worked through Sully, to put order in state finances, encourage education and undertake public works, such as the drainage of swamps, building bridges, canals and roads.[6] He strengthened the French military establishment and it was under his direction that the construction of a great line of defenses on the frontiers was begun. Sully had counseled Henri IV to convert to Roman Catholicism, but refused to become a Roman Catholic himself. He became a model of the honest civil servant, rigorously pursuing revolutionary economic policies well before his time: lowering legal interest, authorizing free exportation of grain and wine, setting up a court to try cases of speculation, forbidding provincial governors to raise money on their own authority and eliminating many abuses of tax-collecting. Yet in spite of his successful administration, Sully was not popular. He was hated by Roman Catholics because he remained a Protestant and by Protestants due to his loyalty to the King. Although his

---

3 "... *si Dieu me donne encore de la vie, je ferai qu'il n'y aura point de laboureur en mon Royaume qui n'ait moyen d'avoir une poule dans son pot,*" extracted from: Hardouin de Péréfixe de Beaumont, *Histoire du roi Henri-le-Grand,* Paris: Antoine-Augustin Renouard, 1816, pp. 459–460.

4 The treaty of Vervins, which brought to an effective end the wars of religion in France, was signed between the representatives of Henri IV of France and Felipe II of Spain, on 2 May 1598, at the town of Vervins in Picardy, northern France.

5 "France goûtait depuis la paix de Vervins une félicité qu'elle n'avait presque jamais connue. Les factions Catholiques & Protestantes étaient contenues par la sagesse de ce Roi qui serait regardé comme un grand politique si sa valeur & sabonté n'avaient pas éclipsé ses autres mérites. Le peuple respirait, les grands étaient moins tîrant, la politique était par-tout encouragée, le commerce commençait à fleurir, les loix reprenaient leur autorité. Les dix dernières années de la vie de ce Prince ont été peut-être les plus heureuses de la monarchie. Il allait changer la face de l'Europe éomme il avait changé celle de la France." Voltaire [François-Marie Arouet], (under the pen name L'abbe Bigay), *Histoire du Parlement de Paris,* (2 volumes), Vol. 2, Amsterdam: Jean Jaques du Fay, 1769, p. 42.

6 For an account of the forceful actions of Sully during Henri IV's reign, see: David Buisseret, *Sully and the growth of centralized government in France 1598–1610,* London: Eyre & Spottiswoode, 1968.

personal style was obstinate and rude, his executive capabilities, sane financial policies and relentless efforts in extirpating administrative corruption made him indispensable to Henri. The King lavishly rewarded him both in honors and materially. A long list of titles was bestowed upon him: superintendent of fortifications and grand master of artillery; governor of Nantes and of Jargeau, captain-general of the Queen's *gens d'armes* [heavy cavalry of the king's court], governor of the Bastille, governor of Poitou; and in 1606 he was made Duke of Sully.[7]

By the late 1590s Henri IV and his advisers saw as their first political imperative to restore Paris to its former grandeur. Henri had his own personal wishes as well: "Henri himself seemed genuinely to love the city and wanted to make it great again for the simple reason that he wanted to live somewhere that was busy and vibrant."[8] When Henri IV created the new office of *Grand Voyer de France* [High Commissioner of Highways and Public Works],[9] in May 1599, tailored for—and probably by—Sully, an impressive boom of roads, bridges and canals construction followed, for which foreign experts were recruited. Some of them were the forefathers of a long lasting French dynasty of civil engineers, surveyors and cartographers.[10] Many cities, including Paris, had an office of chief surveyor, responsible for combating transgression of property limits on public roads by individuals. All of them were put in 1599 under the supervision of the *Grand Voyer*, whose duties were purely administrative and did not include litigation. In 1604, Sully unified the office of *Grand Voyer* with that of the Paris surveyor.

7 Maximilien de Béthune duc de Sully, *Memoirs of Maximilian de Bethune, Duke of Sully, Prime Minister to Henry the Great: Containing the History of the Life and Reign of that Monarch, and His Own Administration Under Him*, (6 volumes), Vol 3, London: J. Rivington [and others], 1778, pp. 163–164.

8 Andrew Hussey, *Paris, the Secret History*, New York: Bloomsbury, 2007, p. 137.

9 This new title represented a drastic change in the way public business was conducted. See: Bernard Barbiche, "Une révolution administrative: la charge de grand voyer de France," in *Pouvoir et institution en Europe au XVIème siècle*, edited by André Stegmann, Paris: Libraire Philosophique J. Vrin, 1987, pp. 283–296.

10 For an in-depth discussion of this subject, see: Gustave Fagniez, *L' économie sociale de la France sous Henri IV, 1589–1610*, Paris: Hachette et Cie., 1897 and Jean-Pierre Babelon, "L'Urbanisme d'Henri IV et Sully à Paris," in *L'Urbanisme de Paris et l'Europe 1600–1800*, edited by Pierre Francastel, Paris: Editions Klinchsieck, 1969, pp. 47–60.

Government and Municipal Authorities—Paris Administration and Development in the Late Sixteenth and Early Seventeenth Centuries

The administration of the kingdom was conducted through a complex web of institutions and officials, which cannot be described completely in the scope of this book.[11] It would, however, be useful to dwell shortly on a few institutions that were in charge of the administration of Paris and of the development of its public works, starting with those officials of various natures who bore the title *prévôt* [provost]. In old French Law this designation had many applications. It could denominate any person placed at the head of a branch of the public service, a position which, according to the old principles, habitually carried with it a right of jurisdiction. There were *prévôts des maréchaux de France* [provost marshals, in charge of public security], *prévôt de l'hôtel du roi* [provost of the royal palace], otherwise known as the *grand prévôt de France* [lord high provost of France], and the *prévôt général* [provost general] or *grand prévôt des monnaies* [high provost of the mint]. The *prévôts royaux* [royal provosts], were entrusted with and carried out local royal authority, including the collection of revenues of the Crown's *domaine* [land belonging to the Crown] and all taxes and duties owed to the king within a provostship jurisdiction. They were also responsible for military defense, with tasks such as raising local contingents for the royal armies.

The head of the Paris municipality was the *prévôt des marchands* [provost of merchants], who was appointed for two years and whose nomination could be renewed. He had to have the status of citizen of the city and to be a born in it. Though nobility was not a formal requirement, he was normally selected from a notable family. He was assisted by four *échevins* (aldermen), who were appointed for two years and renewed by half term. Collectively, these five executive officials were known as *Le Bureau de la ville de Paris*.

The main responsibilities of the *prévôt* and the *échevins* included public works, such as pavement construction in certain public places, conservation of bridges and piers, street cleaning, maintenance of fountains and sewers and the controversial task to inspect building alignment according to regulations. The *prévôt* and the *échevins* appointed, and were assisted by, a City Council (*Conseil de Paris*) composed of twenty four councilors. Although the *Conseil* did not play a regular part in the administration of the city, their advice was

11 For a detailed account of French institutions and specifically those of Paris, see: Gaston Zeller, *Les institutions de la France au XVIe siècle*, Paris: Presses Universitaires de France, 1949.

FIGURE 5.1 *Le Prévôt des marchands et les échevins de la ville de Paris.*[12]

sought in times of financial crisis, and they constituted nearly a third of the electors who elected the five men who actually ran the municipality.[13]

The King was always directly represented in the city by the *prévôt de Paris,* who had two lieutenants: a *lieutenant civil* and a *lieutenant criminel.* The *lieutenant civil,* in charge of the police, was the magistrate who was the real power in the Châtelet, the main tribunal court in France, which had jurisdiction over all disputes and matters of distributive justice.[14] As some responsibilities were not clearly defined occasional conflicts arose between the *Grand Voyer* and *the lieutenant civil.*[15] The appointment of a *prévôt des marchands* or a *échevin* conferred nobility, which was hereditary, provided certain behavior requirements were fulfilled.[16]

12 Philippe de Champaigne, *Le Prévôt des marchands et les échevins de la ville de Paris,* 1648, Oil on canvas, (200×271 cm), Musée du Louvre.

13 For an in-depth account of the workings of this institution, see: Barbara B. Diefendorf, *Paris City Councillors in the Sixteenth Century,* Princeton: Princeton University Press, 1983.

14 Roland Mousnier, *The Institutions of France under the Absolute Monarchy, 1598–1789: The Organs of State and Society,* (2 volumes), Vol. 2, Chicago: University Of Chicago Press, 1984, p. 316.

15 Pierre Lavedan, *Histoire de l'urbanisme a Paris,* Paris: Hachette, 1975, p. 136.

16 Roland Mousnier, *The Institutions of France under the Absolute Monarchy, 1598–1789: Society and the State,* translated by Brian Pearce, (2 volumes), Vol. 1, Chicago: University Of Chicago Press, 1979, p. 584.

Already François I and Henri II had grossly interfered in the municipal elections.[17] Henri IV, like Sully, thought the city authorities should be overseen by a king's loyal man, thus even the elections of the *prévôt des marchands* were more or less subject to his will. The newspaper Le Mercure de France, in 1608, stated that "... in the elections '... all is done by the will of the king.'"[18] Robert Descimon, who studied Henri's interference in Parisian elections, concluded that the king and the municipal authorities did their best to maintain the appearance of free elections, while ensuring that their result would suit the king's preferences. However, while Henri practically chose the city's *prévôt des marchands*, he only seldom interfered in the election of *échevins*. In many instances choices for senior candidates made by electors were rubber-stamped by the king, if he had no strong reasons to oppose them. Descimon emphatically maintains that Henri IV dealt with matters based on the most archaic centralistic political framework.[19] A careful study of the list of the *prévôts des marchands* in the sixteenth and early seventeenth century shows to what extent nepotism pervaded in this elected position, with several surnames returning every several years.[20]

The *Parlement de Paris*, essentially a judicial body, had legislative powers that concerned the realm as a whole, but it also had an important role in the administration of the capital, though without normally interfering with its day

17 Colin Jones, *Paris, Biography of a City*, New York: Penguin Books, 2006, p. 151.

18 "... qu'aux élections 'tout se fait par la volonté du Roi'." Lavedan, *Histoire de l'urbanisme*, p. 137, quoting *Le Mercure de France*, année 1608, I, fol. 114.

19 Robert Descimon, "L'Echevinage Parisien sous Henri IV (1594–1609). Autonomie urbaine, conflits politiques et exclusives socials": in Neithard Bulst and J.-Ph. Genet, (eds), *La Ville la bourgeoisie et la Genese de L'Etat*, Paris: Éditions du Centre National de la Recherche Scientifique, 1988, p. 150.

20 Examples abound: Jean Morin (1524–1525) and (1544–1546); Germain de Marle (1526–1527) and Guillaume de Marle (1560–1564); Jean Luillier (1530–1531), Nicolas Luillier (1576–1578) and Jean Luillier (1592–1594); Augustin de Thou (1538–1540), Christophe de Thou (1552–1554) and Augustin II de Thou (1580–1582); Claude Guyot, who served two terms (1548–1552) and (1564–1568) and Antoine Guyot (1600–1602); Nicolas Perrot (1556–1558) and Christophe Perrot (1641); Martin de Bragelongne (1558–1560) and Martin de Bragelongne (1602–1604); Michel Marteau (1588–1589) and Michel Marteau, seigneur de la Chapelle (1589–1590); François Miron (1604–1609) and his brother Robert Miron (1614–1616); Jacques Sanguin (1609–1612) and Christophe Sanguin (1628–1632); Oudart Le Féron (1638–1641) and Hiérome Le Féron (1646–1650). For the complete list of prévôts, see: Marcellin Berthelot, Camille Dreyfus and others, (eds), *La grande encyclopédie: inventaire raisonné des sciences, des lettres et des arts*, (31 volumes), Vol. 25, Paris: Lamiraulot, 1886, p. 1063.

to day administration.[21] The *Parlement* sprang into action when a crisis or difficult problem threatened the city, leaving the execution of its decisions to the municipal authorities, but closely supervising their actions.[22] The various bodies of Paris's municipal administration, including the *prévôt des marchands*, the *échevins* and the *Conseil du Paris* were considered by the members of the *Parlement* as inferior officials, only representing the bourgeois population of Paris. The relations between the *Parlement* and the municipality varied according to circumstances. As a reflection of the Paris population, the city's *Conseil* was intractable when prosperity predominated, becoming humble and submissive under different circumstances. The *Parlement* skillfully navigated between both extremes.[23]

In late 1599, Henri used to its utmost his influence to have Antoine Guyot, *seigneur de* Charmeaux, who had been president of the *Chambre des Comptes* of Paris [Chamber of Accounts], elected as *prévôt des marchands* of the city in the elections scheduled for August 1600. He wrote directly to members of the *Bureau de la ville de Paris* in May of the same year, pressuring them to approve the nomination[24] Guyot was elected in what represented an additional case of nepotism: his father, Claude Guyot, who had occupied the same position in 1548, was reelected in 1552 and in 1564, a unique case in the annals of the city.[25] Henri counted on Guyot to obey the orders of the newly appointed *Grand Voyer* in order to implement his rebuilding policies, which Guyot and his successors willingly did. The *prévôt des marchands* described the crown's intentions in a declaration to the municipal government in 1601, emphasizing the king's concern for the well-being of Parisians, which, rather surprisingly for the period, he links to drinking pure water, and the restoration of the public fountains.[26]

---

21 For an extensive detail of the authority of the Parlement during the period of our study, see: Voltaire, *Histoire du parlement de Paris* (edited by Charles Palissot de Montenoy), in *Œuvres de Voltaire*, (55 volumes), Vol. 24, Paris: Chez Stoupe, 1792, pp. 175–191.

22 Diefendorf, *Paris City Councillors*, p. 10.

23 Leon de Laborde, *Le Parlement de Paris, sa compétence et les ressources que l'érudition trouvera dans l'inventaire de ses archives. Préface de l'Inventaire des actes du Parlement de Paris*, Paris: H. Plon, 1863, p. liv.

24 Orest Ranum, *Paris in the Age of Absolutism: an Essay*, University Park (PA): Pennsylvania State University Press, 2004, p. 76.

25 Pierre-Victor-Palma Cayet, "Chronologie septenaire ou histoire de la paix entre les roys de France et d'Espagne," in *Suite de la Collection Universelle des Mémoires Particuliers Relatifs à l'Histoire de France*, (72 volumes), Volume 70, Paris: De l'imprimerie et fonderie d'Orizet, 1807, p. 516.

26 Paul Guerin, (ed.), *Registres des Déliberations du Bureau de la Ville de Paris*, (15 volumes), Vol. 12 (1598–1602), Paris: Imprimerie Nationale, 1909, p. 386, 16 March, 1601. English

The reconstruction advanced at a frantic pace, including the Hotel-Dieu, the Hôpital de la Charité, the Arsenal, the Place Royale, the Grande Galerie of the Louvre and, most relevant to our study: the Pont Neuf—connecting the right and left banks of the city.[27] The newspaper *Mercure françois* asserted: "No sooner was Henri IV master of Paris that the masons were at work."[28] He was compelled to do so following the destruction caused by the civil war in Paris, where more than 1,400 houses had been devastated by artillery and fire,[29] out of an estimated overall number of 20,000 houses in the city, a huge upheaval, considering that in that period each house was inhabited by about 20 souls on the average.[30]

The building drive caused also the admiration of foreign visitors, such as Tobie Mathew, an English Member of Parliament and courtier who converted to Roman Catholicism and became a priest. He wrote in 1604 to Dudley Carleton, First Viscount Dorchester, English art collector, diplomat and Secretary of State: "This town is growing much fairer than yow have seen it".[31] A few years later, in October 3, 1608, François de Malherbe enthusiastically wrote to Mr. A. M. de Peiresc, "If you return to Paris in two years, you will not recognize

---

translation extracted from Hilary Ballon, *The Paris of Henri IV—Architecture and Urbanism*, Cambridge (MA) and London: M.I.T. Press, 1991, p. 123.

27 For an account of Henri IV's building drive, see: Reginald Theodore Blomfield, *A history of French architecture from the reign of Charles VIII till the death of Mazarin*, (2 volumes), Vol. 2, London: G. Bell and sons, 1911, chapter XIII, pp. 31–48. For the appearance of the margins of the Seine before the construction of the Pont Neuf, see: François de Belleforest, "Map of Paris," in *La Cosmographie universelle de tout le monde, en laquelle ... sont au vray descriptes toutes les parties habitables et non habitables de la terre et de la mer, ... auteur en partie Münster, mais beaucoup plus augmentée, ornée et enrichie par François de Belle-Forest*, (2 Volumes), Vol. 1, Paris: Michel Sonnius, 1575, Plate 9, between pp. 174–175. Belleforest's book is a translation of Sebastian Münster, *Cosmographiae Universalis*, Basel: Petri, 1550, and his map is based, with additional details, on "Lutetia Parisiorum urbs, toto orbe celeberrima notissimaqúe, caput regni Franciae," included in Münster's book, pp. 88–89.

28 *Le Mercure françois ou la suitte de l'histoire de la paix* (24 volumes), Vol 1 (1605–1610), Paris: Iean Richer, 1612, facsimile published by the Groupe de Recherches Interdisciplinairessur l'Histoire du Littéraire from an original kept at the bibliothèque de L'École des Ponts et Chaussées, p. 475 v.

29 Albert Miron de l'Espinay, *François Miron et l'administration municipale de Paris sous Henri IV de 1604 à 1606*, Paris: E. Plon, Nourrit et Cie, 1885, p. 171.

30 Jean-Pierre Babelon, *Demeures parisiennes sous Henri IV et Louis XIII*, Paris: Hazan, 1991, p. 39.

31 Arnold Harris Mathew and Annette Calthrop, *The Life of Sir Tobie Matthew, Knight*, London: Elkin Mathews, 1907, p. 43.

it"[32] In a short reign Henri IV did more for the capital than any other French king before him, and probably even after him. The king exhorted his subjects to support his building drive; he declared himself 'a friend of buildings' and claimed that he made three things willingly: 'war, love and buildings.'[33]

History of Former Water Supply Undertakings

The ancient city of Lutetia had an abundant supply of water for household needs from wells fed by the Seine. Wells were more numerous in the right bank than in the left bank, as the aquifer has different depths on either side (10 meters in the north against 30 meters in the south).[34] The water drawn was generally of poor quality, heavily laden with minerals.[35] The baths built by the Romans in Cluny were fed by water collected from a vast area and conveyed by the Aqueduct de Rungis, built about CE 360 under the emperor Julian, supplying about 2600 cubic meters per day.[36] Water from springs was also collected at Auteil and brought in earthenware pipes to baths near the present Palais Royal, a daily supply of about 500 cubic meters.[37] By the end of the twelfth century the surplus of the water springs owned by two abbeys in the villages Pré-Saint-Gervais and Belleville was being supplied to the city through lead pipes.

Despite the presence of a major river flowing through it, good water was an extremely scarce commodity in Paris. Worsening the situation, a substantial part of the water flowing to the fountains was diverted for the use of influential people close to the Court, either by their lineage or their wealth. In

32 "Si vous revenez à Paris d'ici à deux ans, vous ne le connoîtrez plus"—François de Malherbe and Nicolas Claude Fabri de Peiresc, *Lettres de Malherbe: ornées du fac-simile de son écriture, dédiées a la ville de Caen avec une vue de cette ville*, Paris: J. J. Blaise, 1822, p. 61.

33 Leonardo Benevolo, *The Arquitecture of the Renaissance*, translated by Judith Landry, (2 volumes), Vol. 2, Boulder (CO): Routledge & Kegan Paul, 1978, p. 693.

34 This discussion draws mainly from: Pierre-Nicolas Bonamy (1694–1770), "Mémoire sur les aqueducs de Paris, comparés à ceux de l'ancienne Rome," in *Mémoires de littérature, tirés des registres de l'Académie royale des inscriptions et belles-lettres depuis l'année MDCCLVIII jusques et compris l'année MDCCLX*, (45 volumes), Vol. 30, Paris: Imprimerie Royale, 1764, pp. 729–755 and Maxime du Camp, "Le service des eaux à Paris," *Revue des Deux Mondes*, Tome 105, (1873), pp. 275–308.

35 Virginia Vidal Lescuyer, "Historia del Uso del Agua en Una Gran Ciudad: Caso de París," *Ambient, Curso 2002–2003*, Barcelona, (July 2003), pp. 1–5, p. 1.

36 Robert James Forbes, *Studies in Ancient Technology*, (9 volumes), Vol. 1, Leyden: E. J. Brill, 1964, p. 186.

37 Bromehead, "The Early History of Water supply (Continued)," p. 185.

1392 Charles VI revoked the rights granted to private persons, except for the Louvre's, his brothers' and uncles' private dwellings, and ordered the piping connections of all others to be destroyed.[38] Most people were forced to draw water from the upstream of the Seine or from three old fountains: the 'Halles', the 'Innocents' and the 'Maubuée' ('bad laundry'- due to the calcareous salts found in the water). The fifteenth century brought little change, besides the construction of three new fountains, all fed by the springs of Pré-Saint-Gervais and Belleville.[39]

The beginning of the sixteenth century was marked by epidemics: whooping cough (*pertussis*) attacked in 1510 and the disease visited the city once again in 1522.[40] In 1553 the public fountains had to serve a population of about two hundred and sixty thousand and delivered only 300 cubic meters, barely over a liter per person per day.[41] During the sixteenth and seventeenth centuries there was no lack of newly established ordinances or by-laws regulating the use and maintenance of 'clean' water in fountains and wells. Various measures were taken, such as the prohibition of raising pigs, rabbits and poultry in Paris, and ordering the digging of pits for the *retraits* in each house, but as usual, effective enforcement was the problem.[42]

Authorities fought, with mixed results, against influential private individuals, who continued making clandestine connections to the conduits carrying water to the public fountains, thus reducing the flow for the general public.[43] By the initiative of King Henri IV many orders were issued to try to improve

38 William Barclay Parsons, *Engineers and Engineering in the Renaissance*, Cambridge (MA) and London: M.I.T. Press, 1976, p. 241. For the complete text of the edict see: Félix Lazare and Louis Lazare, *Dictionnaire historique des rues et monuments de Paris [en] 1855: avec les plans des 48 quartiers*, Paris: Servedit, 1855, p. 108.

39 For an extensive account of the construction and development of the whole network of fountains in Paris, see: Pierre-Simon Girard, *Recherches sur les eaux publiques de Paris, les distributions successives qui ont été proposés pour en augmenter le volume*, Paris: Imprimerie Impériale, 1812, pp. 2–15.

40 This illness caused inextinguishable cough, which Parisians believed they could evade by wrapping the head in a *coqueluche*, a kind of hood which gave its name to the deadly disease.

41 Bromehead, "The Early History of Water supply (Continued)," p. 189.

42 La Mare, *Traité de la police*, p. 548 and ff.

43 See for example: Alexandre Tuetey, (ed.), *Registres des Déliberations du Bureau de la Ville de Paris*, (15 volumes), Volume 11 (1594–1598), Paris: Imprimerie Nationale, 1902, p. 121 and Paul Guerin, (ed.), *Registres des Déliberations du Bureau de la Ville de Paris*, (15 volumes), Vol. 12 (1598–1602), Paris: Imprimerie Nationale, 1909, p. 411. Several additional examples are given in Parsons, *Engineers and Engineering*, pp. 241–246.

the situation, and significant actions to improve the water supply of the capital were implemented. In order to finance the repair of the fountains of the city, among other things, the king imposed in 1594 a tax of "… 7 *sous* and 6 *deniers* for each *muid* of wine carried into Paris."[44] By 1602 the restoration of fountains was almost complete, but this did not suffice to supply enough water to the thirsty city.[45]

La Samaritaine

By the end of the sixteenth century water needs, both of the King's palaces and of Paris huge population, were far from being met. This situation encouraged entrepreneurs to propose innovative methods and machines to lift water from rivers, usually attributing to their inventions extraordinary capabilities. Such was the case when in March 1585 King Henri III issued "letters-patents for a new method for raising water, and other beautiful and practical inventions"[46] to Nicolás Wasseur Hun—a citizen of Basel, Paul Latreille and Jehan Desponde, granting them the exclusive right to employ their invention in France for thirty years. During that period, nobody would be allowed to imitate or interfere with them, under the threat of imprisonment and property confiscation.[47]

The letter-patents lack practical details, such as the quantity of water to be lifted, the means of water distribution or the location of the intended device, but include bizarre claims "… Moreover they have a means to raise more earth in a day with two horses than hitherto has been done with ten pair of oxen,

44 "… *7 sous 6 deniers tournois* chaque *muid* de vin qui entrait dans Paris" Lazare, *Dictionnaire historique*, p. 109. One *livre tournois* contained 20 *sous*, subdivided into 12 *deniers* each—(John Pringle Nichol, *Cyclopædia of the physical sciences: comprising acoustics, astronomy, dynamics, electricity, heat, hydrodynamics, magnetism, philosophy of mathematics, meteorology, optics, pneumatics, statics, &c. &c*, London: Charles Griffin and Company, 1868, p. 173). One *muid* was equivalent to a wagon-load of roughly 2,500 liters—(Greengrass, Mark, *France in the Age of Henri IV, the struggle for stability*, London: Longman, 1984, p. 173).

45 Lazare, *Dictionnaire historique*, p. 109.

46 The complete text of the letters-patents is quoted in M. L. Cimber and F. Danjou, *Archives curieuses de l'histoire de France depuis Louis XI jusqu'à Louis XVIII, ou Collection de pièces rares et intéressantes … publiées d'après les textes conservés à la Bibliothèque royale, et accompagnées de notices et d'éclaircissements; ouvrage destiné à Leber*, Paris: Beauvais, 1836, pp. 419–423.

47 Sophie Crawford Lomas, (ed.), *Calendar State Papers Foreign, Elizabeth*, Volume 19: August 1584–August 1585, London: HMSO, 1916, pp. 228–234, entry: January 18, 1585.

and to draw carts heavily laden more easily with six horses than has commonly been done with thirty."[48] No records exist of what, if at all, was accomplished by the inventors.

In the framework of Henri IV's and Sully's quest to promote agriculture and drain swamps to create productive crop lands, a memorandum was prepared in 1597 for the King about a pump invented by Sire Travah from Metz. The claim was that the pump would be inexpensive and long lasting, and suitable to drain marshes and transform them into arable land and water meadows.[49] These meadows would increase forage production, causing livestock augmentation and as result larger quantities of manure, allowing for a bigger agricultural output.[50] The claims about the pump's prowess were extraordinary: it would make water mills unnecessary, a desirable result due to the materials cost for their construction—mainly wood, and the fact they hindered navigation; it was supposed to supply a whole town and develop enough pressure to reach the higher stories of houses and be able to power bellows employed by ironmasters.[51] This invention too was most probably not implemented, since no records exist of such impressive achievements.

In 1601, Jean Lintlaër,[52] a Fleming, submitted to King Henri IV a proposal to erect in Paris, at the Pont Neuf, upon one of the piers of the bridge, an engine consisting of lifting pumps to be driven by the current of the Seine.[53] The

48 Ibid.

49 Henry Heller, *Labour, Science and Technology in France, 1500–1620*, Cambridge: Cambridge University Press, 2002, p. 159.

50 Ibid., p. 159.

51 Ibid., p. 176.

52 Spelled sometimes as Jehan, and his surname as Linteler. V. J. Guiffrey, "Liste alphabétique des artistes et artisans employés à l'embellissement et à l'entretien des châteaux royaux du Louvre, des Tuileries, de Fontainebleau, de Saint-Germain etc., de 1605 à 1656, avec la mention de leurs gages." in *Nouvelles Archives de l'art français,* pp. 1–54., Paris: F. de Nobele, 1872, p. 35, states the true surname was *Lentlaer*. Edmond Bonnaffe, in his Dictionnaire des amateurs français au XVIIe siècle, Paris: A. Quantin, 1884, states that the French called him *Linclair*. In the text Lintlaër will be used, as Lintlaër himself signed.

53 Pierre-Thomas-Nicolas Hurtaut and Magny, *Dictionnaire historique de la ville de Paris et de ses environs*, (4 volumes), Vol. 3, Paris: Moutard, 1779, p. 82; Germain Brice, *Nouvelle description de la ville de Paris: et de tout ce qu'elle contient de plus remarquable*, (4 volumes) Vol. 4, Paris: Julien-Michel Gandouin, 1725, p. 195, Michel Félibien, *Histoire de la Ville de Paris*, (5 volumes), vol. 4, Paris: Chez Charles-Jean-Bapt. Delespine, 1735, pp. 99–100 and Jean Almar Piganiol de la Force, *Description historique de la ville de Paris et de ses environs*, (10 Tomes), Tome 2, Paris: Les Libraires Associes, 1765, p. 52, attribute to King Henri III the project, which is obviously mistaken, as Henri III died in 1589.

proposal was accepted, and a grant extended on January 2, 1602, detailed the terms:

> The King being at Paris and desiring for the improvement of his gardens of the Tuileries to have water taken from the river Seine to his said park at such height as his Majesty may be benefited, as has been proposed to do by means of a pump of which Jean Lintlaër of Flanders is the inventor and of which he controls the use; and his Majesty, not wishing to lose this opportunity, has ordered that, under the Pont Neuf in the city of Paris and in the arches that the said Lintlaër may select as the best and easiest for the execution of his design, there shall be erected a building on piles ... to hold two pumps of the kind and capacity to raise water in the largest possible quantity and there to lodge the said Lintlaër in order that he may supervise the operation and maintenance ... also that a reservoir be constructed in the cloister of Saint-Germain L'Auxerrois to receive water in abundance."[54]

This machine, which was eventually built and used for the declared purpose, became known as La Samaritaine, taking its name from a group of sculpture that adorned the small building in which it was placed. According to Hilary Ballon, the sculptural relief of the Good Samaritan bore a symbolic message, as it associated the biblical episode of Christian charity with the king's magnanimous effort to provide the city with water.[55]

It is not known whether Lintlaër's project used Hun's, Latreille's and Desmonde's invention. In any case, the latter did not raise any claim against the Fleming's enterprise. On the other hand, his idea was at first strongly opposed by Martin de Bragelongue, the *prévôt des marchands*, and the *échevins* of Paris at the time, who alleged, according to Fernand de Mallevoüe, that the device would interfere with the navigation of the river.[56] They may have been right, though it seems that this opposition was rather motivated by uneasiness about the infringement of the city's prerogatives, as the municipality was

54 Parsons, *Engineers and Engineering*, p. 249.

55 Ballon, *The Paris of Henri IV*, p. 122.

56 Fernand de Mallevoüe, (ed.), *Maximilien de Béthune, duc de Sully, Les actes de Sully passés au nom du roi de 1600 à 1610: par devant Me Simon Fournyer, notaire au Châtelet de Paris, dont les minutes sont conservées en l'étude de Me Henri Motel, notaire à Paris*, Paris: Imprimerie Nationale, 1911, p. 152, footnote No. 3. The holders of the position during that period were: Antoine Guyot (1600–1602) and Martin de Bragelongne (1602–1604). Extracted from Berthelot, *La grande encyclopédie*, p. 1063.

considered 'the queen of the river'.[57] The fact that Lintlaër was a foreigner certainly added to the bad feelings of the Parisian magistrates.[58] Consequently, the municipality ordered Lintlaër to cease his work on this project.[59] Henri, however, discarded this opposition on the ground that construction of the Pont Neuf was largely financed by his own revenues.[60] This was not unusual, as revenues which the city collected from taxes financed only ordinary expenses, such as the maintenance of the city's fortifications. In those cases in which the King granted subsidies for extraordinary expenses, no fixed rules were followed. As was generally the case when the city authorities felt that their privileges were infringed, Henri found a justification for his actions.[61]

In 1600, eager to gain the support of the city's authorities for the water supply project, the king recommended the election as *échevins* of Messrs. Garnier and Champin, while he opposed for some time the candidacy of François Miron to the post of *prévôt des marchands*. But at the climax of the struggle, François Miron[62] was elected as *prévôt* on 16 August, 1604,[63] and subsequently, influenced by Sully and convinced by the evidence in favor of the water supply project, the newly elected *prévôt* finally adopted the King's views in this matter. In order to keep the honor of the municipality, he resorted a few days later to the well known tactic of commissioning an expert opinion. Accordingly, the *Bureau de la Ville* nominated on 27 August, 1604, Pierre Guillain, master of masonry, John Gross, merchant, Mathieu Mascrier and Nicolás Bourguillot, both masters of the bridges of the City, as experts, who eventually submitted a posi-

57 Miron de l'Espinay, *François Miron*, p. 259.

58 Daniel Roche in "Le temps de l'eau rare du Moyen Age a l'époque moderne," *Annales. Histoire, Sciences* Sociales, 39e Année, No. 2 (Mar.–Apr., 1984), pp. 383–399, p. 386, adds that the city was ill-disposed towards the Dutch in general. For an in-depth discussion of the situation in France in this respect see Jules Marthorez, *Les étrangers en France sous l'ancien régime, histoire de la formation de la population française*, (2 Volumes), Paris: E. Champion, 1919–1921.

59 François Boucher, *Le Pont-Neuf*, (2 volumes), Vol. 1, Paris: Chez Le Goupy, 1925, p. 124.

60 On August 23rd, 1604, the king wrote to Sully: "… ledit pont est fait de mes deniers et non des leurs." See: Maximilien de Béthune duc de Sully, "Mémoires des sages et royales oeconomies d'estat, domestiques, politiques et militaires de Henry le Grand…: et des servitudes utiles, obéissances convenables et administrations loyales de Maximilien de Béthune," (2 volumes), Vol. 2, second serie of Joséph François Michaud and J. J. F. Poujoulat, *Nouvelle collection des Mémoires pour servir à l'histoire de France depuis le XIIIe siècle jusqu'à la fin du XVIIIe*, Paris: Editeur du Commentaire analytique du Code Civil, 1837, p. 540.

61 S. Annette Finley-Croswhite, *Henry IV and the Towns: The Pursuit of Legitimacy in French Urban Society, 1589–1610*, Cambridge: Cambridge University Press, 1999, p. 71.

62 For a biography of Miron, see: Miron de l'Espinay, *François Miron*.

63 Boucher, *Le Pont-Neuf*, p. 120.

tive opinion about the water-lifting device. Asked about their recommendation on the best location to install the pumping station, the experts proposed to attach it to the second arch on the right bank.[64]

As was customarily, from 1596 to 1609 Miron also filled the post of *lieutenant civil*. He had first been nominated as *lieutenant civil* by Henri IV in 1596, who is quoted as saying: "… he [Miron] is known for his faithful service, his frugality on behalf of the state, and that in this capacity he would steal less than others…"[65] Although frequently in friction with the king, Henri IV found in Miron a very active collaborator for the realization of his major urbanization projects. Most significant is the admonition to Miron's successor, Jacques Sanguin, in 1606: "I do not say anything to exhort you to your duty, except that you follow *lieutenant* Miron's way, because my City of Paris in his watch has been greatly beautified with buildings for public benefit."[66] Miron and Sanguin became the instrument through which Henri IV fulfilled his desire to rejuvenate Paris.

His contemporary François Eudes de Mézeray glorifies François Miron[67] and so does André Duchêne (1584–1640), who, after Mézeray, confirmed that during his time in office, François Miron employed all income for the embellishment of the city, but unfortunately his example had not been followed since.[68] At his death in 1609, Pierre de L'Estoile paid Miron this tribute: "He

64 Paul Guerin, (ed.), *Registres des Déliberations du Bureau de la Ville de Paris*, (15 volumes), Vol. 13 (1602–1605), Paris, 1905, p. 349.

65 "En parlant de M. Miron, dit à M. de Villeroy qu'il le conoissoit fidéle a son service, point avaricieux, & qui en cet estat des robberoit moins que les autres." Pierre de L'Estoile, *Journal du Règne de Henry IV. Roi de France, et de Navarre: Tiré sur un Manuscrit du tems*, (2 volumes), Vol. 1, Paris: [s.n.], 1732, p. 127.

66 "Je ne vous dirai autre chose pour vous exhorter à votre devoir, sinon que vous en suiviez le liutenant Miron qui vous a devancé dans cette charge, car ma ville de Paris sous sa pré vôté a été de beaucoup embellie de bâtiments pour les commoditiés publiques." Jean-Pierre Babelon, *Henry IV*, Paris: Fayard, 1982, p. 794.

67 "… il n'a point veu de Magistrat qui ait estably une plus exacte police dans la ville, dans les marchez, & sur les ports, qui ait embraslé si courageusement les interests du peuple, & qui ait apporté plus de foin & plus de ménage à faire revenir les biens & les droits de la ville, à acquitter ses debtes, à l'entretenir dans la splendeur où doit estre la capitale du Royaume, à la decorer de divers ornements, & à l'enrichir de toutes les commoditez publiques. Plusieurs rues eslargies, plusieurs pavées de nouveau & accommodées en pente pour écouler les eaux, huit ou neuf places & carrefours ornez de fontaines jallissantes…" François Eudes de Mézeray, *Abrégé chronologique ou extrait de l'histoire de France*, (3 volumes), Vol. 3, Paris: Chez Thomas Iolly, 1667, pp. 1410–1211.

68 François Rittiez, *L'Hôtel de ville et la bourgeoisie de Paris: origines, moeurs, coutumes, institutions municipales depuis les temps les plus reculés jusqu'à 1789*, Paris: Schlesinger Frères, 1863, p. 356.

FIGURE 5.2
Statue of François Miron in the Hôtel de ville, Paris.[69]

honored the state of lieutenant civil, he pursued the satisfaction of the public with more sincerity, diligence probity, and sufficiency, than any of his predecessors, because he had a beautiful spirit, lively, quick, inclined to virtue and all things beautiful, not sordid, not tight-fisted, not corrupt, loving the people and being loved by it."[70]

Miron has been praised lavishly since then for the impetus he imparted to public works, stressing the importance of water availability. The historian Michel Félibien stated Miron's reputation would never die, noting that one of his accomplishments had been the repair of Paris fountains.[71] It is estimated that he doubled the quantity of water available for the population.[72] Robert Laffont asked a rhetorical question: "Has there been a greater administrator? Attentive both to the king and the Parisians…?"[73]

---

69 Henri-Frédéric Iselin (1826–1905), "Statue of François Miron," Hôtel de ville, Paris, 1881.

70 "… M. Miron, Lieutenant-civil, personnage qui honoroit cest estat, lequel il a exercé, au contentement du public, avec autant de sincérité, vigilance et preud'homie et suffisance, qu'aucun autre de ses prédécesseurs; car il avoit un esprit beau, vif, prompt, porté à la vertu, et à toutes choses belles et hautes, point sordide, point avare, point corrompu, aimant le peuple et aimé d'ice…" Pierre de L'Estoile, *Mémoires-journaux, Journal de Henri IV, 1607–1609*, (10 volumes), Vol. 9, Paris: Librairie des bibliophiles, 1881, p. 268.

71 Félibien, *Histoire de la Ville de Paris*, Vol. 4, p. 494.

72 Laure Beaumont-Maillet, *L'eau a Paris*, Paris: Hazan, 1991, p. 71.

73 "Paris a-t-il jamais eu plus grand administrateur? A la fois tres attentive du roi et des Parisiens…," Robert Laffont, *Histoire de Paris et des Parisiens*, Paris: Editions du Pont Royal, 1957, p. 78.

Notwithstanding its power, the role of the *Parlement de Paris* was unusually passive in the whole issue of water supply to the city. The only specific reference to water-raising devices registered in its records relates to a certain Pierre Langevin and his descendants, who were granted in 1598 letter-patents for fifty years to implement and use a movement of his invention to feed fountains in cities and castles located in high places, at any height desired and capable of draining waters to make land arable.[74]

With regard to the Samaritaine, though no documents substantiate the contention, the attitude of the *Parlement* seems to have been positive. At least some of its members might have benefited from the water supply directly, as can be inferred from a later edict (1619) that regulated the distribution of water from the Arcueil and Rungis aqueducts, which benefited them and other influential notables personally.[75] With reference to the last years of Henri's reign, Michel de Waele writes that the relationship between the king and the *Parlement* of Paris was based on help and mutual support. Conflicts that arose quickly faded away, allowing the return to the daily routine of a relationship between two institutions which could not do without one another.[76] The *Parlament's* lack of opposition to the Samaritaine, a project close to Henri IV's heart, was most probably influenced by this collaborative relationship.

In any case Henri ultimately prevailed, and, as he had financed the construction,[77] the building received the status of *maison royale*, depending administratively on the *Surintendance des Batiments Royaux*. Its caretaker was granted the title of *gouverneur*.[78] Extensive notarial records remain, in which it can be seen that the King was deeply involved in the minutiae of construction and operation of the Samaritaine. The documents describe the work to be performed, with extraordinary detail. As a matter of fact, the long texts describe

74 Gérard Jubert, *Ordonnances Enregistrees au Parlement de Paris Sous le Règne de Henri IV. Inventaire analitique des registres X1a 8640 a 8646*, Paris: AN, 1993, pp. 139–140. There is another, later reference, though it is not clear if raising water was the objective of this invention—in November 1608, Letter-Patents extended to Louis Garcin, to build inside the kingdom mills and fountains of his invention for different kinds of milling applications. See Jubert, *Ordonnances Enregistrees*, p. 162.

75 Lazare, *Dictionnaire historique*, p. 109.

76 Michel de Waele, *Les relations entre le parlement de Paris et Henri IV*, Paris: Publisud, 2000, p. 412.

77 Daniel Roche in "Le temps de l'eau rare du Moyen Age a l'époque moderne," *Annales, Histoire, Sciences Sociales*, 39e Année, No. 2 (Mar.–Apr., 1984), pp. 383–399, p. 388 and Jean-Pierre Babelon, *Henry IV*, p. 830, add that one of the sources for this financing was the tax levied on wine (see p. 254–255 above).

78 M. Berger de Xivrey, M., *Recueil des Lettres Missives de Henri IV*, (9 volumes), Vol. 6, 1603–1606, Paris: Imprimirie Impériale, 1853, p. 285, footnote No. 1.

extensively what simple sketches would have depicted much better, while leaving less room for misunderstanding.

On 12 February, 1607, the masonry work for the construction of the big water cistern, ten *toises*[79] long and 4 *toises* wide, which was to be built in the church of Saint-Germain-Auxerrois alongside the wall facing the river, was adjudicated to Pierre Robelin and Clément Metezeau, the latter described as an architect, at a price of 60 *livres* per *toise* of masonry and six *livres* per *toise* of excavation.[80] Only three weeks later, on 5 March, 1607, an agreement was reached between Mr. de Fourcy, superintendent of the buildings of the King, and the clergymen of St. Germain de l'Auxerrois, relating to the assignment of five *toises* depth from a wall for the establishment of another reservoir intended to contain the waters from the Samaritaine.[81] According to a document dated 1653, this tank was placed in the third floor of a building located at the end of the cloister, toward the Seine.[82] On 24 March, 1608, the wood work for the building housing the pumps and artificial 'fountains' of the Pont Neuf, was adjudicated to Antoine Le Redde, for an amount of 38 *livres*.[83] On March 29th, 1608, the masonry work for the construction of the aqueduct that would convey the water to the nursery garden of the Tuileries garden was adjudicated to Martin Boullet, at a price of 13 *livres*, 5 *sols* per *toise*.[84] On May 24th, 1608, the soil flattening and excavation work for the construction of the trenches required to lay the lead piping which would convey the water from the 'artificial fountain and pump' of the Pont Neuf from the gate of L'Escolle St. Germain de l'Auxerrois to the greenhouse located in the garden of the Tuileries Palace, were adjudicated to Pierre Disle, *terrassier*, at a price of thirty five *sols tournois* per cubic *toise*.[85]

---

79 Toise was a French unit of measurement for length (1.949 m. or six *pieds*—one pied is equivalent to 32.4 cm.), area (36 square *pieds* or 3.799 square meters) and volume. In the case of masonry work it is based upon trade customs. Although sometimes the use of the term is confusing, the generally accepted equivalent volume is 216 cubic *pieds* (7.4 cubic meters).

80 Mallevoüe, *Maximilien de Béthune, duc de Sully*, p. 145.

81 Archives Nationales, *Répertoire de la Sous-Série O' (Maison du Roi), Eaux et Fontaines*: O[1] 1597–1, unnumbered folio.

82 Boucher, *Le Pont-Neuf*, p. 126.

83 Mallevoüe, *Maximilien de Béthune, duc de Sully*, p. 148.

84 Ibid., p. 150.

85 Ibid., p. 153.

Although one testimony indicates that the pump itself was already at least partly operational by the end of May 1605,[86] it certainly was fulfilling its objective by 1608. "The water from the pump of Pont-Neuf has reached the Tuileries," wrote François de Malherbe triumphantly to Mr. A. M. de Peiresc on 3 October, 1608, as though the impossible had been accomplished.[87] Exactly as agreed, the water was lifted above the bridge and stored in the cistern built in the cloisters of *Saint-Germain de L'Auxerrrois*, clearly seen in maps drawn in 1609 by Benedit de Vassallieu dit Nicolay and François Quesnel,[88] located in front of the Louvre, and described by J. C. Legrand.[89] From the reservoir, the water was conveyed by underground leaden pipes along the pier up to the corner of the Tuileries palace bordering the Seine, from where a deviation fed the large octagonal pool in the Tuileries gardens.[90]

On 10 July, 1609, the masonry work for the construction of the above-mentioned pond, which had a diameter of six to seven *toises*, to be located in the center of the new garden of the Tuileries to store part of the water coming from the 'artificial fountain and pump' of the Pont Neuf, was adjudicated to Denis Roux, master mason, for an amount of 4300 *livres tournois*.[91] The pond had an important decorative role: "In the *Grand Jardin* of the Tuileries is a sort of small pool, into which water supplied by the *Pompe du Pont Neuf* flows through trenches in the garden, in which are maintained carps and other fish."[92]

---

86 "The pompe or water worke is goinge but castes not enough to cleanse so foule a towne as Paris," Mathew and Calthrop, *The Life of Sir Tobie Matthew*, p. 44.

87 "L'eau de la pompe du Pont-Neuf est aux Tuileries," Malherbe, *Lettres de Malherbe*, p. 61.

88 Very similar maps were drawn in the first two decades of the seventeenth century: Benedit de Vassallieu dit Nicolay, "Portrait de la Ville, Cité et de Paris avec les Faubour[g]s d'icelle," (1609). Southeast is at the top. The map consists of 4 sheets, (114×146 cm), Copperplate, Bibliothèque historique de la Ville de Paris, A 103. A discordance between dates exists in this view and another one, depicted also in 1609 by François Quesnel, "Carte ou description nouvelle de la ville, cité, université et faubourgs de la ville de Paris," (504×384 cm), en 12 feuilles, gravure à l'eau forte, BNF, extracted from Ballon, *The Paris of Henri IV*, pp. 233–44. The drawings are either posterior to their dating, or the representation of the statue was added later, as it was installed only in 1614.

89 J. G. Legrand and Charles Paul Landon, *Description de Paris et de ses édifices*, (2 volumes), Vol. 1, Paris: Treuttel, 1818, pp. 55–60.

90 Beaumont-Maillet, *L'eau a Paris*, p. 74.

91 Mallevoüe, *Maximilien de Béthune, duc de Sully*, p. 154.

92 "Dans le gran jardín des Tuilleries est une espece de petit estang, dont les eaux proviennient de ce que se tire par la Pompe du Pont neuf, qui par canaux se rendent en ce jardín; & dans cet estang se nourrissent carpes, & autres poissons en quantité." Jacques du Breul, *Le Theatre des Antiquitez de Paris; où est traicté de la fondation des Eglises et Chapelles de la Cité, Vniuersité, Ville et Diocèse de Paris; comme aussi de l'institution du Parlement,*

The management of the Samaritaine was given to its constructor as *concièrge and gouverneur*.[93] In 1606, in a deliberation of the *Bureau de la Ville* concerning the ways to protect the Samaritaine building from fishermen and other people mooring their boats to the pylons of the pump building, he is referred to as *Jehan Lintelere, maistre fontenier et artifficier du Roy*.[94] In all documents posterior to 1612 he is referred to as *Maistre de la pompe du Roy*.[95]

Jean Lintlaër left a note in his beautiful handwriting, written c.1607, apparently as a draft for the contract that would be ultimately signed in April 1608, stating the need to effectively perform maintenance to the pumps, the auxiliary equipment, the piping and the clock.[96] He strove to get an annual amount of 1200 *livres*, covering by himself repair expenses of up to 50 *livres*. He specifically indicates that the water for the Tuileries greenhouse should be always clean for the King's pleasure. In the same document he mentions that he had delivered the pump clock to the king. An act dated 22 April, 1608 details the final maintenance requirements for the building, the waterwheel and the four installed pumps.[97] The contract was adjudicated to Jean Lintlaër, *Alemant de nation*, referred to as 'Engineer of pumps and artificial fountains' (*Ingenieur en pompes et fontaines artificielle*), with residence in the Pont Neuf. Lintlaër was required to maintain and continually keep in motion the pumps and auxiliary equipment. He was also to be responsible for all masonry repairs, roofing, carpentry, metal work, glazing and other reparations pertaining to the house, waterwheel and the four pumps. In case of any damage to parts made of iron, copper, wood or lead, either due to a breakdown of the device or abandoned boats or floating debris, Lintlaër had to cover the cost up to 50 *livres tournois*, the balance to be covered by the King. An annual sum of 3,000 *livres tournois* was to be paid to him for these services in quarterly installments. This was an important sum those days; compensation for civil servants and architects ranged from 800 *livres* for the royal architects under Henri IV up to the *surintendant* and *controleur*, who received 6,000 *livres* a year. But, not surpris-

---

fondation de l'Vniversité & Collèges, & autres choses remarquables, (4 livres), *Augmenté en cette edition d'un Supplement, contenant le nombre des Monastères, Eglises, l'agrandissement de la Ville & Fauxbourgs qui s'est faict depuis l'Année 1610, jusques a présent.* Supplement by D.H.I. Advocat en Parlement, Paris: Societé des Imprimeurs, 1639, p. 76 of supplément.

93 Guerin, *Registres des Déliberations*, 1905, p. 349, footnote 3.

94 Léon Le Grand, (ed.), *Reg. B.V.*, (15 volumes), Vol. 14 (1605–1610), Paris: Imprimerie Nationale, 1908, p. 127.

95 See Léon Le Grand, *Reg. B.V.*, (15 volumes), Vol. 15 (1610–1614), Paris: Imprimerie Nationale, 1921, pp. 129, 130 and 154.

96 BNF, Site Richelieu, *Ms. fr. 16652*, fol. 97.

97 Mallevoüe, *Maximilien de Béthune, duc de Sully*, p. 152.

ingly, the salaries for professional work were of a different order of magnitude when measured against the earnings of the great personages of court: during the same period Nicolás de l'Hospital[98] was receiving 120,000 *livres* a year as *Capitaine des Gardes du corps du Roi*[99] and other highest army officers had annual incomes varying between 60,000 and 100,000 *livres*.[100]

According to William Barclay Parsons, in 1606 a water main was laid from the north to deliver water to the first public fountain erected on the Pont au Change. The source of water for this fountain was the Samaritaine, as the King had allowed one *pouce*[101] to be diverted.[102] Other historians give a different timing for this permission, stating that it was only in December 1617 that it was granted.[103]

Even after the construction of La Samaritaine only a privileged few, to whom Henri granted a special allocation, continued to enjoy the convenience of free water supply from other sources. An example of this attitude can be found in a letter from Henri to Sully, dated 18 May, 1606, in which he instructs him to provide the city with funds to complete the fountains at the Portes St. Bernard and du Temple, and at the Palace and the Croix de Tirouer.[104]

---

98 A French military career officer, Duc & Marquis de Vitry & D'Arc, who was later appointed by Louis XIII maréchal de France (1581–1644). Louis Moreri, *Le grand dictionaire historique, ou Le mêlange curieux de l'histoire sacrée et profane*, (10 volumes), Vol. 4, Amsterdam: Chez Brunel, 1740, p. 202.

99 The cavalry unit entrusted with the king's security. See Académie française, *Dictionnaire de l'Académie française*, (2 volumes), Volume 1, Paris: Firmin Didot frères, 1835, p. 822.

100 Benevolo, *The Arquitecture*, p. 693.

101 Pumps output was expressed in *pouces d'eau ou pouces de fontainier*, an ancient unit approximately equivalent to 19,20 cubic meters per day (Paul Gille, "Construction and Building," in Maurice Daumas, (ed.), *A History of Technology & Invention*, (3 volumes), Vol. 2, translated by Eileen B. Hennessy, New York: Crown, 1969. pp. 493–585, p. 517). Measuring the water was practically impossible. In the few known private concessions of this period, only the width of the faucet or the delivery pipe was regulated, their dimensions being ensured by a copper or silver ring of the diameter permitted. The water pressure at the junction was not taken into account, because it was believed that the volume of water supplied by an opening was in proportion solely to the diameter of this opening (Horace Say, *Etudes sur l'administration de la ville de Paris, et du department de la Seine*, Paris: Guillaumin, 1846, footnote No. 1, p. 387).

102 Parsons, *Engineers and Engineering*, p. 251. But according to Bonamy, "Mémoire sur les aqueducs," p. 744, no water at all was left for public use.

103 Girard, *Recherches sur les eaux*, p. 22.

104 M. Berger de Xivrey, *Recueil des Lettres Missives de Henri IV*, (9 volumes), Vol. 6, 1603–1606, Paris: Imprimerie Impériale, 1853, p. 613.

On May 5th, 1612, the *Conseil de Paris* ordered the full payment of what was due to Lintlaër for the construction of the Samaritaine, for a decorative gondola made in 1608 which had been set in the garden pond of Tuileries at a price of 150 *livres* and for a bronze bowl from where water was poured into the pond made in 1609 at an agreed price of six hundred *livres*.[105] These gardens were created and tendered by the best gardeners available: Pierre Le Nôtre, Pierre Tarquin and Etienne du Perac.[106]

Built in a very short time, La Samaritaine was an expensive installation to maintain, as the records in the Archives Nationales collected by Boucher show.[107] Necessary repairs were needed rather early and exceeded forecasts, which explain a new contract signed 12 March, 1622, between the Superintendent and Lintlaër, adding 1,000 *livres* per annum for maintenance of water pipes, in addition to the 3,000 *livres* established in the original maintenance contract.[108] The municipal authorities were constantly concerned about potential damage to the building from vessels navigating the river. Although the pump was the king's property, and therefore the responsibility of the superintendent of royal buildings, the City had a real interest in its conservation, as it prevented the royal palaces from using water from the public fountains.[109]

In all, Lintlaër and his family were generously rewarded by the King. That the Lintlaër pumps were of real value in spite of their inherent low efficiency was recognized by Louis XIII, son of Henri IV, who in a letter dated 16 March, 1619 (confirmed by a second letter dated 8 February, 1620), appointed Robin, the wife of Lintlaër, and her sons Claude (the eldest) and Louis (comptroller of the Royal Buildings) as persons well qualified to maintain the plant after Jean's death.[110] Louis Lintlaër's godfather was Henri IV himself, which shows the deep feelings the king had for Lintlaër.[111] This appointment was made "… on account of the public benefit that had come from the four pumps by which a great quantity of clear water had been delivered by Jean Lintlaër, our 'engineer' [sic.]," and in recognition of his services. They were to receive the

---

105 Archives nationales, E. 35^{B}, fol. 53, extracted from, Léon Le Grand, (ed.), *Registres des Déliberations du Bureau de la Ville de Paris*, (15 volumes), Volume 15 (1610–1614), Paris, 1921. Paul Guerin, (ed.), *Registres des Déliberations du Bureau de la Ville de Paris*, (15 volumes), Vol. 13 (1602–1605), Paris: Imprimerie Nationale, 1905, p. 129, footnote 2.

106 David Buisseret, *Henry IV*, London: Taylor & Francis Group, 1989, p. 130.

107 Boucher, *Le Pont-Neuf*, p. 130.

108 Arch. Nationales, *Répertoire de la Sous-Série O¹ (Maison du Roi), Eaux et Fontaines: O¹ 1597–1*.

109 Beaumont-Maillet, *L'eau a Paris*, p. 74.

110 BNF, *Ms. fr. 21631*, fols. 81/82.

111 Boucher, *Le Pont-Neuf*, p. 130, footnote No. 2.

same pay as Jean Lintlaër had. The family achieved a high social standing—Lintlaër's daughter, Catherine, married Jacques Planson, director of the Salt Tax Authority, and her sister Françoise married Bernard de Foras, Seigneur de Panfou.[112] Success in a technology venture had brought the Lintlaër dynasty not only wealth but also honor.

The ingenuity of Lintlaër's successful solution caused interest, for it was considered one of the mechanical wonders of the age. The French contemporaneous historian and prior of Saint-Germain-des-Prés, Jacques du Breul, praised Henri IV's achievement in building the 'pompe': "… King has employed the most ingenious and daring inventions available, leaving an admirable proof on this bridge, as can be seen."[113]

The Operational Principle of La Samaritaine

No drawings have survived of Lintlaër's machine, though we can infer their operation principle from the engines later erected at Pont Notre Dame around 1669 and at Marly in 1682, which are reported to have been similar to that of the Samaritaine. We shall refer to these other two projects late in this same chapter. The type of pumps used was not new in Northern Europe, as conceptually similar ones had already been used to raise water from mines.[114] Laure Beaumont-Maillet is the only historian that has provided data about the dimensions and construction of the first generation Samaritaine. The eight blades waterwheel had a radius of eight *pieds* (about 2.6 meters) and was mounted on a mechanism that regulated its height in such a way that changes in the water level in the river could be dealt with.[115] This objective was achieved by raising or lowering the whole complex of waterwheel and pumps by means of four large screws, thus maintaining the correct depth of the wheel paddles relative to the water level.[116] Another mechanism regulated the force of the current driving the wheel, and directed its full force towards it. From the axis of the

---

112 Mallevoüe, *Maximilien de Béthune, duc de Sully*, p. 152, footnote No. 1.

113 "Les anciens avoient ignoré l'industrie de faire eslever & remonter les eaux plus haut que leur source: et le Roy a cy devant employé les plus ingenieuses & hardies inventions qui se sont offertes, à en laisser la preuve admirable sur ce pont, telle que nous la voyons." Du Breul, *Le theatre des antiquitez de Paris*, Livre 1, p. 186.

114 Ewbank, *A Descriptive and Historical Description*, p. 277.

115 Beaumont-Maillet, *L'eau a Paris*, p. 73.

116 Ian McNeil, (ed.), *An Encyclopaedia of the History of Technology*, London and New York: Routledge, 1990, p. 235.

wheel two rods drove with a reciprocal motion a group of two pumps, whose wooden pistons were sealed with leather. The water pushed by the pumps was divided into two pipes, a small one which drove the mechanism of the clock on top of the building and the main supply pipe.[117] Beaumont-Maillet does not provide any references for her sources. She probably assumed that these features existed prior to the repairs carried out in the early eighteenth century by Bernard Forest de Bélidor, the famous French hydraulic engineer,[118] whose detailed description of the device is closely followed by her. Paul Gille stated that the output at the time was 915 cubic yards (680 cubic meters) per day,[119] but an anonymous contemporary source refers to a much smaller quantity: "...[the pump] should raise 24 *pouces* of water [approximately 480 cubic meters per day], but the truth is that it has never raised, neither in winter nor in summer, more than 20 *pouces*."[120]

In 1714 Bélidor was put in charge of repairing and restoring the machines and building of the Samaritaine, which had suffered decay and damages from frequent floods. He followed the design of the relatively recently installed Pont Notre Dame pumps, which had been constructed on the basis of the Samaritaine model. Bélidor gives a detailed description of the renovated Samaritaine in his *Architecture Hydraulique*,[121] actually using this project to present an extensive discussion on the general technical theory and practice of all the components of a water-driven pumping plant. In *Engineers and Engineering in the Renaissance*, William Barclay Parsons uses Bélidor's description for a summary of the essential technical characteristics of La Samaritaine. The following discussion and quotes are based on Bélidor's original book and Parsons' summary.[122]

The schematic section in fig. 5.3 shows that the cranks operating a pair of pumps were set opposite each other, one rocker arm was descending when the other was ascending so that while one pump was being filled the other was

117 Beaumont-Maillet, *L'eau a Paris*, p. 74.

118 John Aikin, William Enfield, Mr. Nicholson, Thomas Morgan and William Johnston, (eds), *General Biography: Or, Lives, Critical and Historical, of the Most Eminent Persons of All Ages, Countries, Conditions, and Professions, Arranged According to Alphabetical Order*, (10 volumes), Vol. 2, London: G. G. and J. Robinson, 1801, p. 83.

119 Gille, "Construction and Building," p. 519.

120 Boucher, *Le Pont-Neuf*, p. 128.

121 Bernard Forest de Bélidor, *Architecture hydraulique ou l'art de conduire, d'élever et de ménager les eaux pour les différens besoins de la vie*; (2 volumes), Vol. 2, Paris: Chez L. Cellot, 1782.

122 Parsons, *Engineers and Engineering*, pp. 249–250. The characteristics of the device in Beaumont-Maillet's description are substantially identical to Bélidor's, so we are only adding the features she doesn't include in her account.

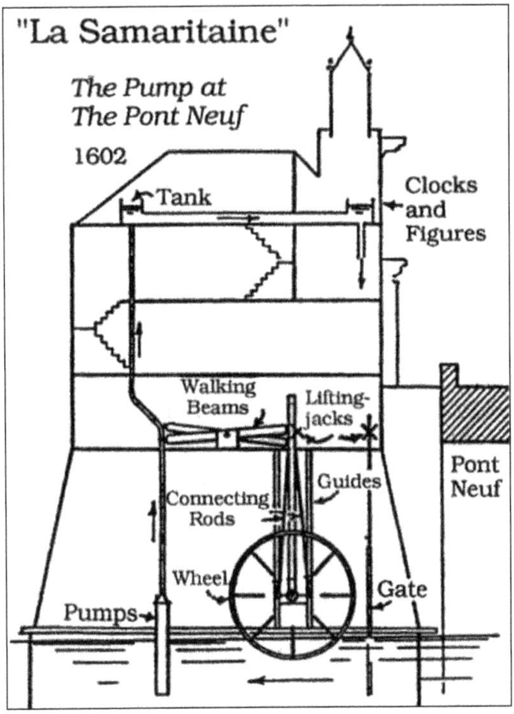

FIGURE 5.3
Schematic section of La Samaritaine.[123]

discharging, thus inducing a continuous flow. The cranks of the two pairs of pumps were set at 90 degrees with respect to each other so that their movements were just half a stroke apart. The rocker beams that actuated the pistons were 20 *pieds* long but non symmetrical—they were longer on the shaft side, and shorter on the pump side, thus giving an increased leverage.

The two fully submerged pairs of pumps, conceptually identical to the future Notre Dame pumps, are represented in fig. 5.4. The internal cylinders diameter was 9 *pouces*,[124] the diameter of the delivery pipe 6 *pouces* and the length of the stroke 3 *pieds*. Two wooden cribs, filled with heavy stones to prevent the flow of the river to displace them, held the shaft of the undershot wheel in place. The cribs shape directed the current towards the wheel, to increase its efficiency. A screen covering the whole width of the device could be lowered to reduce the force of the water flow at times of flood.

123 Ervan G. Garrison, *A History of Engineering and Technology: Artful Methods*, Boca Raton: CRC Press, 2nd edition, 1999, p. 151, figure extracted from James Kip Finch, *The Story of Engineering*, Garden City (NY): Doubleday, 1960, p. 150.

124 The *pouce*, the French 'inch' unit, equal to 1/12 *pied*. Based on the *pied de roi*, the *pouce* equals about 2.707 centimeters. The word *pouce* means thumb in French.

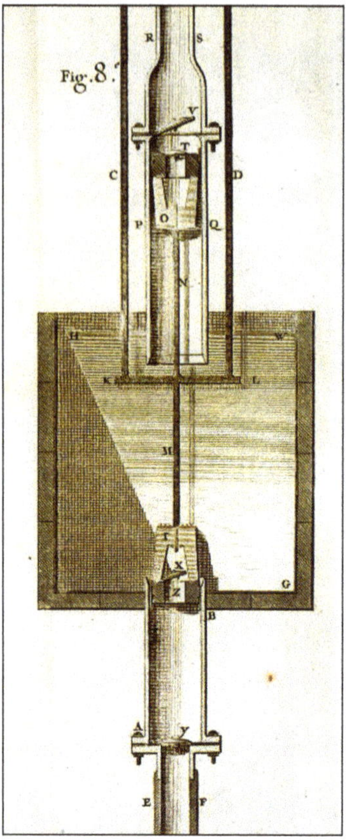

FIGURE 5.4
Sketch of a lifting pump at Pont Notre Dame, similar to the pumps at La Samaritaine.[125]
The contents of the cylinder are lifted up when the piston is raised. The liquid is expelled from the top of the cylinder—it is the water above the piston that is raised. The piston rod in the figures is attached to an iron frame and is suspended to the end of a beam. The valve on the top of the piston, like that at the end of the cylinder opens upwards. When the piston descends (which it does by its own weight and that of the frame) its valve opens and the water enters the upper part of the cylinder, then as soon as it begins to rise its valve closes, and the liquid above it is forced up the ascending pipe. Upon the return of the piston the upper valve is shut by the weight of the column above it, the cylinder is again charged and its contents forced up by a repetition of the movements.

Eugene Belgrand calculated the effective power of the machine at about 2,5 HP, pointing out that, owing to lack of efficiency and friction losses, much of the potential power in the water current was wasted.[126]

Description of the Original Samaritaine Building in Contemporary Travelogues

King Henri's architects had decided that, unlike other bridges in France and other European cities, the Pont Neuf should not have buildings on it, except for

125 Bélidor, *Architecture hydraulique*, chapitre 3, planche 1, right, p. 157. Description based on Ewbank, *A Descriptive and Historical Account*, p. 277.

126 Eugene Bertrand, *Les Travaux souterrains de Paris*, (5 volumes), Vol. 3, Paris: Dunod, 1877, p. 234.

La Samaritaine. This reason probably contributed to later decisions to improve the appearance of the pumping station building, converting it into an attraction by its own merit. The Pont Neuf was "almost finished" in 1608, according to Thomas Coryat,[127] probably meaning that the final decorative touches were about to be completed. The first descriptions of the Samaritaine building itself are found in English travelogues. Peter Mundy's *Travels* (1608) includes a description of Pont Neuf, the Samaritaine building and the clock.[128] François Vinchant, who travelled through France and Italy for several months between September 1609 and February 1610, mentions the "... Pont Neuf, which was built in my time, on which the Samaritaine clock is located, driven by water, which marks the hours, for the amazement of everyone."[129] John Evelyn, noted on the 24 December, 1643[130] that "the landmark clock had a very rare dial of several motions."[131] In June 1650 Evelyn returned to Paris to visit Claude Lintlaër, who was known for his love of art and had succeeded Jean Lintlaër, his father, as governor of La Samaritaine.[132] Claude had considerably improved, at his own cost, the inside of the building, and built an artificial cave, which flabbergasted Evelyn: "... furnished with innumerable rarities both of art and nature; especially the costly grotto, where are the fairest corals, growing out of the very rock, that I have seen ...[133] Edouard Fournier has notes that Lintlaër

127 Thomas Coryat, *Coryat's crudities: hastily gobled up in five moneths travells in France, Savoy, Italy, Rhetia commonly called the Grisons country, Helvetia alias Switzerland, some parts of high Germany and the Netherlands: newly digested in the hungry aire of Odcombe in the county of Somerset, and now dispersed to the nourishment of the travelling members of this kingdome*, (original written in 1608), (2 volumes), Vol 1, Glasgow: James MacLehose and Sons, 1905, p. 171.

128 Peter Mundy, *The travels of Peter Mundy in Europe and Asia, 1608–1667*, edited by Richard Carnac Temple, (5 volumes), Vol. 1, Cambridge: The Hakluyt Society, 1907, p. 125.

129 "... Pont Neuf, qui si faisoit de mon temps, sur lequel est une horloige de la Samaritaine, conduicte par eau, qui fait sonner les heures, avec un estonnement d'un chascun." François Vinchant, *Voyage de François Vinchant en France & en Italie, du 16 septembre 1609 au 18 février 1610*, Bruxelles: Societe Generale D'Imprimirie, 1897, p. 212.

130 John Evelyn, *The Diary of John Evelyn*, edited by William Bray, (2 volumes), Vol. 1, London: Walter Dunne, 1901, pp. 44–45.

131 According to G. Touchard-Lafosse, *Histoire de Paris, Composée sur un Plan Nouveau*, (4 volumes), Vol. 4, Paris: P.-H. Krabbe, 1834, p. 366, the mechanism of the clock was from German and Flemish origin.

132 Musée Carnavalet, *Catalogue—Pont-Neuf, 1578–1978: exposition organisée par le Musée Carnavalet et la Délégation à l'action artistique de la ville de Paris: Mairie annexe du 1er arrondissement, 10 juin–9 juillet 1978, Musée Carnavalet 19 juillet–27 août 1978*, Paris: Le Musée, 1978, p. 20.

133 Evelyn, *The Diary of John Evelyn*, p. 255.

managed to open a passage to the interior of the bridge piers in order to have additional rooms, one of which he converted into a sort of shelter to be used in case of fire. Fournier further gives an extensive but totally incomprehensible description of additional occult building activities by Lintlaër's son in the bridge, which nevertheless makes interest reading.[134] Philippe de Villers, who travelled to Paris in 1657, wrote, amazed by the sophisticated clock, which according to his account marked the hours before noon while going up, and those in the afternoon going down. It also indicated the sun and the moon courses, the months and the twelve signs of the zodiac and sounded the hours, halves and quarters with soft music played by bells.[135] The curious feature of the marking of the hours is confirmed by Francis Mortoft, who visited Paris in 1658–1659: "…[the clock] marks the hours in the forenoone in ascending, and after dinner in descending."[136]

Depictions of the Original Building of La Samaritaine in Maps and Views of Paris

Several extant maps and views of the city, four of which are presented herein chronologically, according to their noted date of publication,[137] allow recreating the appearance of the building of La Samaritaine during the first half of the seventeenth century.[138] Fig. 5.5, by Claude Chastillon, is the first Paris view that includes the Pont Neuf, which must have been depicted between the end of the construction of the Pont Neuf in 1607 and the unveiling of the equestrian statue of Henri IV in 1614. Fig. 5.6, by Benedit Vassallieu is wrongly dated to 1609, as both La Samaritaine and the yet to be installed statue are shown in the

134 Édouard Fournier, *Histoire du Pont-Neuf*, Paris: E. Dentu, 1862, p. 162.

135 "Au-dessus est un horloge fort artificiel qui marque les heures de devant midy en montant, et celles d'après en descendant, avec le cours du Soleil et de la Lune, les mois et les douze signes du Zodiaque et sonne les heures, les demi et les quarts avec une douce musique par le concert de certaines cloches." Philippe de Villers, *Journal d'un voyage à Paris en 1657–1658*, Paris: A. P. Faugère, 1862, p. 44.

136 Francis Mortoft, *Francis Mortoft: His Book, Being His Travels Through France and Italy, 1658–1659*, edited by Malcom Letts, London: The Hakluyt Society, 1925, p. 4.

137 An in-depth comparison between differences in most of the maps shown in this paragraph is presented In the article by Catherine Bousquet-Bressolier, "Matthäus Merian's 1615 Map of Paris: Its Structure, Decoration and Message," *Imago Mundi*, 58: 1, (2006), pp. 48–69.

138 For a comprehensive study of the ancient maps of Paris, including the ones shown in this study, see: Alfred Bonnardot, *Études archéologiques sur les anciens plans de Paris des XVIe, XVIIe et XVIIIe siècles*, Paris: Deflorenne, 1851.

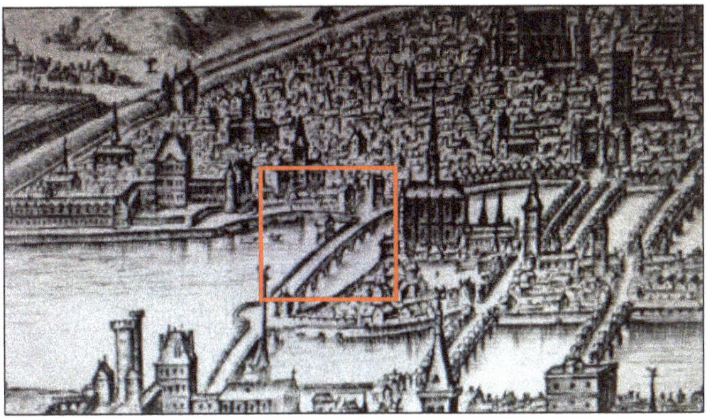

FIGURE 5.5 *Chastillon, Paris (1611).*[139]

FIGURE 5.6 *Vassallieu, Pont Neuf and La Samaritaine, 1609.*[140]

drawing; fig. 5.7, by Matthäus Merian, l'ancien, (1615), shows the equestrian statue turned to the left—Vassallieu drawing skows the statue turned to the right—in both cases most probably to allow viewers to see it fully. The "Map of Paris," engraved by Franciscus Hoiamis [François Van Hoey] in 1619 (fig. 5.8),

139 Claude Chastillon, *Topographie Françoise ou representations de plusieurs villes, bourgs, chasteaux, plans, forteresses, vestiges d'antiquité, maisons modernes et autres du royaume de France*, Paris: Lovys Boissevin, 1655, Plate 58.

140 Benedit Vassallieu (dit Nicolay), "Portrait de la Ville. Cité et Université de Paris avec les Faubourgs dicelle," facsímile edited by Imp. Chardon aîné (Paris), 1881 (77×62 cm), Bibliothèque nationale de France, département Cartes et plans, GE C-10778,

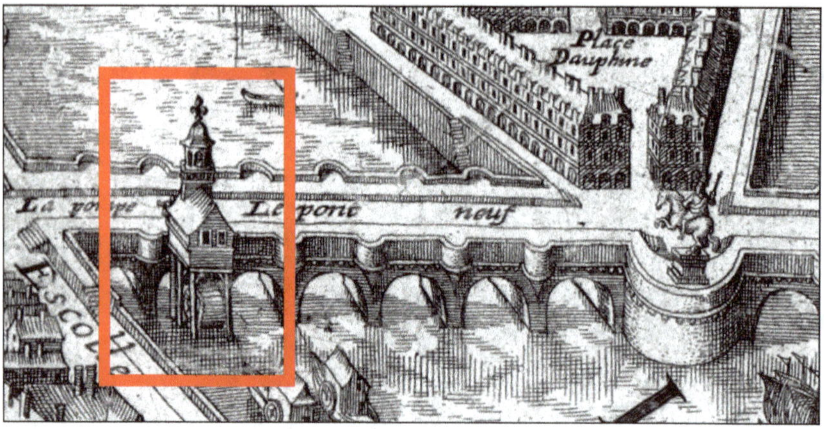

FIGURE 5.7 *Matthäus Merian, l'ancien, (1615).*[141]

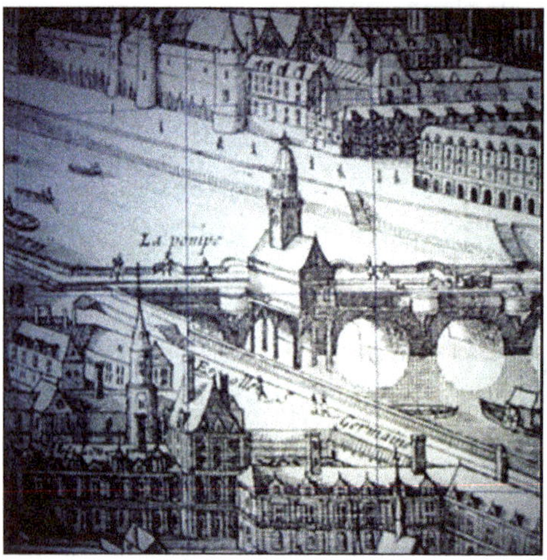

FIGURE 5.8
F. Hoiamis (1619).[142]

---

141 Matthäus Merian, l'ancien, "Le plan de la ville, cite, université et Fauxbourgs de Paris avec la description de son antiquité et singularités," (1615), engraved by Merian and published in Paris by Nicolas de Mathoniére. East-southeast is at the top, (50×101 cm), Bibliothéque historique de la Ville de Paris, catalogued under A 104 f. A later similar view, dated 1630: Jean Sauvé, "Paris," 1630, carte, image fixe, (39.5×50.5 cm), date of publication: 1666–1690, BNF, Appartient à: Collection d'Anville; 00810, Bibliothèque nationale de France, Appartient à: Collection d'Anville; 00810 shows the equestrian statue wrongly placed in an opposite direction, allowing the viewer to appreciate it in its entirety.

142 Franciscus Hoiamis, "Map of Paris," detail, engraved by François Van Hoey, B.N. Est. Va 212, 4 planches, (41×212 cm) assemblées, Paris, 1619. Photo by the author from a micro-

FIGURE 5.9 *La Pompe de La Samaritaine, en 1635.*[143]

confirms Merian's perspective, though it is not its copy. In all these descriptive drawings floating mills can be seen, similar to those described in chapter 2 above.

An anonymous picture, "View of Paris, close to the Pont Neuf," (fig. 5.9) depicts La Samaritaine as a rather elaborated building. Even though Eudoxe Soulié and A. Bonnardot maintained that the painting represents the real likehood of the Samaritaine at about 1635, we believe that the much more modest building, as shown in the 1640 image by Abraham Bosse (a small background detail in a much bigger picture), is closer to its true appearance at that time (fig. 5.10).[144]

film in BNF, Site Richelieu (july 2010). Shown in Werner Hennings, Uwe Horst, Jürgen Kramer, *Die Stadt als Bühne: Macht und Herrschaft im öffentlichen Raum von Rom, Paris und London im 17. Jahrhundert*, Bielefeld: transcript Verlag, 2016, p. 202.

143 Anonymus, "View of Paris, close to the Pont Neuf, detail (c.1635)," BNF, Site Richelieu, Microfilm A 19957, photo by the author. The original painting (2.06×1.15 cm) is kept in the Musée Impérial de Versailles. For details, see: Eudoxe Soulié, *Notice du Musée Impérial de Versailles* (3 volumes), Vol. 1, Paris: Charles de Mourgues frères, 1859, catalogued as No. 780, p. 215. For an extensive description of the painting see: A. Bonnardot, *Iconographie du vieux Paris*, Paris: [s.n.], 1856, pp. 301–307.

144 Supporting the theory of a fancier building appearance in the 1630s is a very similar picture: Félix Godefroy, d'après Pernot, "La Pompe de La Samaritaine, en 1635," Musée Carnavalet—Histoire de Paris, XIXe siècle, Lithographie, G 39161.

FIGURE 5.10
Abraham Bosse, "Le marchand de bois," c.1640, detail.[145]

The Samaritaine Building after its Renovation in 1714

The building, which was partially renovated towards 1665,[146] decayed to such a, extent by the beginning of the 1700s that it became necessary to totally rebuild it,[147] a task done under the direction of Robert de Cotte in 1714.[148] The famous engineer Bernard Forest de Bélidor designed both the new pump plant and

145 Abraham Bosse, "Le marchand de bois," c.1640, estampe, BNF, département Estampes et photographie. Collection Michel Hennin. Estampes relatives à l'Histoire de France. Tome 33, Pièces 2964, p. 379.

146 Musée Carnavalet, *Catalogue—Pont-Neuf*, p. 9.

147 George-Louis Le Rouge, *Les Curiositez de Paris, de Versailles, de Marly, de Vincennes, de S. Cloud, et des envirions: avec les antiquitez justes & précises fur chaque sujet*, seconde edition, (2 volumes), Vol. 1, Paris: Chez Saugrain l'aine, 1718, p. 60.

148 Two of the mid-seventeenth century drawings now preserved in the BNF seem to have been relevant to the renovation, and copied by de Cotte to this end: Anonymus, "Plan des arches du Pont Neuf, plan de la Samaritaine, quay de l'Escolle," aquarelle, (36,7×53,9 cm), 1645, BNF, département Estampes et photographie, FOL-VA-224(B) and Anonymus, "Elévation et coupe transversale de l'ancienne pompe Samaritaine," aquarelle, (41.8×30.5 cm), 1645, BNF, département Estampes et photographie, FOL-VA-224(B). Photos taken by the author in the BNF are kept in his possession.

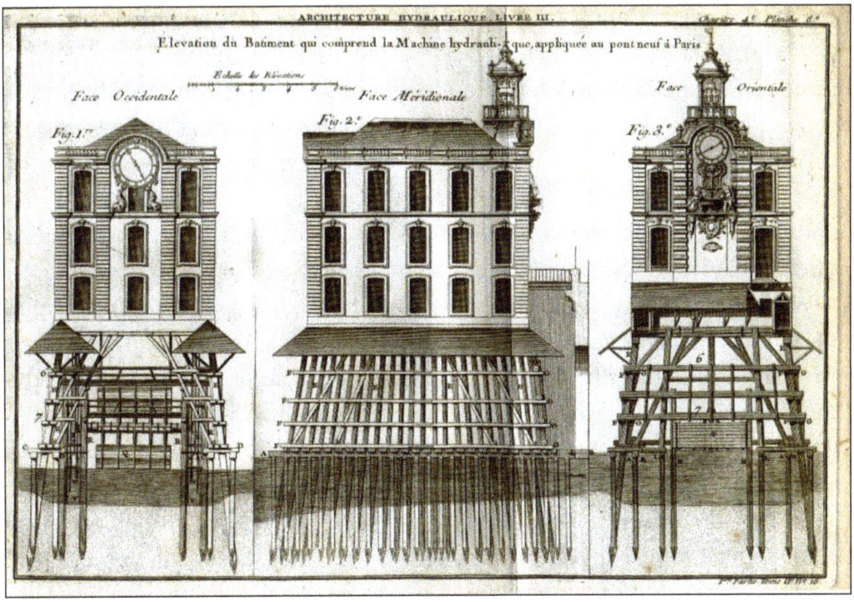

FIGURE 5.11 *Views from the east, west and south.*[149]

the building which replaced the original one, as extensively described in his *Architecture Hydraulique*. The different facades of the Samaritaine are shown in fig. 5.11, where it can be appreciated that the building was much more impressive than the relatively simple original structure, which nevertheless was comfortable enough to serve as Lintlaër's dwelling.

The new building consisted of three stories, the second of which was level with the bridge. A new group of sculptures, based on the same theme that gave the pumping station its name, included a statue representing Jesus receiving water from the woman of Samaria at Jacob's well. The well was represented by a basin, into which a sheet of water descended from a shell. Below the two figures was the inscription, taken from the New Testament,[150] which indicated the subject of the group, and the destination of the building: *Fons Hortorum, Puteus aquarum viventium* [The Fountain of the Gardens, The Well of the

149 Bélidor, *Architecture hydraulique*, p. 263. The plan of the floors is shown in Ibid., p. 268.

150 Luke 10:25–37. For the context and commentary of this passage, see: L. R. Arul Sam, *The Love Commandment of Jesus Christ in the Gospel of Luke and Its Implication in the Indian Context*, Delhi (India): ISPCK, 2008, pp. 57–58.

Living Waters].[151] Controversy exists concerning the authorship of the sculptures and the clock during the different periods of the existence of the pumps.[152] François Boucher, the only historian who refers to the sculptors of the statues in the original building, states that unnamed 'ancient authors' attributed them to Germain Pilon 'le fils',[153] and adds that payments were made in December 1665 to Jeanne and Etienne Blanchard for lead figures they produced for the Samaritaine.[154] This date coincides with the first renovation of the building, which included a new clock, made in 1665–1667.[155]

With regard to the sculptures carried out in 1714, it is agreed among Piganiol de la Force, François and Noel and L. J. M. Carpentier, Augustus John Cuthbert Hare and Germain Brice that the one representing Jesus is to be attributed to Philippe Bertrand and the one of the Samaritan woman—to Rene Fremin.[156] The artistic value of these two sculptures is also a matter of discussion, as they have been considered "d'une exécution mediocre."[157] The clock that filled the rounded space above the group was also renovated by Claude Lory at the same time with the statues,[158] though apparently not successfully, as it seldom marched properly.[159] Between 1748 and 1749 the clock was repaired by Tuilier and the carillon of bells actuated by the clock renovated by an organ maker, Richard Pere, in 1752. Finally, after this work was performed, the carillon started

151 A. and W. Galignani, *The History of Paris from the Earliest Period to the Present Day: Containing a Description of Its Antiquities, Public Buildings, Civil, Religious, Scientific, and Commercial Institutions*, (3 volumes), Vol. 3, Paris: A. and W. Galignani, 1832, p. 101.

152 This controversy is presented by Jean Babelon, *Germain Pilon*, Paris: Les Beaux-Arts, Edition d'etudes et de documents, 1927, p. 72.

153 Boucher, *Le Pont-Neuf*, p. 133.

154 Ibid., p. 133.

155 F. J. Britten, *Old clocks and watches & their makers, being an historical and descriptive account of the different styles of clocks and watches of the past, in England and abroad, to which is added a list of ten thousand makers*, London: Batsford, 1904, p. 583, gives Coignet as the name of the clockmaker. For the appearance of the clock, see: Joannes de Ram after Gabriel Perelle, "La pompe de la Samaritaine," paper, etching, c.1670–1695, (12.5×16.5 cm), BM, DPD, Registration number 1981,U.3308.

156 Jean Almar Piganiol de la Force, *Description historique de la ville de Paris et de ses environs*, (10 volumes), Vol. 2, Paris: Les Libraires Associes, 1765, p. 52; François Noel and L. J. M. Carpentier, *Dictionnaire des inventions*, Bruxelles: J. P. Meline, Cans et Compagnie, 1837, p. 476; Augustus John Cuthbert Hare, *Paris*, (2 volumes), Volume 2, London: George Allen, 1887, p. 4 and Brice, *Nouvelle description de la ville*, p. 196.

157 G. Touchard-Lafosse, *Histoire de Paris, Composée sur un Plan Nouveau*, (4 volumes), Vol. 4, Paris: P. H. Krabbe, 1834, p. 365.

158 Boucher, *Le Pont-Neuf*, p. 136.

159 Boucher, *Le Pont-Neuf*, p. 137.

playing day and night every quarter of an hour, charming and entertaining the people.[160] A second dial, facing toward the West, usefully enabled the population to see the time from far away.[161] The accuracy of the clock was legendary, as Alexander Dumas states in *Les Trois Mousquetaires*: "Il était donc ponctuel comme la Samaritaine."[162] The Samaritaine became a popular meeting place, the parapets next to the building became a sort of popular book store, where pamphlets were sold and traded: "The pious books were banned, the praying books unknown..."[163]

Several well-known artists, such as the engraver Jean Chaufourier (see fig. 5.12) and painter Jean-Baptiste Nicolás Raguenet, took the Samaritaine as a worthwhile object for many of their works, such as the iconic "Le Pont Neuf et la pompe de la Samaritaine, vus du quai de la Megisserie."[164]

Falling again into decay, the Samaritaine was rebuilt in 1772, only to be again mutilated in 1789 by the Revolution, when the statues of Christ and the Samaritan woman were defaced and the building damaged.[165] The statue group was finally destroyed in July 1792—"Il rappelait trop l'Evangile..."[166] The machine itself was removed under orders of Napoleon I, at the time when he was keen on beautifying Paris.[167]

---

160 Virginia Scott, *Molière: A Theatrical Life*, Cambridge: Cambridge University Press, 2002, p. 24.

161 Le Rouge, *Les Curiositez de Paris*, p. 61.

162 Alexandre Dumas, *Les trois mousquetaires*, Paris: Michel Lévy frères, 1860, p. 52.

163 Nicolas Mercier de Poissy, *La France et les royaumes ruinez par les favoris et les Reines amoureuses*, Paris: Jean Musnier, 1649, quoted in Saint-Julien, *Les courriers de la fronde en vers burlesques*, (2 volumes), Vol. 1, reviewed and with notes by Celestine Moreau, Paris: Chez P. Jannet, 1857, p. 116. For a description of life around the Pont-Neuf at the beginning of he seventeenth century see: Gabriel Hanotaux, *La France en 1614; La France et la Royauté avant Richelieu*, Londres, Edimbourg et New-York: Nelson, 1913, specially pp. 67–69.

164 Jean-Baptiste-Nicolas Raguenet, "Le Pont Neuf et la pompe de la Samaritaine, vus du quai de la Megisserie," Huile sur toile (46 × 83,2 cm), 1777, Musée Carnavalet. The Musée, where all relevant Raguenet paintings are exhibited, preserves a model of the little Chateau d'eau in which the pumping machine was housed: Pasquier, Robin et Bourron, "Maquette de la pompe de la Samaritaine," Wood and bronze, painted, (98×75×56 cm), 1772.

165 Catherine Hannah Charlotte Elliot Jackson, *Old Paris: its Court and Literary Salons*, (2 volumes), Vol. 1, Boston: Joséph Knight Company, 1895, p. 29.

166 Cuthbert Hare, *Paris*, p. 258, quoting from an unnamed source.

167 John Bloundelle-Burton, *The Fate of Henry of Navarre: A True Account of how He was Slain, with a Description of the Paris of the Time and Some of the Leading Personages*, London: Everett, 1911, p. 59.

FIGURE 5.12 *The pump de la Samaritaine built in an arche of the pont-Neuf.*[168]

Additional Initiatives to Solve the Water Problem in the Seventeenth Century

Hardly any benefit resulted to the population of Paris at large from La Samaritaine, as almost all the water it raised was used by the Louvre and Tuileries palaces and gardens. But the technical success achieved by this venture led several entrepreneurs to plan the construction of similar devices in different parts of the Seine. On September 9, 1626, two natives from Lyon, Girard Desargues and François Villette, submitted to the *prévôt des marchands* and the *échevins* a proposal to build a machine, driven by the river flow, which would continuously raise water to a height of 40 *pieds*, at a low annual maintenance expense of three hundred *livres*. They offered the City the choice between acquiring their invention or granting them the franchise to sell the water raised by their machines. They obtained the right to test one machine, at their own expense, under the condition that navigation in the river would not be impaired.[169] As the records of the deliberations of the office of the city of

168 Germain Brice, *Nouvelle description de la ville de Paris: et de tout ce qu'elle contient de plus remarquable*, (4 volumes) Vol. 4, Paris: Julien-Michel Gandouin, 1725, p. 195.

169 Girard, *Recherches sur les eaux publiques* p. 26.

Paris contain no further mention of the proposal, it seems that the partners did not implement the authorization that had been granted to them.[170]

Henri IV and Sully, aware that the general public almost did not benefit from Lintlaër's pump, persevered in their efforts to increase water supplies, especially to the neglected left bank—the areas of St. Germain and St. Marceau and the inhabitants of the University area, which didn't have wells in their neighborhoods, suffered the most.[171] Hilary Ballon has a plausible explanation for the acceptance of this fact by the population: "Nevertheless, royal publicists effectively promoted the image of a populist king because his urban policies, however much they served the interest of the crown, were ultimately directed to a broader citizenry."[172]

The king and his minister planned to build the aqueduct of Arcueil, but Henri was assassinated before the project was realized. Water conveyed by it was to be distributed in equal parts between the palace, 14 fountains on the left bank and the last third at the discretion of the contractor. The project was encouraged by Marie de Medici, who, tired of living in the Palace of Louvre, intended to build a palace on the left bank on land purchased from the Duke François de Luxembourg. Construction began in 1613, under Louis XIII, and water started flowing in 1623, and was delivered to public fountains a year later.[173]

King Louis XIII inherited from Henri IV's first wife, Marguerite de Valois, her estate and her debts. The king found himself forced to sell the Hostel de la Reyne Marguerite, the huge property on the left bank of the Seine, which she had built in 1609 (fig. 5.13), to a consortium of speculators and businessmen, the most important of whom was Louis Le Barbier, General Controller of the forests in the Ile de France, the chief financier of the project.

In order to promote sales of land lots in the area, the investors sought to enhance their property by facilitating its communication to the right bank. Le Barbier negotiated the construction of a wooden bridge, in itself a profit motivated enterprise, whose income would be tolls paid by its users.[174] Initially called Pont Barbier, after the entrepreneur who built it, it was renamed Pont St. Anne in honor of Anne of Austria, and later called Pont des Tuileries due to its location, but popularly called Pont Rouge, because of the red painting

170 Arthur Birembaut, "Quelques documents sur Desargues," *Revue d'histoire des sciences et de leurs applications*. 1961, Tome 14 No. 3–4. pp. 193–204, pp. 194–196.

171 Bonamy, "Mémoire sur les aqueducs," p. 744.

172 Ballon, *The Paris of Henri IV*, p. 123.

173 Bromehead, "The Early History of Water supply (Continued)," p. 189.

174 Henri Sauval, *Histoire et recherches des antiquités de la ville de Paris*, written c.1660 (3 volumes), Vol. 1, Paris: G. Moette, 1724, p. 240.

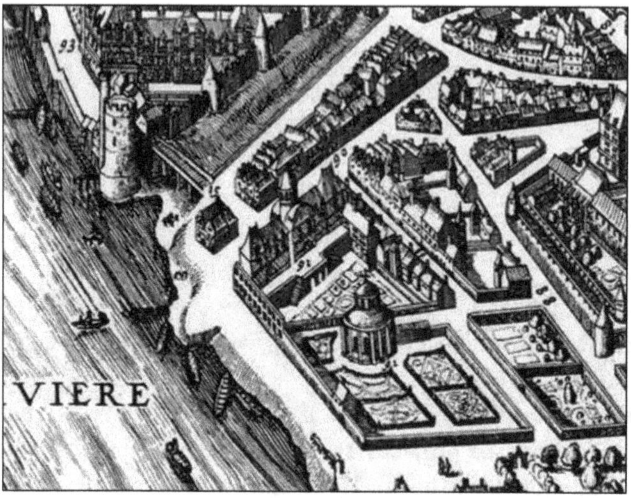

FIGURE 5.13 *Hostel de la Reyne Margueritte, Visscher, 1618.*[175]

which covered it. It was built between 1631 and 1634 to replace the ferry linking both sides of the river. The bridge was destroyed by a flood in 1642, rebuilt in 1651, damaged by fire in 1654, destroyed once again by another flood in 1656, rebuilt in 1660, and finally carried away by a flood in 1684. It can be seen in all Paris maps and drawings produced until its destruction, of which the earliest map was drawn by Melchor Tavernier[176] and the last one by Caspar Merian (fig. 5.14).[177]

175 Claes Jansz Visscher, "Lutetia Parisiorum urbs, toto orbe celeberrima notissimaque, caput regni Franciae," (50×75 cm), Paris: [s.n.], 1618. Original version: [s.l.]: [s.n.], 1618, BNF, département Cartes et plans, GE D-21287. This view is a copy of Merian's 1615 map, with minor changes.

176 Melchior Tavernier, "Le Plan de la Ville, Cite, Université, Faubourgs de Paris avec la Description de son Antiquité," 1628 edition, copperplate, map alone (2 sheets) (55×73 cm), Bibliothéque historique de la Ville de Paris; A 109 A. The map is wrongly dated, as by 1630 the bridge did not exist.

177 Caspar Merian, "View of Paris," *Topographia Galliae*, (4 volumes), Vol. 4, (33.1×19.7 cm), Frankfurt: Verlag Caspar Merians, 1665. The bridge was reconstructed after one of its destructions with 15 supports instead of the original 12. The bridge and the pump are represented as well in Jean Boisseau, "Plan general de la ville, cité, université isles et faubourgs de Paris, dedié a messieurs les gouverneur prevost des marchands et eschevins de la dite ville/ par leurs tres humble et tres affectione serviteur Jean Boisseau, enlumineur du roy pour les cartes geographiques," Paris, 1654, extracted from: Adolphe Alphand, *Atlas des anciens plans de Paris*, (2 volumes), Vol. 1, Paris: Imprimerie nationale, 1880, pl. XVII bis &

FIGURE 5.14 *Caspar Merian, Pont Rouge, Topographia Galliae, detail (c.1665).*[178]

On the bridge Le Barbier also had a pump built to supply water from the Seine to the residents of the area.[179] The pump was not reliable, and had to be repaired frequently. It was burnt, apparently, by a servant who threw a lit torch inside a boat loaded with hay, which was tied to the building housing the pump.[180] No data is available about the capacity of the pump nor about the area it served, and the lack of any reference about its contribution to the water supply of the city is in itself an indication of its negligible impact. Further support for this assumption is found in John Evelyn's account on his travels to Paris in 1643: "… only the Pont St. Anne, landing the suburbs of St. Germains at the Tuileries, is built of wood, having likewise a water house in the midst of it … but much inferior to the Samaritan."[181]

In 1685 a new stone bridge was built replacing the old Pont Rouge, named Pont Royal, as it was totally financed by Louis XIV.[182] No water supplying device was installed in the new bridge.

Mathurin de Moncheny, a veteran *échevin*, proposed in 1655 to raise a certain volume of water by means of a hydraulic machine to be placed adjacent

---

XVII ter and in Johannes Janssonius, "Lutetia Parisiorum, Vulgo Paris," *Theatrum Urbium Germaniae Superioris*, (8 volumes), Vol. 1, (43×52.5 cm), coloured, Amsterdam, 1657.

178 Caspar Merian, "View of Paris," *Topographia Galliae*, (4 volumes), Vol. 4, (33.1×19.7 cm), Frankfurt: Verlag Caspar Merians, 1665. The bridge was reconstructed after one of its destructions with 15 supports instead of the original 12.

179 William O. Goode, "Moving West: Three French Queens and the Urban History of Paris," *The French Review*, Vol. 73, No. 6 (May, 2000), pp. 1116–1129, p. 1222.

180 Pierre-Thomas-Nicolas Hurtaut and Magny, *Dictionnaire historique de la ville de Paris et de ses environs*, (4 volumes), Vol. 4, Paris: Chez Moutard, 1779, p. 108.

181 Evelyn, *The Diary of John Evelyn*, pp. 43–44.

182 Sauval, *Histoire et recherches*, p. 240.

to the Arsenal. He proposed as well to examine additional ways to increase the flow of water provided by the Belleville, Pré-Saint-Gervais and Arcueil aqueducts. The project was accepted and Moncheny received letter-patents to build his machine and begin new excavations in the three villages. After further deliberations, the municipality revoked the permission to build the device out of its constant concern for impediments to navigation. The entrepreneur was also required to submit a detailed plan for his digging work in the villages, in order to ensure that they would not adversely influence the existing springs. The detailed plan was not submitted, and no further work was undertaken.[183]

A new *quartier*, later known as Faubourg Saint Germain, developed in the second half of the seventeenth century, inhabited by well-to-do families. This represented a new market for Seine water, and entrepreneurs rushed to satisfy the demand. A proposal to build a hydraulic machine to lift water from the branch of the Seine that separated the Pré-aux-Clercs from the Île des Cygnes was presented to the King Louis XIV in May 1666, by Claude Regnault de la Fontaine. The commercial purpose was to distribute the water to paying customers, in accordance with regulations to be established. Regnault was to remain the owner of the device, be responsible for its maintenance and he committed himself to purchase the land in which the distribution piping would be laid. The project was approved in September 1666,[184] under conditions extremely favorable for the entrepreneur.[185] The city authorities were then asked for their position, before the grant became effective. Notwithstanding the harsh drought of the years 1658 and 1659, which had caused great discomfort to the city's inhabitants and should have motivated the *prévôt des marchands* and the *échevins* to seek the means of increasing the water supply to Paris,[186] they forbade Regnault and his associates in March 1667 to install their machines on the current of the river, fearing once again that they would interfere with navigation. Consequently, the project had to be abandoned.[187]

In 1676 Jean-Baptiste Colbert, who served as the Minister of Finance of France from 1665 to 1683 under the rule of King Louis XIV, planned to dig a canal to bring the water from the Ourcq River from Lizy to Paris. Louis XIV granted letters-patent to Pierre-Paul Riquet (1609–1680) and Jacques Demance for the construction of the Ourcq canal. Unfortunately Colbert and Riquet

183 Michel Félibien, *Histoire de la ville de Paris*, reveue, augm. et mise au jour par B. Guy-Alexis Lobineau, Paris: chez Guillaume Desprez et Jean Desessartz, 1725, pp. 150–151.

184 Sauval, *Histoire et recherches*, p. 211.

185 *Ms. fr. 21631*, fos. 87–90, BNF, site Richelieu.

186 Sauval, *Histoire et recherches*, p. 211.

187 Girard, *Mémoire pour servir d'introduction*, p. 33.

died, and the *Parlement de Paris*, lacking the funds, aborted the project in 1684. But Demance had carefully preserved all the plans, maps and titles. His widow Catherine Talon, custodian of her husband's assets at his death, presented a memorandum to the Duke of Orleans to continue the enterprise.

Pumps at Pont Notre Dame

On 29 November 1669, Daniel Jolly,[188] at that time *gouverneur* of the pump of Samaritaine, proposed to the municipality to establish, close to the bridge Notre-Dame, a device similar to that whose administration was entrusted to him, offering to raise thirty *pouces* (approximately 600 cubic meters) daily to a height of 80 *pieds* (25 meters) above the level of the river, at the price of 20,000 *livres tournois*. Jolly committed himself to complete his work within four months from the acceptance of his proposal.[189] His initiative was accepted by a decree published on 27 February 1671, and permission was given to erect at the Pont Notre-Dame an installation similar in principle and in details of design to that of the Samaritaine at the Pont Neuf.[190] The waterwheel used was taken from the smaller of two wheat grain mills operating already under the bridge. The work was executed by Jolly, but the plant was purchased by the city soon after its completion and was thereafter operated as a municipal undertaking.[191]

A similar project was submitted at the same time to the *Conseil* by Jacques Demance, who intended to raise 55 *pouces* (approximately 1,050 cubic meters) daily of water by means of a new hydraulic machine, at a price of 40,000 *livres tournois*. This device would take advantage of the waterwheel of the biggest of the wheat mills, adjacent to Jolly's device.[192] His offer was accepted as well.

---

188 Sometimes spelled Joly, Joli or Jolly.

189 Pierre-Simon Girard, *Mémoire pour servir d'introduction au devis général ... pour la distribution des eaux du canal de l'ourcq dans l'intérieur de Paris*, Paris: Imprimerie Impériale, 1812, p. 37. Although no information could be found about the payment terms, the short time due to elapse until the commissioning of the device probably made the financing of this important amount relatively easier.

190 Jacques Antoine Dulaure, *Histoire de Paris, (Continuée jusqu' à nos jours par Camille Leynadier)*, (5 volumes), Vol. 4, Paris: H. Boisgard, 1857, p. 53.

191 Bonamy, "Mémoire sur les aqueducs," p. 748.

192 The first Paris map showing the pumping station is Pierre Bullet et Nicolas Blondel, "Plan de Paris, levé par l'ordre du Roy & par les soins de Messieurs les Prevost des Marchands & Eschevins en l'année 1676," Paris: Blondel, 1676. For details about the map see: Alfred Franklin, *Les anciens plans de Paris, notices historiques et topographiques*, (2 volumes), Vol. 2, Paris: Willem, 1878, pp. 64–114.

Both undertakings worked successfully, and fed several new fountains built for use by the city population.[193]

From the point of view of technical development, the machines in themselves did not represent any progress, as they were but copies of Lintlaër's device. Nevertheless, an important improvement—a raising/lowering mechanism, which according to Germain Brice was inexistent in the 'first generation' Samaritaine engine,[194] enabled the supply of the same quantities of water at all time of the year, regardless of the river level.[195] However, in harsh winters ice disrupted supply by blocking both the paddle wheels of the pumps of Pont Notre-Dame and those of the Samaritaine. Water floods also caused heavy damage. When supply was interrupted, the fear of fires stressed authorities and firefighters.[196]

Henri Sauval (who wrote around 1660) praised the Notre Dame pumps for their reliable operation, continuously lifting water through two pipes with a diameter of six *pouces* to a 22 m. high tower. The whole building can clearly be seen in Turgot's map of Paris (fig. 5.15). From the reservoir located on top of the

FIGURE 5.15 *Paris, Turgot's plan, detail (1739).*[197]

193 Dulaure, *Histoire de Paris*, p. 53.

194 As has been already noted, there are historians contending this type of mechanism existed already in Lintlaër's device.

195 Brice, *Nouvelle description de la ville*, p. 358.

196 Daniel Roche, "Le temps de l'eau rare du Moyen Age a l'époque moderne," *Annales. Économies, Sociétés, Civilisations*, 39e Année, No. 2 (Mar.–Apr., 1984), pp. 383–399, p. 390.

197 Louis Bretez et Michel-Etienne Turgot, *Plan de Paris commencé l'Année 1734* (269×325 cm), levé et dessiné par Louis Bretez, gravé par Claude Lucas, 1739.

tower, conduits carried out the water raised from the Seine towards the fountains of the city.[198] In contrast, Sauval had veiled criticism for La Samaritaine, when he remarks: "This pump provides water only to the Louvre, the Tuileries gardens and the Palais Royal, and amuses passers-by…"[199] Denis Diderot and Jean le Rond d'Alembert, in their monumental *Encyclopédie*, gave a detailed description of the Pont Notre Dame pumping plant,[200] which somehow in a circular argument, confirms the close similarity between the La Samaritaine, renovated by Bélidor and the pumps at Pont Notre Dame. The engineering drawings that they presented are of a high quality (fig. 5.16).

FIGURE 5.16 *Pumping plant of Pont Notre Dame.*[201]

198 Sauval, *Histoire et recherches*, p. 212.

199 Ibid., p. 212.

200 Denis Diderot and Jean le Rond d'Alembert, *Encyclopédie, ou Dictionnaire raisonné des sciences, des arts et des métiers*, (17 volumes—text), Vol. 13, Paris: Briasson, David, Le Breton, Durand, 1751–1772, pp. 66–67.

201 Denis Diderot, *Recueil De Planches Sur Les Sciences, Les Arts Liberaux Et Les Arts Méchaniques: Avec Leur Explication*, (11 volumes—illustrations), Vol. 2, Paris: Briasson, David, Le Breton et Durand, 1763, Planche XXXVII, p. 311.

The pumps, placed on piles, were contained in a building whose door, designed by Pierre Bullet, was decorated with two bas-relief masterpieces by Jean Goujon. It carried a medallion of Louis XV with an inscription in Latin verse by Santeuil, translated to the French by P. Corneille.[202] Towards the demolition of the whole building housing the pump shortly after a new bridge was built, in an article published in 1857 Fulgence Girard marvelled about the beauty of the recently built quays, steps, esplanades, locks and ports in Paris, noting that the waterworks were the only remaining ugly structure breaking the harmony of the improved river sights.[203] Several months later, a fellow reporter for the same magazine, Francois Lacour, concurred with Gerard.[204] Not everybody agreed with them—the graphic representation in fig. 5.17 proves that 'Beauty is in the eyes of the beholder.'

Water pumped from the Seine continued to be critical for Parisians for many years: The "Plan of Bullet and Blondel" (1710 edition, revised by Jaillot) gave as water sources: the Belleville aqueduct—8 *pouces* of water; Pre-Saint-Gervais—20 *pouces*; sources of Rungis—83 *pouces*, from which the city re-

FIGURE 5.17 *"La joute des mariniers entre le pont Notre-Dame et le pont au Change (1756)"*.[205]

202 Fulgence Girard, "La pompe du pont Notre-Dame," *Le Monde illustré,* no 10, 20/06/1857, p. 11.

203 Ibid.

204 François Lacour, "La pompe du pont Notre-Dame," *Le Monde illustré,* no 48, 13/3/1858, pp. 166–167.

205 Jean Baptiste Nicolas Raguenet, "La joute des mariniers entre le pont Notre-Dame et le pont au Change," oil on canvas laid on panel, (59×96 cm), 1756, Musée Carnavalet.

ceived 23 *pouces*; and the waters of the Seine raised by pumps, 80 *pouces*. The Seine provided to the city a third more than all other sources combined.[206] But the unsatisfactory result was that by the last decades of the seventeenth century, Paris provided for a population of 460,000 inhabitants, living in 23,000 houses, a total water volume of 3300 m^3/day, or less than 7 liters per inhabitant per day.[207] As could be expected, much water was diverted by the nobility, and as the pumps installed in the Seine were often out of order, by the end of the century the public supply per capita for the mass of the population was estimated to be only 3.5 liters per day.[208]

The Machine at Marly[209]

Louis XIV began in 1678 to gradually move his court to Versailles. The court was officially established there on 6 May 1682. In order to supply the public gardens at the Marly and Versailles palaces from the Seine the most elaborate and famous water-operated machine ever constructed for raising water was planned and built in the 1680's at Marly-le-Roi, situated at about 17 km. west of the center of Paris. According to legend, Louis asked his finance minister, Jean-Baptiste Colbert, and his architect, Jules Hardouin-Mansart, to organize a competition to solve the problem of irrigating the gardens: "Ask the wise men of France when there will be a machine that can raise the waters of the Seine."[210]

---

206 Alfred Franklin, "Plan of Bullet and Blondel (1710 edition, revised by Jaillot)," in *Les anciens plans de Paris*, p. 112.

207 Daniel Roche, *The people of Paris: an essay in popular culture in the 18th century*, Berkeley and Los Angeles: University of California Press, 1987, p. 20.

208 Bromehead, "The Early History of Water Supply (Continued)," p. 189.

209 The discussion about Marly's water works is based partly on: an article authored by Prof. Nicolas Maurice Dehousse—*"The Marly waterworks,"* Université de Liège, Belgium, 1996. http://www.hec.be/~evyncke/marly, accesed 24 August, 2009; Pierre Rousseau, *Histoire des techniques et des inventions*, Paris: Fayard, 1958, pp.185–186; Paul Gille, "Construction and Building," in Maurice Daumas, (ed.), *A History of Technology & Invention*, (3 volumes), Vol. 2, translated by Eileen B. Hennessy, New York: Crown, 1969. pp. 493–585; Friedrich Klemm, *A History of Western Technology*, translated by Dorothea Waley Singer, Cambridge (Mass.): M.I.T. Press, 1991, pp. 205–208; Ian Thompson, *The Sun King's Garden: Louis XIV, Andre le Notre and the Creation of the Gardens of Versailles*, New York: Bloomsbury Publishing USA, 2006 and Joséph-Adrien Le Roi, *Curiosités historiques sur Louis xiii, Louis xiv, Louis xv, Mme de Maintenon, Mme de Pompadour, Mme du Barry, etc.*, Paris: Henry Plon, 1864, chapter VI, pp. 115–199.

210 Thompson, *The Sun King's Garden*, p. 247. For previous attempts to pump river water to Versailles, see Chandra Mukerji, *Territorial Ambitions and the Gardens of Versailles*, Cambridge University Press, 1997, pp. 187–191.

The winning proposal was submitted in 1678 by a 25 years-old Liegeois, Arnold de Ville, who subcontracted a carpenter of Liège, Rennequin Sualem,[211] to be responsible for the plant that was erected at the riverside of an arm of the Seine.[212] Sualem was a Walloon carpenter born in 1645, in the mining area of Liège in Belgium, along the Meuse. Trained by his father, a master carpenter, he had already built pumping devices for the drainage of coal mines. Moreover, he had produced a water-lifting machine in the Hoyoux River, based on the existing Palfour mill (fig. 5.18),[213] for the Count de Marchin's chateau de Modave, located about 100 km. southeast of Brussels, where the height to overcome was 50 m.[214]

FIGURE 5.18 *Model of the Palfour water-raising device.*[215]

211 According to John Gorton, *A general biographical dictionary*, (4 volumes), Vol. 3, London: Henry G. Bohn, 1851, (pages not numbered, equivalent to p. 28), his proper name was Swalm Kenkin. For a portrait of Rennequin, see: Edward Foucaud, *The book of illustrious mechanics, of Europe and America*, Hartford (CN): W. J. Hamersley, 1845, p. 38.

212 Brandstetter, "The Most Wonderful Piece of Machinery, p. 206.

213 François Boulet, *Leçon d'histoire de France: Saint-Germain-en-Laye: des antiquités nationales à une ville international*, Paris: Presses franciliennes, 2006, p. 91 who cites as a source: Pierre Nickler, "La machine hydraulique de Palfour sous la Terrasse de Saint Germain-en-Laye. Des ingénieurs Liégeois à Sir Samuel Morland," in "André Le Nôtre," *Bulletin des Amis du Vieux Saint-Germain*, no 38, 2001, pp. 25–39.

214 Rousseau, *Histoire des techniques*, p. 185.

215 Website of the Chateau de Moldave: http://www.modave-castle.be/.

According to the proposal, in order to raise the water from the Seine to the plateau of Versailles, several conditions were required: a wooden dam across the river (20 km downstream from Paris), a navigable branch to bypass the dam, a three-stage pumping system along the steep river bank and an aqueduct, at the top of the hill, to convey the water to the Versailles complex. The plant took seven years to be built (1681–1688) by a workforce of 700 people, mostly originating from the Liège area, at a huge cost.[216] The theoretical discharge was to be 5.200 m^3 per day,[217] though the real water volume supplied was much less, due to losses in the valves and the energy lost by friction of the huge number of moving parts.[218]

In the Seine, moved by the current created through the dam, 14 wheels (12 m. in diameter) operated pumps. A first series of 64 pumps pushed the water into a first reservoir, at 48.50 m. above the Seine level. In this reservoir, a second series of 69 pumps pushed the water into a second reservoir, located 56.50 m. higher. Finally a third series of 78 pumps pushed the water 57 m. uphill to the top of the aqueduct, which delivered the water by gravity flow to Versailles, some five kilometers away. It was estimated that the complex delivered at least 75 horsepower.[219] Long iron transmission rods were used to operate the first and second reservoir pumps. These rods were powered from the Seine by crank arms fixed on the wheels' axles. The smallest ones extended 200 m. from the Seine to the first reservoir. The longest ones covered 650 m., measured from the Seine to the second reservoir.

King Louis entrusted the famous scientist Edmé Mariotte to solve the problem of finding suitable pipe materials to withstand the high pressure. Mariotte experimented with different materials, including glass, before selecting cast iron, in an early example of an ambitious undertaking rendering no real benefit in itself, but advancing technology.[220] As a consequence, this was the first device that used cast iron pipes, applied until then only for cannons due to

216 For an extensive detail of the expenses, see: Joséph-Adrien Le Roi, *Curiosités historiques sur Louis xiii, Louis xiv, Louis xv, Mme de Maintenon, Mme de Pompadour, Mme du Barry, etc.*, Paris: Henry Plon, 1864, chapter VI, pp. 138–175.

217 John Theophilus Desaguliers, *A Course of Experimental Philosophy*, (2 volumes), vol. 2, London: W. Innys, 1744., pp. 530–531.

218 Michel Baridon, *A History of the Gardens of Versailles*, Philadelphia: University of Pennsylvania Press, 2008, p. 104.

219 Abbott Payson Usher, *A History of Mechanical Inventions*, Cambridge, (MA): Harvard University Press, 1962, p. 336.

220 Antaki, *Piping and Pipeline Engineering*, p. 3.

their high cost.[221] Louis XIV evidently thought it was worth doing anything for the sake of the royal gardens.[222] Figs. 5.19 and 5.20 illustrate the size of the machine and its complexity.

The expertise acquired by Rennequin was passed over to his son Gervais, who also performed extensive maintenance work on both La Samaritaine and the Notre Dame pumps, as can be seen in documents dated 1716, kept in the French National Archives.[223]

The Marly machine was considered one of the most extraordinary achievements of the seventeenth century, and it was visited at that time as a tourist attraction. It was kept in operation for over century.

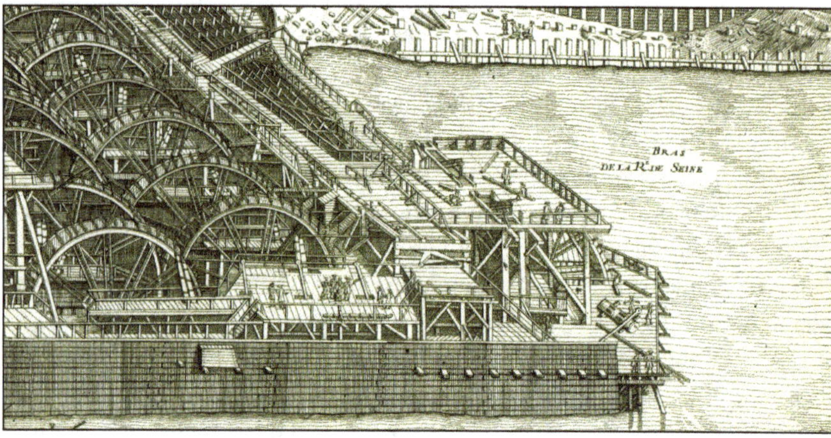

FIGURE 5.19 *Machine de Marly engraving, detail, showing part of the water-wheels.*[224]

221 Most of these pipes were excavated in 1939, and found to be still in good condition—E. Stewart Saunders, "Louis XIV: Patron of Science and Technology," *Purdue University Libraries Research Publications*, Paper 46 (1984). http://docs.lib.purdue.edu/lib_research/46, accessed 14 April 2016.

222 Kirby, *Engineering in History*, p. 150.

223 Archives Nationales, *Répertoire de la Sous-Série O^1 (Maison du Roi), Eaux et Fontaines*: 1597—fol. 23.

224 Gaspard Baillieul, publisher, "Vue de la fameuse machine de Marly qui élève l'eau de la Seyne 535 pieds de haut servant à faire jouer les fontaines du Palais-Royal, de Versailles et de Marly," estampe, (52.8×63.1 cm), c.1700, BNF, département Estampes et photographie, Paris, Collection Michel Hennin, catalogued under No. 5257.

FIGURE 5.20 *View of the Marly Machine and the Aqueduct at Louveciennes, 1722.*[225]

Land Reclamation and Sewerage in France

In medieval Northwest France, several free prosperous peasants undertook land reclamations, consolidating enclosed blocks of land.[226] Henri III granted in 1583 an entitlement to drain the lake Pujaut, in southern France. But it was the rise to the throne of Henri IV that started a comprehensive policy of land reclamation, whose implementation peaked between 1599 and the end of the 1650s. This period represents a turning point in the history of drainage, caused by the concerted efforts of French aristocracy and Flemish and Brabant leading members of the merchant elites of Amsterdam. At least 12 swamps were drained and about 260 square kilometres of wetlands were converted to arable

225 Pierre-Denis Martin, "View of the Marly Machine and the Aqueduct at Louveciennes, 1722," oil on canvas (115×165 cm), Musée Promenade, Marly-le-Roi, France.

226 Daniel R. Curtis and Michele Campopiano, "Medieval land reclamation and the creation of new societies: comparing Holland and the Po Valley, c.800–c.1500," *Journal of Historical Geography*, Vol. 44, (2014), pp. 93–108. For medieval Paris area undertakings, see Charles Higounet, *Défrichements et villeneuves du Bassin parisien: XIe-XIVe siècles*, Paris: Éd. du Centre national de la recherche scientifique, 1990.

lands. For more than thirty years, the draining of the French marshes relied on the association of two different groups; French aristocrats and foreign capitalists, mainly from the southern Netherlands.[227]

One of the major figures instrumental in promoting and implementing land reclamation both in England and in France in the last years of the sixteenth century and the beginning of the seventeenth was Humphrey Bradley, the son of an English wool merchant, born in the Netherlands.[228] Bradley, who initiated several drainage projects during the ten years he spent in England, suggested that the greatest impediments to all his projects were not technical, but were caused by the selfish interests of individual landowners and counties, which paralyzed the authorities.[229] Funding represented another limitation, thus, the rejection of Bradley's proposal for a general drainage scheme in 1593 was due not to its technical defects but to his failure to obtain financial backing when other groups of potential English investors had a rival interest in fen reclamation. After his return to the Netherlands in 1594, Bradley and two other experts were sent to France, charged to advice and assist king Henri IV in land reclamation and drainage assignments for strategic military purposes.[230] He was appointed on 1 January 1599 *maître des digues du royaume* (Master of the dams of the Kingdom), a title that conferred upon him a practical monopoly of land drainage throughout the country. During his tenure of office, France launched projects of land reclamation that would then continue, with variations in intensity, for many years to come. Sully saw the draining of low-lying lands not only as a economic issue but more as a hygienic measure—in the marshes of Bordeaux malaria claimed many victims every year.[231]

In Paris, from ancient times the wastewater was poured onto fields or unpaved streets, and finally reached back into the Seine.[232] The first sewer system

227 Raphaël Morera, "Environmental change and globalization in seventeenth century France: Dutch traders and draining of French wetlands (Arles, Petit Poitou)," *International Review of Social History*, No. 55 (2010), Supplement, pp. 79–101.

228 For Bradley's biography, see: Lawrence Ernest Harris, *The Two Netherlanders: Humphrey Bradley and Cornelius Drebbel*, Leiden: E. J. Brill, 1961.

229 Ibid., p. 62.

230 Holmes, "Drainage Projects," p. 90.

231 Harris, *The Two Netherlanders*, pp. 92–93.

232 For extensive accounts of the history of sewage in Paris, see: Donald Reid, *Paris Sewers and Sewermen: Realities and Representations*, Cambridge (MA): Harvard University Press, 1993 and A. and W. Galignani, *The History of Paris: From the Earliest Period to the Present Day; Containing a Description of Its Antiquities, Public Buildings, Civil, Religious, Scientific and Commercial Institutions ... To which is Added, an Appendix: Containing a Notice of the*

was built in the 1200's by Phillipe Auguste—open troughs that ran down the center of each cobblestone paved road. It was neither effective nor healthful and contributed to the spread of epidemic diseases. In 1370 Hughes Aubriot, a *Provost* of Paris under Charles v, had a covered, stone walled sewer built in the *rue Montmartre*, a street traced in 1137, when the Marché des Halles was created. This sewer collected wastewater and drained it into the *ruisseau de Menilmontant* (Menilmontant brook) a tributary of the Seine River. The increase in the Parisian population brought about the evolution of this little river into the *Grand Egout* (main sewer) with a two-meter width in the seventeenth century. However, the wastewater was still drained in the open air. The sewers were expanded slowly throughout the next years. Nevertheless, due to their widespread disrepair they remained a problem for the city. In the Late Middle Ages the human concentration inside the walls of the city caused appalling hygiene conditions: contents of chamber pots were thrown out of the windows, so the open sewers carried, when having an adequate slope, filth, spoiled food and excrement to the Seine. Discharges from tanneries, dyeing plants, refuse cast by poultry merchants, carcasses of slaughtered animals and other garbage aggravated what was already an intolerable situation. Many dry pits, popularly called 'stinking holes', were used for human waste, resulting in pollution of water wells. This extreme lack of hygiene is likely to have aggravated the results of the catastrophe of the Black Death (1347–1349), which left 80,000 dead in Paris alone.[233] In 1531 a law mandated landlords to provide a latrine and a cesspool for every house, but it doesn't seem to have been enforced.[234]

Violations of sanitation were prevalent, even in the Louvre people defecated without restraint in the courtyards, behind doors, on the stairs and balconies, while the palace attendants did not seem to bother. Not only the lower classes were guilty of such gross violations of hygiene and decorum, they had excellent examples in royalty members: the same day in 1606 that an order was given prohibiting any resident of the palace of Saint Germain from relieving themselves in the property "the King's son urinated against the wall

Church of Saint Denis; an Account of the Violation of the Royal Tombs; Important Statistical Tables, Paris: A. and W. Galignani, 1825.

233 For a report on the contemporaneous (1348) explanations of the cause of the disease by the Paris Medical Faculty, see: Rosemay Horrox, *The Black Death*, Manchester: Manchester University Press, 1994, pp. 158–183.

234 Harold Farnsworth Gray, "Sewerage in Ancient and Medieval Times," *Sewage Works Journal*, Volume 12, No. 5 (1940), pp. 939–946, p. 943.

of his room."[235] Sanitary facilities were almost inexistent—in 1636, a report mentioned 24 sewers in Paris, of which only 6 were covered, and all of them were clogged or ruined.[236]

In the framework of the public works undertaken by Louis XIV, a large sewer was built on the banks of the Seine, conducing water to the Bièvre River, which flows to the left bank. Diverse legislation at times made the police responsible for street cleaning, at other times this responsibility fell upon citizens to clean the part of the street adjacent to the house they inhabited. By the time of the French Revolution privies were widespread in Paris, but unkept and filthy to a degree that people favored relieving themselves in open places, such as the terrace of the Tuileries. Authorities decided to install a latrine and charge an admission fee for its use. The public was enraged, and began 'frequenting' the Royal Palace grounds, leading the Duke d'Orleans to construct a dozen privies, better maintained. A law was enacted in 1721, by which property owners had to "pay for the cleaning of the covered sewers beneath their building." In 1736 and 1755 additional legislation was put in place, preventing the dump of waste into the covered sewers.[237]

---

235 Ibid.

236 M. Gandy, "The Paris Sewers and the Rationalization of Urban Space," *Transactions of the Institute of British Geographers*, 24, (1999), pp. 23–44, p. 39.

237 De Feo and others, "The Historical Development of Sewers Worldwide," p. 3960.

CHAPTER 6

Conclusions

Urban Identities

The development of Toledo, London and Paris during the sixteenth and seventeenth centuries responded to distinct characteristics. Frederick Jack Fisher, in an essay written in 1948, analyzed the reasons for the growth of the population of London at a rate far higher than the growth of the number of England's inhabitants as a whole.[1] Fisher applied to the city principles described by Giovanni Botero in *A Treatise Concerning the Causes of the Magnificencie and Greatnes of Cities* (1588).[2] We are taking the liberty of adapting his conclusions to the case of Paris, and we shall see to what extent, if at all, they are applicable to Toledo.

The growth of London was viewed by city authorities with pride on the one hand, and apprehension on the other. The accelerated increase in the population was accompanied by housing shortages, public health hazards, high unemployment, totally inadequate water supplies and other grave problems. Nevertheless, London complied with the four rules, almost necessary conditions, which Botero firmly set out for towns to transform themselves into large cities. First, it was built on the banks of a navigable river and its economy was based on trade.[3] But the impressive growth of the city did not merely result from the development of trade and industry, but also from being an important center of consumption, a status it reached by fulfilling the other three preconditions that Botero enumerated in his book: London served as the seat of

1 Frederick Jack Fisher, "The Development of London as a Centre for Conspicuous Consumption in the Sixteenth and Seventeenth centuries," in Frederick Jack Fisher, *London and the English Economy 1500–1700*, edited by P.J. Corfield and N.B. Harte, London: Hambledon Press, 1990, pp. 105–118.

2 Giovanni Botero, *A treatise, concerning the causes of the magnificencie and greatnes of cities, devided into three books by Sig: Giouanni Botero, in the Italian tongue; now done into English by Robert Peterson*, London: Printed by T. Purfoot for Richard Ockould and Henry Tomes, 1606.

3 Botero extensively describes the advantages of rivers in: Botero, *A treatise*, First Book, chapter x, pp. 17–29.

 | DOI 10.1163/9789004312425_008

government,[4] the main courts of justice were located in it[5] and the majority of the nobility resided in the city.[6]

Paris fulfilled Botero's requirements as well. However, Pierre Lavedan, who has thoroughly investigated the development of urbanism in Paris, attributes the success of the Paris renovation in the beginning of the seventeenth century to four specific factors.[7] Most importantly, he credits King Henri IV's deep involvement with the well-being of the Parisians and the beautification of the capital, supported by Sully's outstanding performance as 'minister of urbanization', taking advantage of his title of *grand voyer de France*. Secondly, though the municipality had its own organs and acted as a counterbalance to the king and his staff, its head, the *prévôt des marchands*, was in practice appointed by the king, before the city authorities went through the motions of electing him. Moreover, the *prévôt des marchands* saw his position as just one step in his administrative career, so his cooperation with the king was assured. Thirdly, adequate financial and legal conditions for expropriation were established, as operational planning required money and land. The last factor Lavidan identified was the existence of speculators motivated by profit, who acted as agents of change and thrived in a situation of expansion and transformation. The most influential were great entrepreneurs, who were granted land to develop in exchange for what they presented as a public service. The policy of reliance on private capital was justified by Miron in a letter to Sully (14 October 1604), whose arguments sound familiar to our modern ears: "The building industry is better nurtured with money from property owners or promoters, in preference to that of the city …"[8]

Toledo had the opportunity of developing as London and Paris did, but this potential was not realized. Already the capital city of Castile, with Carlos I's accession to the throne in 1519 it became one of the most important cities in the world, home of the court of the most powerful man in Europe at that time. This period, although short, brought to Toledo an era of grandeur and demographic expansion accompanied by important works carried out under the patronage of the Crown and of the archbishop. Thus, by the middle of the sixteenth century Toledo was at the peak of its imperial greatness, which paradoxically marked its decline, because the topographical layout prevented

4 Ibid., p. 65.

5 Ibid., p. 45.

6 Ibid., p. 63.

7 Lavedan, *Histoire de l'urbanisme*, pp. 177–184.

8 Miron de l'Espinay, *François Miron*, p. 269.

it from expanding into a great metropolis able to fulfill the needs of an imperial bureaucracy. Although Toledo continued being the ecclesiastical capital and kept its artistic and cultural importance after the transfer of the capital to Madrid in 1561, Felipe's desertion of Toledo for a new administrative center for the nation negatively affected the development of the city. Thus Toledo ceased to comply with even one of Botero's or Lavedan's previously noted conditions, and suffered a drastic decay in its importance. Within a few decades it was transformed into a small, provincial town.

Water Supply as a Specific Problem

Water supply, well developed in Roman times, decayed after the fall of the Empire. The consequences were dire: a decrease in the quality of life of the population and inadequacy to cope with the hygienic standards required to prevent epidemics. During most of the Middle Ages, in some specific locations—what Adam Lucas called 'pockets of innovation'[9]—advances were achieved over the methods used during previous centuries: drawing water from springs, digging wells, collecting rain water in cisterns and conveying it from faraway sources by aqueducts. It is doubtful if the water scarcity was felt gravely by the majority of the population, since personal hygiene and civic water usage was not perceived as having the same importance as it had had for Romans. In the Late Middle Ages, the contribution derived from gravity-based technologies of water supply, often improved by monasteries, permeated to society at large, where it was further developed. However the evidence is fragmentary and the reconstruction of the complete picture of what was really achieved in a broader scale is difficult to achieve. Even less is known about the diffusion of knowledge through the outer boundaries of Western Christendom, and in what way the sophisticated Islamic know-how in this field influenced Western Europe.

In late medieval times, the operation of urban-scale *norias* represented an advance in lifting water to relatively low heights, and its importance rested in the concept of exploiting the energy of a river flow to obtain this lift. However, it was only during the Renaissance that the first recorded attempts were made to use complex machines and auxiliary equipment to raise water for water supply. The engineers who created and built these sophisticated devices were among the precursors of the inventors of the Industrial Revolution.

9 Lucas, *Wind, Water, Work*, p. 231.

Their profession became coveted by all sorts of people, who strove to work as engineers for governments and rulers, a position which conferred, at least in some cases, social prestige, power and rewarding income. This symbiotic relationship enhanced the essential role of the engineer as an individual agent of change, mastering both the theoretical and practical aspects of the water-powered machines, in combination with other instrumental agents of change, such as financiers and state authorities. Moreover, while seeking rewarding work proposals, engineers strove to elude restrictions and move between cities, enabling kings and their institutions to draw foreign talent to carry out public works that necessitated complex technologies. Cities and states competed to recruit the best brains, but seldom took upon themselves the financial risk that technical failure could imply. However, as we have seen, the potential efficiency of centralized water supply to larger cities represented a powerful motivation to entrepreneurs to offer their inventions to city authorities.

Land Reclamation Motivation vs. Waste Disposal Disregard

We are able to discern patterns of similarity between urban water supply undertakings and large-scale land reclamation enterprises during the sixteenth and seventeenth centuries. Moreover, both the history of land drainage and that of water supply are not only the histories of technological interaction between those disciplines and their development, but they demonstrate the importance of economic, financial and political factors. Prices, climate, migration, political and social circumstances affected, and in turn were affected by, these public works projects. Obviously, some of the technologies in drainage and water lifting are common, even though the order of magnitude of the height to which the water had to be raised was different—much higher in water supply. The majority of the patents requested were granted for devices capable both to lift water and drain flooded areas. The 'modus operandi' of the entrepreneurs fit the economic and political context in which they acted: Bradley in England resembles Peter Morris, Bradley in France resembles Lintlaër. No figure in Spain plays a role in drainage similar to Juanelo's in water supply, which is not surprising as land reclamation was strongly resisted and no central authority took upon itself the task of imposing these development schemes.

The relatively positive attitude of authorities towards water supply and land drainage projects contrasted sharply with their aloof approach to waste disposal and hygiene subjects in general. Since the late Middle Ages the hygiene and cleanliness of London and Paris had become one of the main declared objectives of the municipal administrations, as was as well the case with Madrid but not with Toledo. These stated intentions were not backed by the authorities with a proportionate response to the real needs required by the new scale of the cities. This response should have included the deployment of adequate technical, human and financial means to undertake effective cleaning tasks, paving of streets and disposing of waste. But they did not allocate the budgets these tasks demanded. They placed the burden squarely on the shoulders of the population. However, even given these limitations, the authorities invested efforts to find ways to deal with a major part of the cleanup, as was the evacuation of sludge and silt from the streets for the purpose of facilitating transit on public roads, lessening the harmful effects of putrid odors and preventing excessive moisture from deteriorating basements, foundations, walls of houses and other urban infrastructure. But even when efforts were invested, practical results were disappointing. In the late sixteenth century evacuating waste, mud and cleaning the huge urban centers of London, Paris and Madrid—capital city of one of the greatest empires in history, had become an extremely difficult task to perform. The construction of separate sewer systems for the evacuation of sludge, storm water, excess irrigation and domestic rubbish was expensive and complex to implement. Neither municipal nor royal authorities recognized its absolute necessity, since the scientific knowledge at the time did not fully understand that the lack of sanitary sewerage was one of the main cause of pestilence and a vehicle of contamination. Nevertheless, according to some modern historians, medieval Europeans were aware that there was a link between pollution and disease, even if they ignored the existence of pathogenic organisms. The uncontrolled growth of the population worsened the sanitation problems, polluting the sources of clean water with human wastes.

A crucial aspect of any attempt to clean-up or waste disposal is legislation and, above all, enforcement. Numerous laws were passed prohibiting various types of littering and pollution under the punishment of fines, which were supposed to be levied when they were violated. But, although some of the medieval laws were surprisingly advanced, litigation was rare and influential people disregarded the writs and provisions that regulated waste disposal. The abundance of legal proclamations creates the impression that a lot was said but almost nothing done to enforce their contents.

The Entrepreneurs

The veneration of foreign knowledge in all the undertakings that are the subject of this book enhanced the role of foreign engineers/entrepreneurs, who basically acted as cross-border 'knowledge carriers'. The 'climate' that ruling authorities in general created for foreigners was a determining factor in the success of the enterprises studied, as can be seen in the positive attitude Henri IV and Sully had towards industry experts in general and Lintlaër in particular. Peter Morris was warmly welcome as well, thanks to his proven technological competence. Fanatic Spain and its authorities, though apparently well disposed to Juanelo the foreigner, actually ruined him. Notwithstanding the carrot-and-stick approach used by cities with their local experts in order to prevent specific know-how from being transferred to other areas, Juanelo was granted freedom to roam between Italy and Spain and performed work in both, with the consent of Felipe II, who desired to preserve his close relationship with the Pope. The efforts of authorities did not always succeed in preventing knowledge theft. Such was the case of Pedro de Zubiaurre, the protagonist of what was to become an early example of industrial espionage.[10]

The three entrepreneurs in Toledo, London and Paris had in common being largely self-taught, or products of the apprentice system; none of them had received formal training, which did not yet exist at that period for the study of engineering. The only one who seems to have also been engaged in artistic activities was Lintlaër, as indicated by his commission to make decorative sculptures for the Tuileries pond. Juanelo displayed multifarious technical knowledge of hydraulics, transportation, practical mechanics, and many other fields, depending on the needs of his successive patrons. But he also undoubtedly possessed advanced mathematical capabilities, as shown in his clock-making innovations and his contribution to the Gregorian calendar project. Peter Morris was the ultimate expert in one technical area: water-raising, either for drainage or water supply. He obviously possessed business acumen, but nothing is known about other technologies in which he might have been involved. Juanelo either invented, or was the first to implement, an extremely complex solution. Morris as an entrepreneur did not hesitate to apply a well-known technology, water drainage, to water supply. So did Lintlaër.

Peter Morris developed to a new dimension a business niche started earlier in Central Europe—selling water for a fee. He lobbied the London 'establishment' through powerful friends, founded a company, which marketed and serviced his device and pipelines, and seems to have done all that simply for the

10 Curiously, Zubiaurre's actions as a military daredevil are recorded by British historians, though his technology plagiarism is not mentioned by them at all. Zubiaurre's name is named 60 times in the digital library of the Institute of Historical Research, British History, (http://www.british-history.ac.uk/), but only referring to his diplomatic, spying and military actions.

sake of profit. Lintlaër worked under the auspices and protection of an aggressive developer-king in a country administrated by an effective finance minister. In sharp contrast, Spain, though it had received huge amounts of silver and gold from South America in the sixteenth century, allowed a great part of the empire's trade to be carried out by foreigners and neglected local industry. This set of priorities of the local elites meant that in reality almost no financial support by the government was forthcoming for public enterprises and none for the entrepreneurs themselves. In London the intercession of Bernard Randolph enabled a smooth payment stream to Morris, while the Spanish authorities promised future profits to the Toledo entrepreneurs and shamelessly ignored their commitments. Lintlaër seems to have received his salary and compensation for specific commissions on time, as no record has been found about delay or curtailment of monies due to him. Juanelo and Castillo operated as 'one man armies', counting only on promises of payment by a financially starved city and unreliable kings. Moreover, though Juanelo was enthusiastically received by the king and municipal authorities, no guarantee was offered to him to protect either his intellectual property or any advance payment forthcoming. A few years after the initial enthusiastic reception of their work by their contemporaries, Juanelo and Castillo entered into oblivion, and the lack of motivated followers prevented advanced engineering developments from taking place in Spain.

The Contemporary Image of the 'Engineer'

The admiring attitude of the literary giants of the Spanish Golden Century towards Juanelo's *Artificio* illustrates the fact that it was one of the first machines considered by its contemporaries to be an object of historic and artistic value, worthy of an admiration comparable to the best art works. This remains true despite the fact that the *Artificio* did not provide sufficient water to meet the needs of both the king and the population. From the beginning of Juanelo's project, a veil of secrecy covered all phases of construction and the inner workings of the machine. The quest for prestige pervaded every action and in this sense it succeeded, becoming almost immediately after its inauguration a technological myth. The fame of the machine overtook its operational capabilities. Sadly, an advanced piece of technology that was ahead of its time ended up as a museum showpiece.

The will of French writers to glorify technological achievements by members of their own nation sometimes brought farfetched arguments into play, such as the controversial contention that it was the French Louis de Foix who was the mastermind behind the construction of the *Artificio* of Toledo. Lintlaër did enjoy a close relationship with Henri IV and Sully, and was consequently well accepted by the king's entourage. However, as the water raised by La Samaritaine did not alleviate the citizens' plight for water, the extant records refer mainly to the wonderful clock in the building as being popular, without identifying its creator.

Morris did not achieve the popularity that Juanelo did, his device being less spectacular, though much more effective. His success, achieved in spite of bureaucratic impediments, allowed contemporary opponents of the city functionaries to mock their incapacity and their unwillingness to take risks in order to develop London. Similarly, Sir Hugh Myddelton received from no other than the famous chronicler John Stow, a passionate eulogy upon the successful completion of the New River project, in which the opponents of the project were ridiculized.

Juanelo, Morris and Lintlaër, like many others in other fields, imagined a project, designed it, managed the team that carried out the construction, and were appreciated for their combination of skill and knowledge. They belonged to a new breed of engineers, catalyzers of a new positive attitude toward engineering, technology and the engineers themselves, who became famous, no longer suffering disdain by humanists of the era,[11] but on the contrary, being praised for their technological prowess. Thomaso Gorzoni shows no bounds in his admiration for architects and engineers in his widely disseminated *La piazza universale di tutte le professioni del mondo* (1585)—a collection of 155 '*discorsi*' (elaborated texts), each dedicated to one or more professions or trades, for a total of 400 activities.[12] Further proof of the official and public disregard towards sewerage and waste disposal issues is the fact that all the engineers or technicians who dealt with this type of undertakings have remained anonymous in all of the cities we have researched.

11 For an account of the attitude by humanists towards the mechanical arts and craftsmen, see: Edgar Zilsel, *The Social Origins of Modern Science*, Dordrecht: The Netherlands, Springer, 2003, pp. 7–21, especially pp. 12–14.

12 Thomaso Garzoni da Bagnacauallo, "De gli Architetti in universale, overo, maestri d'edificij, e fortificatori di fortezze, e maestri di machine, & mecanici in commune overo Ingegnieri," in *La piazza universale di tutte le professioni del mondo*, Venice: Gio. Battista Somasco, 1588, pp. 756–764, p. 762.

Financial Aspects of the Enterprises

The unwillingness of the authorities to take risks impaired, on the one hand, the transfer of know-how, but on the other helped to create innovative legal mechanisms developed to motivate entrepreneurs to finance by themselves the projects, under the promise of future profits. They represented a continuation of an old pattern of partnership between the state, local and private contractors to carry and operate roads, bridges, canals or water supply, all works or services that are usually described as public, but which are historically often the result of private initiative.[13] Thus, in London, both Morris and Myddleton, the profit motivated entrepreneurs who made the decisions concerning the long-term investments required for their water supply projects, often adopted a short-term approach, looking into immediate savings and not considering the not-so-distant future. This attitude was exemplified in the choices they made when laying the pipe networks to distribute the water raised or conveyed. Wooden pipes had on the average a useful life of about 14 years,[14] but despite their limited life span, they were used in the London Bridge Waterworks and in the New River.[15] It seems to have been financially easier to raise funds at regular intervals for the lower-cost wood pipes than to make a much higher initial investment in metal pipes, which had a longer life span.

Though data on the income (or, as was frequently the case, lack of income) of the entrepreneurs in the three cities is available, very little information on their investments has survived, except a few elements related to quantities of materials employed to build the *Ingenio* in Toledo. Thus, it is impossible to reach a substantiated evaluation of the investments necessary for building and maintaining each of the three systems. The *Artificio* project represented a highly risky, major capital investment for Juanelo, since the expected revenues were exclusively dependent on the success of his device in fulfilling the provisions established in the contract. Almost four years elapsed from the signing of the contract with the city until the inauguration of the *Ingenio*. It is difficult to

---

13 See: Xavier Bezançon, *Les services publics en France: du Moyen Age a la Revolution*, Paris: Presses de l'École Nationale des Ponts et Chaussées, 1995. For examples of several hydraulic public works contracted in France during the sixteenth century, see: Xavier Bezançon, *2000 ans d'histoire du partenariat public-privé: Pour la réalisation des équipements et services collectifs*, Paris: Presses de l'École Nationale des Ponts et Chaussées, 2004.

14 Geoffrey F. Read, "The Development of Public Health Engineering," in Geoffrey F. Read and Ian Vickridge, (eds), *Sewers. Rehabilitation and New Construction. Repair and Renovation*, Burlington (MA): Butterworth-Heinemann, 1996, pp. 1–21, p. 9.

15 Ian Douglas, *Cities: An Environmental History*, London: I. B.Tauris, 2013, p. 110.

assess the actual time invested in building Castillo's *ingenio*. Though it fit in the existing building and took advantage of the materials of Juanelo's first device, thus saving time and money, these resources were anyhow wasted by a long and tortuous process of approval.

Fewer personal risks were involved in Peter Morris' project, as he invested other people's money. When he approached the City Corporation of London for permission to build his waterwheel-driven engine, the Corporation welcomed the possibility of expanding its public water supply. They granted Morris a long lease on one London Bridge arch and loans to purchase materials and labor. This attitude of long-term planning and the loan made the City's commitment to his project credible and encouraged Morris to undertake his venture. Thus, the development of London's water supply provides a very early example of privatization of a service commonly thought of as requiring government regulation, or even ownership. Water supply increased dramatically with the new systems, though at the price of forsaking the old ethic, by which water was freely guaranteed to all classes of the population. It is obvious that governments at all levels most probably did not feel, and certainly did not act, as if waste disposal should also be provided to the inhabitants of their cities.

The only concern of owners and shareholders of the water companies was to satisfy paying customers. Even when public conduits remained operational after private water services were installed, their maintenance became a lower priority to the well-to-do citizens, who considered them traffic obstructions.[16] We do know for certain that Morris became rich by exploiting his invention and commercializing the water lifted. The construction of sewers was not a lucrative proposition, and it was awarded the lowest priority in public works.

Lintlaër's project, which took about four years to complete, was a risk-free proposition as far as the construction was concerned, as it responded to demand created and paid for by the Court. Lintlaër only took a gamble when he committed himself to cover the annual costs of maintenance for a fixed fee. Nevertheless, his was the less risky venture, as behind him blew a strong tail wind generated by Henri IV and Sully. These leaders behaved honestly with him, and were generous in their rewards for his proficiency. Lintlaër led a very comfortable life, the fruit of his ingenuity and the king's fairness, achieving a high social standing and receiving lavish praise from his patron monarch. Juanelo and Castillo were never able to collect the sums due for their labor. Their enterprises caused them poverty and heavy debts.

16 Magnusson, *Water Technology*, p. 171.

Politics

The issue which few historians have referred to, namely "... the extent to which technological innovation was itself circumscribed by the political and administrative limitations of the age,"[17] is well reflected in the three cases examined here: London and Paris seem to have had effective bureaucracies which were up to the task. Notwithstanding its huge administrative apparatus, Toledo showed itself on the whole incapable of either initiating water supply works or even of fairly compensating entrepreneurs who had the requisite know-how, drive and risk-taking attitude. Thus, future private initiatives were discouraged. A late echo of this incapacity is the fact that the ruins of the *Artificio* were demolished (by blowing them up with dynamite!) in 1868, causing the creation of several academic and official commissions which investigated how such a decision could have been taken by the authorities at the municipal level.[18] In contrast, the emphasis in London was on the commercial side: the cost of the water to be supplied and the price to be levied for it, the entrepreneurial and financial forecast results and the length of the franchise period granted by the authorities. The entrepreneur negotiated with the City aldermen, and not with the king. Mutual convenience appears to be the main motivation for this undertaking—an inkling of the future Anglo-Saxon capitalism.

The municipal authorities' attitudes differed towards the three enterprises: when the Great Fire of London in 1666 destroyed both the public Great Conduit system and Morris' private waterworks, the waterworks were promptly restored. In Toledo, city officials who supposedly cared about solving the water supply problem, actually interfered as much as they could to prevent Juanelo to make progress when building the second *Artificio*. In Paris conditions were, once again, different. Before the Samaritaine was inaugurated, a long thread of corruption, water theft and favoritism had caused the failure of previous undertakings. As could be expected from the autocratic Felipe II,

17 Peter Elmer, "The Early Modern City," in Chant, *The Pre-Industrial Cities*, pp. 186–241, p. 189.

18 A huge number of documents was generated by this incident, a record of which can be seen in Jorge Maier, "Catálogo e Índices," *Comisión de Antigüedades de la Real Academia de la Historia Castilla-La Mancha,* Madrid: Publicaciones del Gabinete de Antigüedades de la Real Academia de la Historia, 1999, pp. 97–100 and in Francisco García Martín, *La Comisión de Monumentos de Toledo (1836–1875)*, Toledo: Ledoria, 2008, pp. 230–237. See also Daniel Crespo Delgado, "Un episodio de la historia de la conservación del patrimonio tecnológico de España: La destrucción del artificio de Juanelo en 1868," in *Fundación Juanelo Turriano 1987–2012. 25 años*, Madrid: Fundación Juanelo Turriano, 2012, pp. 57–67.

but not from Henri IV, whose acts and words had contributed to his image as a ruler concerned with the wellbeing of his people, the French monarch did not hesitate to keep all the water raised by the Samaritaine for his Paris palace gardens, instead of alleviating the shortage felt by his subjects. Notwithstanding the opposition of the city authorities, the king, who was the source of the required funds, felt he had the right to do as he pleased.

Louis XIV set an even more extreme example of disregard with the Marly project. Its sole purpose was to deliver the water necessary for the fountains of the king's parks and gardens at Marly and Versailles. At no time did Louis XIV claim that the general public would benefit from the water raised, and indeed not a drop was diverted for the inhabitants in the areas of Marly and Versailles.[19] The magnificent machine was obviously a means of highlighting the sovereign's power, as Thomas Brandstetter puts it: "The evident disproportionality between the edifice's ends, the amusement of the noble guests, and the enormous efforts to put it into action was central to absolutist representation."[20] This sort of attitude obviously affected, as it does today, the will of the people to support or resist the economic projects developed by their governments.

For the entrepreneurs, political ability was an asset at least as pivotal as good engineering—having or lacking the correct 'friends' who could influence the authorities proved critical for all of them. In all three cities the importance of the individual entrepreneur is obvious, but another necessary condition for success was to have a 'champion of the cause', like François Miron in Paris and Bernard Randolph in London. Unfortunately such a figure, who might have been able to influence conservative institutions, was missing in Toledo. The overcoming of technological and economic considerations and constraints was obviously essential for the success of the private companies marketing water in London, but other challenges had to be mastered as well: managing the competing desires and interests of all who claimed to have access to the same sources, and maintaining the confidence of all constituencies—customers, city authorities and shareholders.

In England, all major figures instrumental in promoting, implementing and managing the water supply and drainage projects closely fit the 'expert mediators' profile, in accordance to Eric H. Ash's thesis on the crucial role

19 This was not surprising, as the output of the Marly machine was deemed to be insufficient for the needs of the Versailles complex. See: Henri Martin, *Martin's history of France: the age of Louis XIV*, translated by Mary L. Booth (2 volumes), Vol. 2, Boston: Walker, Wise, 1865, p. 5.

20 Brandstetter, "The Most Wonderful Piece of Machinery," p. 206.

played by people who could on the one hand skillfully manage the financial and technical resources required for implementation of projects, and on the other hand act as a bridge with authorities in the realm.[21] Such was the case of Hugh Myddelton, and Bevis Bulmer before him, both skilled in mining of precious metals, which gave them both insight into hydraulics and the managerial acumen to run complex projects. Peter Morris had the drainage expertise and showed himself capable of tackling a new application—water supply—in a large scale, however, he could only breakthrough with Sir Bernard Randolph's 'mediation,' which in his case included direct political influence as well. Humphrey Bradley had to overcome not technical impediments, but resistance from landowners and counties, so he had to develop negotiating skills in order to placate the selfish interests of individual landowners and counties.

In Rome, the restoration of the Acqua Vergine, the Acqua Felice and the Acqua Paola and the distribution of their waters to public and private fountains at the end of the sixteenth century were major engineering feats. While various popes used the administration of the aqueduct restoration to wrest privileges away from the Roman Council and as a means to restore its temporal power, the Council sought to uphold its authority to maintain its prerogative to choose contractors and supervise construction. Both the papal administration and the Roman Council strove to control the construction process and assure the attribution of its ensuing 'glory'. These two powers were often rivals with very different political agendas and strategies, yet their competitive relationship actually contributed in a vital way to the preservation of the ancient past. Above all, scholars see these contribuitions to the *renovatio Romae* as part of an overall papal strategy to restore the prestige and moral authority of the Roman Catholic Church in the face of challenges posed by the Reformation and the Ottoman Empire.

Technology and Knowledge Transfer

The Toledo, London and Paris water supply installations were complex, as they dealt not only with the invention of a single machine, but with water-distribution systems, quantity allocation between the ruling monarch and the general population, payment for the service, maintenance and other issues.

21 Eric H. Ash, *Power, knowledge, and expertise in Elizabethan England*, Baltimore: Johns Hopkins University Press, 2004, pp. 1–18.

Besides, objective parameters influenced the outcome of the projects; London and Paris had existing bridges, under their arches waterwheels could be installed, and water needed to be raised only up to a reasonable height above the river level. In Toledo, the challenge was of a completely different magnitude. There, 100 meters height and 300 meters sideways had to be overcome. The specific location of the devices responded to operational and economic logic, but only in Paris were visual elements of image building considered, as shown in the lavish construction of La Samaritaine. In Toledo, when the available technologies of the period were not capable of providing the materials or the tools, an alternative unconventional way to solve the water supply problem had to be found. The causes of critical technical problems were recognized and solved by Juanelo, albeit on an insufficient scale.

A quantitative comparison between the amounts of water the London Bridge Waterworks, la Samaritaine and the *Artificio* would have supplied to each inhabitant of the respective cities, had this water been distributed evenly to all of them, has no real validity, as we have shown that water had never been supplied uniformly to all city dwellers. However, we can estimate virtually the impact of the undertakings by making a rough calculation of the volume of water raised daily by each of them immediately after the beginning of their operation: the *Artificio* de Toledo supplied in 1571 about 17 cubic meters, the London Bridge Waterworks in 1582 about 200 cubic meters and La Samaritaine in 1604 about 540 cubic meters.[22]

In a relatively short time, urban needs demanded an ever-increasing capability on the part of engineers, managers and financiers to plan new systems and expand old ones. These qualities were present in London in the 1580s and in Paris only about ninety years later, where water supply systems developed as a logical consequence of their degree of success. Expansion was a result of a virtuous circle: the better the supply, the greater the attraction to form population clusters where water was abundant, which in itself fostered the expansion of these systems—in the case of London, a fivefold increase in the waterworks capacity; in the case of Paris by the construction of the Notre

22 The data on the *Artificio* extracted from Llaguno y Amirola, *Noticias de los arquitectos*, p. 103, The capacities for the First Morris' waterworks (in 1582—one waterwheel) and La Samaritaine (1607—as per Beaumont-Maillet's description) calculated by the method used by Henry Beighton in his study of the London bridge waterworks (1721). For a rough estimation of the power of the waterwheels installed in the London Bridge Waterworks, La Samaritaine, the Notre Dame pumps and the Marly Machine for the period 1700–1800, see Reynolds, *Stronger than a hundred men*, pp. 176–178.

Dame pumps. No such spirit existed in Toledo, and its decline was probably influenced by this fact.

Kostas Gavroglu, Manolis Patiniotis and others have explored the notion that in any transmission of science or technology from one place to another, progress in linear expanding or deepening of knowledge is achieved.[23] The latter assumption cannot always be corroborated in the period covered here. At times, after one step forward, two or three were taken backward, probably due to lack of effective communication between inventors and their disciples or plain misunderstanding by plagiarists. As an example, Agricola's and Ramelli's designs greatly surpassed much later ones by other treatises authors, both in sophistication and practicality. Not surprisingly, machines installed in important political centers were hailed as great inventions, even though similar ones had operated much earlier in smaller towns for long periods.

We perceived distinct cases of a historiographic tendency to analyze the past through a "perspective of technocentric and progressivist notions about the dominance of one form of technology over another." This thinking embodies retroactive expectations on the way technological change was supposed to have occurred in western industrial societies—new and more efficient technologies 'should' have replaced older ones. Social relevance and suitability to specific needs are neglected, thus reaching distorted conclusions.[24]

All extant texts on the subject of water supply refer to closely kept working models as the means to explain and convince authorities about the merits of a technical proposal. Although in the early sixteenth century some city authorities, like those in Venice, demanded a working model in order for machinery to be patented, difficulties arose when scaling up to the intended size. The process of knowledge circulation of water-raising devices, as exemplified in the three enterprises that form the core of the present study, was that neither drawings nor formulas or mathematic calculations were transferred, but only sheer experience. An exception to the rule that only practical know-how and experience were transmitted, and not abstract mathematical knowledge, was the application of theoretical know-how to the deployment of water supply piping networks. These networks expanded rapidly in the sixteenth century, and their construction necessitated a level of skill not unlike the engineering know-how required to build pumping stations.

We have shown that in the three cases studied, know-how was passed from the entrepreneur to other members of his family, who were assigned the task

23 Kostas Gavroglu et al., "Science and Technology in the European Periphery: Some Historiographical Reflections," *History of Science*, Vol. 46, 2008, p. 153–175.

24 Rynne, "Technological Continuity," p. 104. See also Lucas, *Wind, water, work*, p. 37.

of maintaining the devices after the creator's death, being considered the ones who acquired from him the specific expertise. These are not cases of nepotism, but a logical outcome of the fact the entrepreneurs and their families were intimately involved with the operation of the devices, sometimes, as Lintlaër, literally living inside them. In an iterative process, know-how developed for one objective migrated into different avenues, as was the case of the metallurgical know-how first developed when Louis XIV ordered the high pressure water pumps for Marly. By the beginning of the eighteenth century, the use of cast iron pipes in European waterworks was widespread.

The documentation pertaining to the Spanish and French projects treated in this study shows that it took several pages of tightly packaged text to define the work to be performed, whereas a simple sketch would have sufficed. For the present research, the information obtained from circulating graphic sources such as maps, city views and paintings have proved to be essential—not less than the textual documents—in the investigation of the operation of the water supply devices. These iconographical sources are practically the only tools available in order to track external changes in the devices, from which we can infer the improvements made during their existence, improvements that are not recorded in texts that have reached us. This method has proved to be most useful, despite the often inaccurate nature of sixteenth and seventeenth-century graphic presentations, which more often portray a conceptual or artistic vision than an accurate reproduction of reality. If the author wanted to stress the importance of a building, he represented it in a relatively bigger size, or if its location did not allow its inclusion in the picture, he changed its position to enable its depiction.[25] In our study, all city-views dating from the sixteenth and seventeenth centuries show these intentional distortions, including those in Toledo, London and to a lesser degree in Paris after 1610. It is only by the beginning of the eighteenth century that scaling becomes real, such as in Palomeque's "View of Toledo," (c.1720) and Turgot's "Bird's view of Paris, (1739). Not surprisingly, no sign, drawing or notes referring to sewers or waste disposal conduits are shown in the contemporaneous views of the cities.

We have found cases in which the engineers refer positively to the relation between theoretical knowledge and practical implementation. Such is the mention of the potential improvements that mathematicians and architects could contribute to the effectiveness of a new pump submitted to Henri IV in

25 Newman, "Towards a Topographic Imaginary," p. 62.

1597.[26] Juanelo himself, when commenting on Ambrosio de Morales's amazement at the proportion and accuracy of the *Artificio* motion, gave as an explanation for this perfection: "… I have seen others who are more knowledgeable of Astronomy and Geometry but none who knows Arithmetic as I do …"[27]

As shown in the discussion of the three main projects, historical research that has dealt in one way or the other with these issues has often proven to be inaccurate. In many cases, technical and time sequence descriptions of the three undertakings studied are erroneous. Much of the blame rests in the researcher's almost exclusive reliance on textual sources, usually written in his/her native language. The circulation of knowledge among present historians proved to be rather imperfect.

Cultural Factors and Technological Clusters

A hypothesis, according to which technological development is based on the Christian ideal of the individual, has been articulated by Richard Shelton Kirby:

> The Christian ideal of the infinite worth of man and the correlative aversion to submitting men to work which required no intelligence or judgment were among the principal factors which incited the evolution of mechanical power as a replacement for human muscles.[28]

This thesis has not been supported by the present research. No record at all has been found in any of the sources in which a reference is made, either by authorities or by entrepreneurs, of alleviating the burden of somebody who performed the grueling labor of manually lifting water before the devices were built. On the contrary, evidence abounds for the opposition that these water supply systems generated, by potentially cutting off the possibility to earn a living by carrying and distributing water. The exception was the clergy, whom technology could free from labor to devote more time to contemplation and

26 See chapter v, p. 244.

27 Morales, *Las antigüedades*, p. 334, quoting Juanelo Turriano's own words.

28 Kirby, *Engineering in History*, p. 96.

prayer,[29] in Jacques Le Goff's words: "... a Judeo Christian legacy which emphasized the primacy of the contemplative life ..."[30]

Furthermore, from a technological perspective, no basis has been found in this study to substantiate the widely diffused notion of Protestantism as conducive to increased awareness of the importance of scientific and technological knowledge, as opposed to Catholic backwardness. This 'thesis', formulated in 1936 by Robert Merton, holds that the rise of science in early modern England could be linked to Protestant values, which encouraged efforts to identify God's influence on the world, thus providing religious justification for scientific research.[31] Nevertheless, Protestantism may have been a factor in providing a context that gave birth to water supply as a business, which had not been the case in Western Europe before London's project.

A pattern of parallel improvement of technologies emerges from the present research, such as the simultaneous development in far-apart European cities of quasi-identical complex machinery based upon ctesibic pumps. This finding supports Carlo Cipolla's assertion that "... innovations show a remarkable tendency to cluster in time and space,... attention should also be extended to the anonymous forces of the environment."[32] Cipolla elaborates on the meaning of "the anonymous forces of the environment," noting that countries in which intolerance and fanaticism prevailed lost precious brain power, which migrated to more open-minded countries. He attributes the success stories of England, Holland, and Sweden in the sixteenth and seventeenth centuries to this quality.[33] As the Spanish historian Manuel Silva Suárez has stated, these

---

29 Jaques Le Goff, *Time, Work, and Culture in the Middle-Ages*, translated by Arthur Goldhammer, Chicago and London: University of Chicago Press, 1980, p. 80.

30 Ibid., p. 90.

31 Robert King Merton, "Science, Technology and Society in Seventeenth Century England," *Osiris*, Vol. 4. (1938), pp. 360–632 and Steven Shapin, "Understanding the Merton Thesis," *Isis*, Vol. 79, No. 4 (Dec., 1988), pp. 594–605. For additional extensive discussions on this subject, see: Wietse de Boer, "A Comparative Approach to Social Discipline," in John W. O'Malley, Kathleen M. Comerford and Hilmar M. Pabel, (eds), *Early Modern Catholicism: Essays in Honour of John W. O'Malley, S.J.*, Toronto: University of Toronto Press, 2001, pp. 84–96 and C. Scott Dixon, "Introduction," in C. Scott Dixon, Dagmar Freist, Mark Greengrass, (eds), *Living with Religious Diversity in Early-modern Europe*, Farnham (UK): Ashgate Publishing, Ltd., 2009, pp. 1–20.

32 Cipolla, "The Diffusion," p. 46.

33 Ibid., p. 52.

changes support the underlying idea that "... engineering is not only culture, but a strong catalyst of cultural change."[34]

In London, the pioneering enterprise established by Morris was followed by a series of initiatives by entrepreneurial engineers and investors, such as Bevis Bulmer and Sir Edward Ford, who installed additional pumps to raise water at other places along the Thames through private companies that built, owned and operated pumps, pipe networks and reservoirs to supply water to homes, public taps and businesses. Undoubtedly, Morris' success spurred Myddleton to undertake the New River project. In contrast to the three main projects treated in the present research, which were based on mechanical devices, the 'New River" marked a temporary return to the ancient Roman approach, namely conveying water by gravitation through aqueducts or canals, from a supply source at a sufficient altitude. These various initiatives helped to turn London into one of the best water-supplied cities in Europe. Ultimately, the technical improvements in waterworks mechanisms found their way into English looms and later into steam engines.

In Paris, the technical success of La Samaritaine motivated other entrepreneurs in the first half of the seventeenth century to propose schemes to lift water from the Seine, of which only one, the pump at the Pont St. Anne, was implemented. During Henri IV reign, the alternative path to water supply—aqueducts and canals, was explored as well, though the projects crystallized only after Henri's assassination. The breakthrough in water-raising from the Seine came sixty years later, with the construction of the Pont Notre Dame highly reliable pumping station, which brought a substantial addition to the water supplied to the city for almost 200 years. Its success influenced the overly ambitious project of La Machine de Marly, which undoubtedly pushed the limits of water lifting technology, but at a huge cost.

No technological cluster was born in Toledo and no attempt to imitate Juanelo's invention is known, either in Spain or elsewhere. Even what at the time were already 'conventional' pumps were successfully installed by Fernández del Castillo, the technological thread was not pursued, thus contributing to the stalling of the city and Spain's progress.

It has always been surmised that technology advances have influenced society, sometimes radically so. The reverse notion has not been studied until

34 Manuel Silva Suárez, "Introduccion," in Manuel Silva Suárez, (ed.), *Técnica e Ingeniería en España I: El Renacimiento*, (6 volumes), Vol. 1, Zaragoza: Real Academia de Ingeniería, Institución Fernando El Católico y Prensas Universitarias de Zaragoza, 2004, pp. 9–22, p. 16.

recent years with the same depth. This study hypothesized that technology solutions are decisively influenced by the cultural, social, economic and political milieus in which they are implemented.[35] The case of urban water supply to European cities in the later part of the sixteenth and the early years of the seventeenth centuries illustrates the fundamental importance of external conditions and attitudes on the adoption of different technological solutions. The influence of the cultural and economic contexts helps to explain the lack of interest in investing seriously in maintaining the cities clean. Municipal and royal governments could achieve 'glory' by supplying water and reclaiming land. No 'glory' could be found in removing filth.

When political will and administrative ability existed, as in France and England, the projects succeeded. When these conditions were lacking, failure ultimately overcame what could have been a successful enterprise. Similar as well was the kind of opposition to the projects by interested parties. Moreover, there is a common denominator in the behavior of the early modern elites involved in the three undertakings: human nature and motivations have not changed with time. Each of the players was driven by a mixture which included profit motivation, pride, favoritism, suspicion, need for 'influential protectors', royal whims and currency-starvation at city levels. Not too much has changed over time. Each of the projects achieved a different result: resounding success in London, accomplishment of ambitious objectives in Paris, and failure of utilitarian goals in Toledo. The evolution of water supply systems and specifically water lifting devices has acted as an indicator of change in the patterns of urban development and municipal public works. Not only the technical characteristics of each one of these enterprises were different, the differences between these projects resulted from external conditions not directly related to the water supply problem. The personal experiences of the entrepreneurs and engineers who built them and the different social and political contexts in which they were carried out, and, above all, the will and capabilities of the authorities were the determining factors in their success or failure.

---

35 In 1987 a ground-breaking book: Wiebe E. Bijker, Thomas P. Hughes and Trevor Pinch, *The Social Construction of Technological Systems, New Directions in the Sociology and History of Technology*, Cambridge (MA): MIT Press, 1987, introduced a method of inquiry—social construction of technology, that became a key part of the wider discipline of science and technology studies. See also John Law and Vicky Singleton, "Performing Technology's Stories: On Social Constructivism, Performance, and Performativity," *Technology and Culture*, Vol. 41, No. 4 (2000), pp. 765–775. Adam Robert Lucas is presently the foremost advocate of this discipline for medieval milling studies.

The water supply undertakings in London in the late sixteenth century distinctly fit into John Ulric Nef's concept of an early industrial revolution.[36] Not only did the output of consumable water as a valuable commodity significantly increase, but radical technical improvements and changes in organization took place. Nef work suggested that the rise of industrialism in Great Britain can be more properly regarded as a long process stretching back to the middle of the sixteenth century and coming down to the final triumph of the industrial state towards the end of the nineteenth. On Nef's opinion, the projects supported by the French crown were not an early example of industrial capitalism. He saw them as artificial schemes, often doomed to fail, and based on the assumption that industry could be created at the will of the state. Nef's ideas about France have been confirmed on what concerns the water supply enterprises—neither *La Samaritaine* nor the Notre Dame projects inaugurated an era in which water became a commodity in France. Toledo suffered even more than London and Paris from lack of water. The king and the city authorities strove to solve the problem, but placing the risk and the financial burden on the shoulders of the technically gifted men who undertook the water supply ventures. However, even when partial success was achieved, the disregard of the king and the municipal officials towards the entrepreneurs became a precedent that negatively influenced the will of others to try to solve Toledo's water problem in the future.

When analyzing the broader results of the construction of the three projects, only the case of London allows us to perceive a turning point in the control over water, dramatically changing the relationship between men and water as a precious resource. Since the construction of the first conduits in the thirteenth century, the feeling had been that water was a collective resource and solidarity dictated its universal free provision; that access to water should not be restricted and should never become an object of commerce. This was to be no longer. In the cities of England, and especially in London, water had

36 For Nef's concepts in this respect, see: John Ulric Nef, "The Industrial Revolution Reconsidered," *The Journal of Economic History*, Vol. 3, No. 1 (May, 1943), pp. 1–31; John Ulric Nef, *The Rise of the British Coal Industry*, Vol. 1, Oxford: Routledge, 1966, p. 165; John Ulric Nef, "The Progress of Technology and the Growth of Large-Scale Industry in Great Britain, 1540–1640," *The Economic History Review*, Vol. 5, No. 1 (October 1934), pp. 3–24, p. 22 and Jeff E. Biddle, Ross B. Emmett and Warren J. Samuels, (eds), "*Further University of Wisconsin Materials: Further Documents of F. Taylor Ostrander*," in *Research in the History of Economic Thought and Methodology*, Volume 23-C, pp. 59–78, p. 63.

become an article of commerce, a good example of 'commoditization', in the Marxist sense of the word.[37]

A broad perspective is crucial to appreciate the development patterns of water supply systems. The comparison offered in this study will hopefully also help to improve comprehension of the evolvement of these systems, and the different ways in which they influenced, or failed to influence, the demographic, social and economic development of each city. From the technical treatises published before the three enterprises were carried out or contemporaneous with their execution, we learn that the same basic technology could apparently have been known at the time in all three cases. However, the geographical, managerial, engineering, entrepreneurial and cultural nature of each region differed and so did the technological choices they made, which influenced costs, reliability and effectiveness of the solutions adopted.

---

37 Jenner, "L'eau changée en argent?" p. 643.

Sources and Bibliography

Archives, Libraries and Museums

Spain

Archivo General de Palacio (AGP), Madrid
Cédulas Reales (*C. R.*): tomos 2, 3, 4, 6, 8, 12 and 13; Caja 10.977/7
Archivo General de Simancas (AGS), Simancas
Cámara de Castilla: libros de cédulas 119, 126 and 131.
Cédulas Reales (*C. R.*): tomo 2, cédulas nos. 119, 126, 131
Casas y Sitios Reales, (*C & S. R.*): legajos 261, 271/2, 303/4, 321, 327/8, 331/2.
Mapas, Planos y Dibujos: Drawing XXVII-3
Archivo Histórico de Protocolos (AHP), Madrid
Protocolos notariales: 262 12
Archivo Municipal de Toledo (AMT), Toledo
Libro de actas: 12
Archivo secreto: cajón 6
Biblioteca Nacional de España (BNM), Madrid
Colección de manuscritos: catalogued under R-26692, MSS 3372/3376.
Archivo Histórico Provincial, Toledo
Protocolo notarial: Juan Sánchez de Canales, libro 1594.
Museo del Greco, Toledo
El Greco paintings
Centro Cultural San Marcos, Toledo.
Models of the Artificio

United Kingdom

British Museum, London
Department of Prints and Drawings (BM, DPD): Registration numbers: 1859,1210.1004; Q,6.58; 1880,1113.1120; 1880,1113.1124.1–4; 1880,1113.1125; 1880,1113.1132; 1864,0611.438; Q,6.136; 1868,0808.13614; 1981,U.3308 and Crace Collection: Museum number: I,7.65–76.
London Metropolitan Archives (LMA) (Includes the records of the Corporation of London Records Office (CLRO), London
The Repertories of the Court of Aldermen, 1495–1835.

France

Archives du Bureau de la Ville de Paris, Paris
Repertoire numerique de la sous-serie H^2

Archives Nationales, Site de Paris, Le Caran—Centre d'accueil et de recherche des Archives nationales, Paris
Répertoire de la Sous-Série O' (Maison du Roi)
Bibliothèque historique de la Ville de Paris, Paris
Documents cartographiques, cartes de Paris (see A.III, under the individual headings) A-104, A-103, 104, 109,
Bibliothèque nationale de France (BNF), Site Richelieu, Paris
Département Estampes et photographies (see A.III, under the individual headings)
Collection d'Anville; 00810
Collection Michel Hennin. Estampes relatives à l'Histoire de France, Site Richelieu, FOL-VA-224(B). and FOL-VA-224(B)
Département des manuscrits, manuscrits français 21631 Ms. fr. 16652
Département Cartes et plans, GE D-21287
Musée Carnavalet, Paris.
Maps, views and paintings of Paris

USA

The Metropolitan Museum, Havemeyer Collection, New York
El Greco paintings

Published Sources

Agricola, Georgius (Georg Bauer), *De re Metallica*, translated from the first Latin edition of 1556 by Herbert Clark Hoover and Lou Henry Hoover, New York: Courier Dover Publications, 1950.

Agricolae, Georgii, *De re metallica Libri XII: quibus officia, instrumenta, machinae, ac omnia denique ad Metallicam spectantia, non modò luculentissimè describuntur, sed et per effigies, suis locis insertas, adiunctis Latinis, Germanicisque appellationibus ita ob oculus ponuntur,... Eiusdem De Animantibus subterraneis liber, ab autore recognitus: cum indicibus diversis ...*, Basileae: Froben, 1561.

Al-Idrisi, Muhammad, *Geografía de España* (unattributed translation of the c.1140 original), Valencia: Anubar, 1974.

Al-Idrisi, Muhammad, *Geographie d'Edrisi*, trad. Amédée Jaubert, Paris: La Societe de Geographie, 1840.

Ambroise, *L'Estoire de la Guerre sainte: histoire en vers de la troisième croisade (1190–1192)*, publ. et trad. d'après le manuscrit unique du Vatican et accompagnée d'une introd., d'un glossaire et d'une table des noms propres par Gaston Paris: Imprimerie nationale, 1897.

Anonymous, "Le monologue du bon vigneron sortant de sa vigne, et retournant soupper en aison," in Charles Moiset, (ed.), "La poesie auxerroise au XVIe siècle," *Annuaire historique du département de l'Yonne*, xxi (1857), pp. 73–83.

Bacon, Francis, *The Advancement of Learning: With a Brief Memoir of the Author*, Whitefish (MT): Kessinger Publishing, 1994.

Bate, John, *The Mysteries of Nature and Art*, 3rd. edition, London: R. Bishop for Andrew Crook, 1654.

Beawes, Wyndham, *A Civil, Commercial, Political, and Literary History of Spain and Portugal: In Two volumes* (2 volumes), Vol. 1, London: Faulder, 1793.

Beighton, Henry, "A Description of the Water-Works at London-Bridge, Explaining the Draught of T a b. I," *Philosophical Transactions*, Vol. 37 (1731–1732), pp. 5–15.

Berger de Xivrey, M., *Recueil des Lettres Missives de Henri IV* (9 volumes), Vol. 6, 1603–1606, Paris: Imprimirie Imperiale, 1853.

Besson, Jacques, *L'art et science de trouver les eaux et fontaines cachées soubs terre autrement que par les moyens vulgaires des Agriculteurs & Architectes*, Orléans: P. Trepperel, 1569.

Besson, Jacques, *Theatrum instrumentorum et machinarum*, Lugduni: Apud Barth, 1578.

Biringuccio, Vannoccio, *De la pirotechnia libri X dove ampiamente si tratta non solo di ogni sorte e diuersita di miniere, ma anchora quanto ricerca intorno à la prattica di quelle cose di quel che si appartiene à l'arte de la fusione ouer gitto de metalli come d'ogni altra cosa simile*, Veneto: V. Roffinello for C. Navo, 1540.

Böckler, Georg Andreas, *Theatrum Machinarum Novum: Neu-vermehrter Schauplatz der mechanischen Künsten, handelt von allerhand Wasser-Wind-Ross-Gewicht- und Hand-Mühlen, wie dieselbige zu dem Frucht-Mahlen, Papyr- Pulver- Stampff-Segen- Bohren- Walcken-Mangen, und dergleichen anzuordnen*, Nürnberg: Paulus Fürsten, 1661.

Bonamy, Pierre-Nicolás, "Mémoire sur les aqueducs de Paris, comparés à ceux de l'ancienne Rome," in *Mémoires de littérature, tirés des registres de l'Académie royale des inscriptions et belles-lettres depuis l'année MDCCLVIII jusques et compris l'année MDCCLX* (45 volumes), Vol. 30, Paris: Imprimerie Royale, 1764, pp. 729–755.

Boorde, Andrew, *The fyrst boke of the introduction of knowledge*, (originally written c.1541), edited by C. J. Furnivall, London: N. T. Trübner & Co., 1870.

Botero Benese, Giovanni, *A treatise, concerning the causes of the magnificencie and greatnes of cities, devided into three books by Sig. Giouanni Botero, in the Italian tongue; now done into English by Robert Peterson*, London: R. Ockould and H. Tomes, 1606.

Branca, Giovanni, *Le machine: volume nuovo et di molto artificio da fare effeta maravigliosi tanto spiritali quanto di Animale operatione arichito di bellissime figure con le dichiarationi a ciascuna di esse in lingua volgare et latina*, Roma: Iacomo Mascardi, 1629.

Braun, Georg and Frans Hogenberg, *Civitates Orbis Terrarum* (6 volumes), Vol. 1, Coloniae Agrippinae: Typis Theodori Graminae, 1572.

Braun, Georg and Frans Hogenberg, *Civitates Orbis Terrarum*, (Ausgabe *Beschreibung vnd Contrafactur der vornembster Stät der Welt*), (6 volumes), Vol. I, Köln: Kempen, 1582.

Braun, Georg and Frans Hogenberg, *Civitates Orbis Terrarum* (6 volumes), Vol. 3, Coloniae Agrippinae: Apud Godefridum Kempensem, 1581.

Braun, Georg und Franz Hogenberg, *Civitates orbis terrarium* (6 volumes), Vol. 5, Cologne: [s.n.], 1598.

Brice, Germain, *Nouvelle description de la ville de Paris: et de tout ce qu'elle contient de plus remarquable*, (4 volumes) Vol. 4, Paris: Julien-Michel Gandouin, 1725.

Bullein, William, *The government of health: a treatise written by William Bullein, for the especiall good and healthfull preservation of mans bodie from all noisome diseases, proceeding by the excesse of evill diet, and other infirmities of nature: full of excellent medicines, and wise counsels, for conservation of health, in men, women, and children. Both pleasant and profitable to the industrious reader*, London: Valentine Sims, 1595.

Burton, Robert, *The anatomy of melancholy what it is. With all the kindes, causes, symptomes, prognostickes, and seuerall cures of it. In three maine partitions with their seuerall sections, members, and subsections. Philosophically, medicinally, historically, opened and cut up. By Democritus Iunior. With a satyricall preface, conducing to the following discourse*, Oxford: Iohn Lichfield and Iames Short, for Henry Cripps, 1621.

Caldecott, Thomas, *Report of Cases Relative to the Duty and Office of a Justice of the Peace from Michaelmas Term 1776, inclusive to Trinity Term 1785, inclusive*, London: His Majesty's Law Printers, 1786.

Campo, Antonio, *Cremona fedelissima citta et nobilissima colonia de Romani rappresentata in disegno col suo contato, et illustrata d'una breve historia delle cose piv notabili appartenenti ad essa, et dei ritratti naturali de duchi et duchesse di Milano, e compendio delle lor vite da Antonio Campo*, Milano: Bidelli, 1645 (originally written in 1585).

Cardano, Girolamo, *Hieronymi Cardani Medici Mediolanensis De svbtilitate*, Libri XXI, Norimbergae: apud Ioh. Petreium, 1550.

Cardano, Girolamo, *Hieronymi Cardani Medici Mediolanensis De svbtilitate*, Libri XXI, Basilae: Per Ludovicum Lucium, 1554

Cardano, Girolamo, *Hieronymi Cardani Mediolanensis, De Subtilitate*, Libri XXI, Lugduni: apud Stephanum Michaelem, 1580.

Castelli, Benedetto, *Della Misura dell'acque correnti*, Bologna: H. H. del Dozza, 1660.

Castillo de Bobadilla, Jerónimo, *Política para corregidores y señores de vasallos en tiempo de paz y de guerra y para Iuezes eclesiásticos y seglares y de sacas, aduanas y de residencias y sus Oficiales y para Regidores y Abogados y del valor de los corregimientos y Goviernos Realengos y de las Órdenes* (2 volumes), Vol. 2, Madrid: Luis Sanchez, 1597.

Caus, Isaac de, *New and Rare Inventions of Water-works Shewing the Easiest Waies to Raise Water Higher Then the Spring: By which Invention the Perpetual Motion is Proposed: Many Hard Labours Performed: and Varieties of Notions and Sounds Produced: a Work Both Usefull Profitable and Delightfull for All Sorts of People*, translated into English from the French by John Leak, London: Joséph Moxon, 1659.

Caus, Isaac de, *Nouvelle invention de lever l'eau plus haut que sa source avec quelques machines movantes par le moyen de l'eau, et un discours de la conduit d'icelle*, Londre: [s.n.], 1644.

Caus, Salomon de, *Les raisons des forces mouvantes, avec diverses machines tant utiles que plaisantes, auxquelles sont adjoints plusieurs dessings de grotes & fontaines*, Francfort: Jan Norten, 1615.

Ceredi, Giuseppe, *Tre discorsi sopra il modo d'alzar aque da' luoghi bassi*, Parma: Appresso Seth Viotti, 1567.

Cervantes Saavedra, Miguel de, *El Ingenioso Hidalgo Don Quijote de la Mancha*, edited by Florencio Sevilla Arroyo and Antonio Rey Hazas, Madrid: Centro de Estudios Cervantinos, 1994.

Cervantes Saavedra, Miguel de, "La ilustre fregona," in *Novelas exemplares de Miguel de Cervantes Saavedra*, Madrid: Imprenta de Villapando, 1799, pp. 3–147.

Chastillon, Claude, *Topographie Françoise ou representations de plusieurs villes, bourgs, chasteaux, plans, forteresses, vestiges d'antiquité, maisons modernes et autres du royaume de France*, Paris: Louys Boissevin, 1655.

Chaucer, Geoffrey, *The Canterbury Tales* (originally published in 1475), edited by D. Laing Purves, Auckland (New Zealand): The Floating Press, 2012.

Chavannes, Ernest, "Extraits des manuaux du Conseil de Lausanne, 1536–64," in *Mémoire et documents publiés par la Société d'histoire de la Suisse Romande, 2d ser. 1*, Lausanne, 1887.

Cimber, M. L. and F. Danjou, *Archives curieuses de l'histoire de France depuis Louis XI jusqu'à Louis XVIII, ou Collection de pièces rares et intéressantes ... publiées d'après les textes conservés à la Bibliothèque royale, et accompagnées de notices et d'éclaircissements; ouvrage destiné à Leber*, Paris: Beauvais, 1836.

Codagnellus, Johannes, *Annates placentini*, edited by O.-Holder-Egger, Hannover: Lipsiae, Impansis Bibliopolii Hahniani, 1901.

Collingridge J. H., and R. B. Wernham, (eds), *Calendar of the Patent Rolls Preserved in the Public Record Office: Elizabeth [I]: 1560–1563*, London: His Majesty's Stationery Office, 1948.

Coryat, Thomas, *Coryat's crudities: hastily gobled up in five moneths travells in France, Savoy, Italy, Rhetia commonly called the Grisons country, Helvetia alias Switzerland, some parts of high Germany and the Netherlands: newly digested in the hungry aire of Odcombe in the county of Somerset, and now dispersed to the nourishment of the travelling members of this kingdome*, (2 volumes), Vol 1, Glasgow: James MacLehose and Sons, 1905.

Crawford Lomas, Sophie, (ed.), Public Record Office, *Calendar of State Papers Foreign, Elizabeth*, Vol. 19: August 1584.–.August 1585, London: His Majesty's Stationery Office, 1916.

D'Acres, R., *The art of water-drawing, or A compendious abstract of all sorts of water-machins, or gins, practiced in the world*, London: Henry Brome, 1660.

Dasent, Jean Roche, (ed.), *Acts of the Privy Council of England, new series, Volume XII, CE 1580–1581*, London: Her Majesty's Stationery Office, 1896.

De Alcocer, Pedro, *Hystoria, o descripción de la Imperial cibdad de Toledo: con todas las cosas acontecidas en ella desde su principio, y fundación*, Toledo: Iuan Ferrer, 1554.

De Bathe, Henry, *The charter of Romney-Marsh, or, The laws and customs of Romney Marsh, very useful for all professors of the law, and also for all lords of towns.../ framed and contrived by the Venerable Justice Henry de Bathe*, London: Printed by S. R. for Samuel Keble, 1686.

De Bélidor, Bernard Forest, *Architecture hydraulique ou l' art de conduire, d' élever et de ménager les eaux pour les différens besoins de la vie* (2 volumes), Vol. 2, Paris: Chez L. Cellot, 1782.

De Béthune, Maximilien (duc de Sully), "Mémoires des sages et royales oeconomies d'estat, domestiques, politiques et militaires de Henry le Grand ...: et des servitudes utiles, obéissances convenables et administrations loyales de Maximilien de Béthune," volume 2 of the deuxième serie of: Michaud, Joséph François and J. J. F. Poujoulat, *Nouvelle collection des Mémoires pour servir à l'histoire de France depuis le XIIIe siècle jusqu'à la fin du XVIIIe*, Paris: Editeur du Commentaire analytique du Code Civil, 1837.

De Castilla y de Aguayo, Juan, *El Perfecto Regidor*, Salamanca: Cornelio Bonardo, 1586.

De Céspedes, Andrés, *Libro de instrumentos nuevos de geometria, muy necessarios para medir distancias y alturas, sin que interuengan numeros, como se demuestra en la practica: demas desto se ponen otros tratados, como es uno, de conduzir aguas, y otro una question de artilleria, en donde se ponen algunas demostraciones curiosas*, Madrid: Juan de la Cuesta, 1606.

De Colmenar, Juan Alvarez, *Les Délices de l'Espagne et du Portugal*, (5 volumes), Vol. 2, Leide: Pierre van der Aa, 1707.

De Colmenares, Diego, *Historia de la insigne ciudad de Segovia, y Conpendio de las Historias de Castilla. Segunda inpresion*, Madrid: Diego Diez, 1640.

De Covarrubias Orozco, Sebastian, *Tesoro de la Lengua Castellana o Española*, original written in 1611, edited by Felipe C. R. Maldonado, revised by Manuel Camarero, Madrid: Editorial Castalia, 1995.

De Garibay y Camalloa, Esteban, "Memorias de Garibay," in *Memorial Histórico español, Colección de documentos, opúsculos y antigüedades, que publica La Real Academia de la Historia* (11 volumes), Vol. 7, Madrid: Imprenta José Rodríguez, 1854.

De Garibay y Camalloa, Esteban, *Compendio Historial de las Chronicas y Vniversal Historia de Todos los Reynos de España, Donde se Poden en Suma los Condes, Señores de Aragon, con los Reyes del Mesmo Reyno: y Condes de Barcelona, y Reyes de Napóles y Sicilia*, (4 volumes), Vol. 4, Barcelona: Sebastian de Cormellas, 1628.

De Grunenbergh, Carlos y Fernando, "Perspectiva de la Ciudad de Toledo vista de la otra parte de la Puente del camino de Sevilla," (the Colonels' plan), in *Memorial que los Coroneles Carlos y Fernando de Grunenbergh han dado a la Reina Maríana de Austria, sobre la proposición realizada de hacer navegable el río Manzanares desde el Pardo hasta Toledo: se manifiestan los motivos que tuvieron para haber hecho dicha proposición, los fundamentos y razones que hay para la ejecución y facilidad de su fábrica, las utilidades que de ella resultan, las condiciones para su realización y la demostración que han hecho de las corrientes del dicho río y del Jarama*, Madrid: [s.n.], 1668.

De L'Estoile, Pierre, *Journal du Règne de Henry IV. Roi de France, et de Navarre: Tiré sur un Manuscrit du tems* (2 volumes), Vol. 1, 1732 [place and publisher's name unknown].

De L'Estoile, Pierre, *Mémoires-journaux, Journal de Henri IV, 1607–1609*, (10 volumes), Vol. 9, Paris: Librairie des bibliophiles, 1881.

De Malherbe, François and Nicolás Claude Fabri de Peiresc, *Lettres de Malherbe: ornées du fac-simile de son écriture, dédiées a la ville de Caen avec une vue de cette ville*, Paris: J. J. Blaise, 1822.

De Mallevoüe, Fernand, (ed.), *Maximilien de Béthune, duc de Sully, Les actes de Sully passés au nom du roi de 1600 à 1610: par devant Me Simon Fournyer, notaire au Châtelet de Paris, dont les minutes sont conservées en l'étude de Me Henri Motel, notaire à Paris*, Paris: Imprimerie Nationale, 1911.

De Mézeray, François Eudes, *Abrégé chronologique ou extrait de l'histoire de France* (6 volumes), Vol. 3, Paris: Chez Denys Thierry, Guignard et Barbin, 1690.

De Monconys, Balthasar, *Journal des Voyages de Monsieur de Monconys* (3 volumes), Vol. 2, Lyon: Horace Boissat & George Remevs, 1666.

De Pisa, Francisco, *Descripción de la imperial ciudad de Toledo i historia de sus antiguedades i sus grandezas, i cosas memorables, los reies que la an señoreado, o gouernado, i sus Arzobispos mas celebrados* (Dos partes: primera y apuntamientos para la segunda parte), primera parte, Toledo: Diego Rodríguez, 1617.

De Rojas, Pedro (Conde de Mora), *Historia de la imperial, nobilísima, inclita y esclarecida ciudad de Toledo: Fundación, antigüedades, grandezas y principio de la Religión Católica en ella y de su Santa Iglesia ... Vidas de sus arçobispos y Santos y Cosas memorables*, Madrid: Diego Diaz de la Carrera, 1654.

De Rojas Villandrando, Agustín, *El viaje entretenido* (originally written in 1603) (2 volumes), Vol. 1, Madrid: B. Rodríguez Serra, 1901.

De Salazar, Ambrosio, *Almoneda general de las más curiosas recopilaciones de los reynos de España*, Paris: Antonio du Brueil, 1612.

De Thou, Jacques Auguste, *Jaobi Augusti Thuani Historiarum sui temporis tomus primus* [*-septimus*], (first published in 1604), Londini: excudi curavit Samuel Buckley, 1733.

Di Giorgio Martini, Francesco, *Trattati di architettura, ingegneria e arte militare* (c.1480), Codice Saluzziano 148, original manuscript in Biblioteca Reale, Torino (Italy).

Di Iacopo, Mariano (detto il Taccola), *Liber tertius de ingeneis ac edifitiis non usitatis*, Ms. Palatino 766, Biblioteca Nazionale Centrale, Florence, 1430.

Diderot, Denis and Jean le Rond d'Alembert, *Encyclopédie, ou Dictionnaire raisonné des sciences, des arts et des métiers* (17 volumes—text), Vol. 13, Paris: Briasson, David, Le Breton, Durand, 1751–1772.

Díez Daza, Alonso, *Libro de los provechos y daños que provienen con la sola bebida del agua; cómo se ha de escojer la mejor y rectificar la que no es tal, y cómo se ha de beber fria en tiempo de calor sin que haga daño* (2 volumes), Vol. 1, Sevilla: Alonso de la Barrera, 1576.

Digby, Sir Kennelm, *Two treatises, in the one of which the nature of bodies, in the other, the nature of mans soule is looked into in way of discovery of the immortality of reasonable soules*, Paris: Giles Blazot, 1644.

Dodsley, Robert and James Dodsley, *London and Its Environs Described: Containing an Account of Whatever is Most Remarkable for Grandeur, Elegance, Curiosity, Or Use, in the City and in the Country Twenty Miles Round It. Comprehending Also Whatever is Most Material in the History and Antiquities of this Great Metropolis*, (6 volumes), Vol. 4, London: Printed for R. & J. Dodsley, 1761.

Du Breul, Jacques, *Le Theatre des Antiquitez de Paris; où est traicté de la fondation des Eglises et Chapelles de la Cité, Vniuersité, Ville et Diocèse de Paris; comme aussi de l'institution du Parlement, fondation de l'Vniversité & Collèges, & autres choses remarquables*, (4 livres), *Augmenté en cette edition d'un Supplement, contenant le nombre des Monastères, Eglises, l'agrandissement de la Ville & Fauxbourgs qui s'est faict depuis l'Année 1610, jusques a présent.* Supplement by D.H.I. Advocat en Parlement, Paris: Societé des Imprimeurs, 1639.

Errard de Bar-le-Duc, Jean, *Le premier livre des instruments mathematiques mechaniques*, Nancy: J. Janson, 1584. Reproduction en fac-similé de l'édition de 1584 avec une introduction d'Albert France-Lanord, Paris: Berger-Levrault, 1979.

Evelyn, John, *The Diary of John Evelyn*, edited by William Bray (2 volumes), Vol. 1, London: Walter Dunne, 1901.

Everett Green, Mary Ann, (ed.), *Calendar State Papers Domestic of The Reign of Elizabeth, 1591–1594* (11 volumes), Vol. 3, London: Longman, Green, Reader and Dyer, 1867.

Everett Green, Mary Anne, (ed.), *Calendar State Papers Domestic of the reign of Charles II, 1664–1665* (11 volumes), Vol. 4, London: Longman, Green, Longman, & Roberts, 1863.

Everett Green, Mary Anne, (ed.), *Calendar of State Papers, Domestic, Charles II, 1667–8*, (11 volumes), Vol. 8, London: Public Record Office, 1893,

Everett Green, Mary Anne, (ed.), *Calendar State Papers Domestic of The Reign of Elizabeth, 1601–1603, with Addenda 1547–1565*, London: Longmans & Co., 1870.

Felibien, Michel, *Histoire de la Ville de Paris* (5 volumes), Vol. 4, Paris: Chez Charles-Jean-Bapt. Delespine, 1735.

Félibien, Michel, *Histoire de la ville de Paris, reveue, augm. et mise au jour par B. Guy-Alexis Lobineau*, Paris: chez Guillaume Desprez et Jean Desessartz, 1725.

Ficino, Marsilio, *Tratado de la peste*, edited by Fabián Alejandro Campagne, original published in Pamplona: Mathias Mares, 1598, Biblioteca Nacional, Madrid, catalogued under R-26692. The title of the first Italian edition is: *Consiglio contro la peste*, Florence: Apud Sanctum Jacobum de Ripolis [Convent of San Jacopo di Ripoli], 1481.

Fitzstephen, William, *FitzStephen's Description of the City of London, Newly Translated from the Latin Original; With a Necessary Commentary. A dissertation on the Author, Ascertaining the Exact Year of the Production, is prefixed: and to the Whole is Subjoined, a Correct Edition of the Original, with the Various Readings, and Some Useful Annotations*, by an Antiquary, London: B. White, 1772.

Fortescue, Sir John, *Commendation of the laws of England: the translation into English of "De laudibus legum Angliæ"* (originally written c.1470) translated by Francis Grigor, London: Sweet & Maxwell, 1917.

Friedrich I (Duke of Württemberg), "Travel Journal" (1592), in William Benchley Rye (editor and translator), *England as seen by foreigners in the days of Elizabeth and James the First*, London: John Russell Smith, 1865, pp. 1–53.

Fuero Juzgó o Libro de los Jueces, Cotejado con los más antiguos y preciosos códices por la Real Academia Española, Valladolid: Lex Nova, 1980.

Gabriel Hanotaux, *La France en 1614; La France et la Royauté avant Richelieu*, Londres, Edimbourg et New-York: Nelson, 1913.

Garzoni da Bagnacauallo, Thomaso, *La piazza universale di tutte le professioni del mondo*, Venetia: apresso Gio. Battista Somasco, 1588.

Grollier de Servière, Nicolás and Gaspard Grollier de Servière, *Recueil d'ouvrages curieux de mathematique et de mecanique, ou description du cabinet*, Paris: Chez David Forey, 1719.

Guerin, Paul, (ed.), *Registres des Déliberations du Bureau de la Ville de Paris* (15 volumes), Vol. 12 (1598–1602), Paris: Imprimerie Nationale, 1909.

Guerin, Paul, (ed.), *Registres des Déliberations du Bureau de la Ville de Paris* (15 volumes), Vol. 13 (1602–1605), Paris: Imprimerie Nationale, 1905.

Harrison, William, *The description of England: the classic contemporary account of Tudor social life* (1577), edited by Georges Edelen, New York: Dover Publications, 1994.

Hartlib, Samuel, "Ephemerides 1651," in *The Hartlib Papers (a Complete Text and Image Database of the Papers of Samuel Hartlib)*, Sheffield (UK): Humanities Research Institute, University of Sheffield, 2002.

Hatton, Edward, *A new view of London: or, an Ample Account of that City, In Eight Sections* (2 volumes), Vol. 2, London: J. Nicholson, 1708.

Holinshed, Raphael, *The Chronicles of England, Scotland & Ireland* (6 volumes), Vol. 4, London: J. Johnson, F.C. and J. Rivington and others, 1808.

Howell, James, *Familiar letters on important subjects wrote from the year 1618 to 1650*, Aberdeen: F. Douglass and W. Murray, 1753.

Hurtado de Toledo, Luis, "Memorial de algunas cosas memorables que tiene la imperial ciudad de Toledo dirigido a la C. R. M. del rey D. Phelipe de Austria, monarca de las Españas," (1576) in Carmelo Viñas y Ramón Paz, (eds), *Relaciones de los Pueblos de España Ordenadas por Felipe II, Reino de Toledo*, tercera parte, Madrid: Instituto Balmes de Sociología, Instituto Juan Sebastián Elcano de Geografía, Consejo Superior de Investigaciones Científicas, 1963.

Hurtaut, Pierre-Thomas-Nicolás and Magny, *Dictionnaire historique de la ville de Paris et de ses environs* (4 volumes), Vol. 3 and Vol. 4, Paris: Chez Moutard, 1779.

Jean-François, Père de la Compagnie de Jésus, *L'art des fontaines, c'est à dire pour trouver, esprouver, assembler, mesurer, distribuer & conduire les sources dans les lieux publics & particuliers; d'en rendre la conduite perpétuelle; et de donner par art des eaux courantes aux lieux, où elles manquent par nature*, Rennes: Chez Hallaudars, 1665.

Jonson, Ben, *Everyman in his humour*, written in 1598, first acted in 1601, first published in London: William Stanley, 1616.

King, Gregory, "Natural and Politicall Observations and Conclusions upon the State and Condition of England, 1696," an appendix to George Chalmers, *An estimate of the comparative strength of Britain during the present and four preceding reigns; and of the losses of her trade from every war since the Revolution*, London: John Stockdale, 1804, pp. 29–73. King's original manuscript is the Harleian Manuscripts, British Museum.

Kyeser, Conradus, *Bellifortis*, Ms. germ. qu. 15, (originally written c.1405), Elsaß, copy from 1460, kept in the Universitätsbibliothek Johann Christian Senckenberg, Frankfurt am Main, extracted from a facsimile reproduction from the Universitätsbibliothek Johann Christian Senckenberg, Goethe-Universität.

La Mare, Nicolás de, *Traité de la police, où l'on trouvera l'histoire de son établissement toutes les loix et tous les règlemens qui la concernent* (5 volumes), Vol.1, Paris: Jean et Pierre Cot, 1705.

Lastanosa, Pedro Juan de [or Pseudo Juanelo Turriano], *Los veinte y un libros de los yngenios y máquinas de Juanelo, los cuales le mando escribir y demostrar el catholico rei d. Felipe Segundo Rey de Hespanas y nuebo mundo*, Madrid: BNM, MSS 3372/3376.

Le Grand, Léon, (ed.), *Registres des Déliberations du Bureau de la Ville de Paris* (15 volumes), Vol. 14 (1605–1610), Paris: Imprimerie Nationale, 1908.

Le Grand, Léon, (ed.), *Registres des Déliberations du Bureau de la Ville de Paris* (15 volumes), Vol. 15 (1610–1614), Paris: Imprimerie Nationale, 1921.

Le Mercure françois ou la suitte de l'histoire de la paix (24 volumes), Vol 1 (1605–1610), Paris: Iean Richer, 1612. Facsimile published by the Groupe de Recherches Interdisciplinairessur l'Histoire du Littéraire from an original kept at the bibliothèque de L'École des Ponts et Chaussèes.

Le Rouge, George-Louis,[1] *Les Curiositez de Paris, de Versailles, de Marly, de Vincennes, de S. Cloud, et des envirions: avec les antiquitez justes & précises fur chaque sujet*, seconde edition (2 volumes), Vol. 1, Paris: Chez Saugrain l'aine, 1718.

Lebeuf, [Abbé] Jean, *Histoire de la ville et de tout le diocèse de Paris*, (15 volumes), Paris: Auguste Durandl (volumes 1 & 2); Prault (volumes 3–15), 1754–1758.

Lemon, Robert, (ed.), *Calendar State Papers Domestic, Elizabeth, 1581–90*, London: Longman, Green, Longman, Roberts, and Green, 1865.

Lemon, Robert, (ed.), *Calendar State Papers Domestic, of the Reigns of Edward VI, Mary, Elizabeth, 1547–1580*, London: Longman, Brown, Green, Longmans, and Roberts, 1856.

Leupold, Jacob, *Theatri Machinarum Hydraulicarum, Tomus I: oder, Schau-Platz der Wasser-Künste, erster[-anderer] Theil ... ein Werck, so nicht nur Künstlern, Künstmeistern, Berg-Leuthen und Künst-Steigern ... sondern auch Architectis, Ingenieurs, Commissarien, Beamten* (2 volumes), Vol. 1, Leipzig: C. Zunkel, 1724.

Leupold, Jacob, *Theatri Machinarum Hydraulicarum, Tomus II: oder, Schau-Platz der Wasser-Künste, erster[-anderer] Theil ... ein Werck, so nicht nur Künstlern, Künstmeistern, Berg-Leuthen und Künst-Steigern ... sondern auch Architectis, Ingenieurs, Commissarien, Beamten* (2 volumes), Vol. 2, Leipzig: C. Zunkel, 1725

Lhermite, Jehan, *Le Passetemps*, written between 1597–1602 (2 volumes), Vol. 1, Antwerpen: Cr. Ruelens, 1890.

Lupton, Donald, *London and the Countrey Carbonadoed and Quartred into Severall Characters*, London: Nicholas Okes, 1632, in James Haliwell, (ed.), *Books of Characters, illustrating the Habits and Manners of Englishmen from the Reign of James I. to the Restoration*, London: J. E. Adlard, 1857, pp. 257–320.

Lyle, J. V., (ed.), *Acts of the Privy Council of England 1616–1617* (46 volumes), Vol. 35, London: His Majesty's Stationery Office, 1927.

Maitland, William, *The History and Survey of London from its Foundation to the Present Time*, (2 volumes), Vol. 1, London: T, Osborne and J. Shipton, 1756.

Mc.Connell, Anita, "L. F. Marsigli's Voyage to London and Holland, 1721–1722," *Notes and Records of the Royal Society of London*, Vol. 41, No. 1 (1986), pp. 39–76.

1 Alternatively attributed to Claude Marin Saugrain.

Mercier de Poissy, Nicolás, *La France et les royaumes ruinez par les favoris et les Reines amoureuses*, Paris: Jean Musnier, 1649.

Micó, Francesc, *Alivio de los sedientos: en el qual se trata la necesidad que tenemos de beber frio y refrescado con nieve, y las condiciones que para esto son menester, y quales cuerpos lo pueden libremente soportar*, Barcelona: Diego Galán, 1576.

Morales, Ambrosio de, *Las antigüedades de las ciudades de España*, (First edition: Alcalá de Henares: Casa de Iuan Iñiguez de Lequeríca, 1575), Madrid: Oficina de Benito Cano, 1792.

Mortoft, Francis, *Francis Mortoft: His Book, Being His Travels Through France and Italy, 1658–1659*, edited by Malcom Letts, London: The Hakluyt Society, 1925.

Moryson, Fynes, *An itinerary containing his ten yeeres travell through the twelve dominions of Germany, Bohmerland, Sweitzerland, Netherland, Denmarke, Poland, Italy, Turky, France, England, Scotland & Ireland* (4 volumes), Vol. 4, Glasgow: MacLehose, 1907 (original from 1617).

Mundy, Peter, *The travels of Peter Mundy in Europe and Asia, 1608–1667*, edited by Richard Carnac Temple (5 volumes), Vol. 1, Cambridge: The Hakluyt Society, 1907.

Münster, Sebastian, *Cosmographiae Universalis*, Basel: H. Petri, 1550.

Navajero, Micer Andrés, "Viajes por España," in Antonio María Fabié (ed. and trans.) *Viajes por España de Jorge de Einghen: del Baron Leon de Rosmithal de Blatna, de Francisco Guicciardini y de Andrés Navajero, con una introducción*, Madrid: Librería de los Bibliófilos, 1879, pp. 231–352. The original published in Italian: Andrea Navagero (1483–1529), *Il viaggio fatto in Spagna, et in Francia, dal magnifico m. Andrea Navagiero, fu oratore dell'illustrissimo Senato Veneto, alla cesarea maesta di Carlo V. Con la descrittione particolare delli luochi, et costumi delli popoli di quelle provincia*, Venezia: Domenico Farri, 1563.

Noorthouck, John, *A New History of London: Including Westminster and Southwark* (2 books), Book 1, London: R. Baldwin, 1773.

Overall, W. H. and Overall H. C., (eds), *Analytical index to the series of records known as the Remembrancia, Preserved among the Archives of the City of London, CE 1579–1664*, London: E. J. Francis & Co., 1878.

Pérez de Herrera, Cristóbal, *Texto y Concordancias de la Defensa de las criatvuras de tierna edad*, Valladolid: Luis Sánchez, 1604 (original in Biblioteca Nacional, Madrid), edited by Andrea María Bau, Nueva York: Hispanic Seminary of Medieval Studies, 2004.

Pickering, Danby, *The Statutes at Large from the Fifth Year of K. William and Q. Mary to the Eighth Year of King William III* (46 volumes), Vol. 9, London: Joseph Bentam, 1764.

Pickering, Danby, *The Statutes at Large from the Twelfth Year of King Charles II to the Last Year of James II* (46 volumes), Vol. 8, London: Joseph Bentham, 1763.

Piganiol de la Force, Jean Almar, *Description historique de la ville de Paris et de ses environs* (10 Tomes), Tome 2, Paris: Les Libraires Associes, 1765.

Pinheiro da Veiga, Tomé, *Fastiginia, Vida Cotidiana en la Corte de Valladolid*, (originally written in Portuguese in 1605), translated into Spanish by Narciso Alonso Cortés, Madrid: Ambito—Fundacion municipal de cultura, 1989.

Plat, Hugh, *A briefe apologie of certaine new inventions, whereof there hath bene a publicke viewe taken in London by some of her Maiesties priuie Counsell, and diuerse other gentlemen and citizens of good worship and account, with the authors excuse for not publishing the same, to the generall good of his country*, (3 sheets), sheet 1, London: Richard Field, 1593.

Pliny the Elder, *Natural History*, translated and annotated by John Bostock and H. T. Riley, (eds) (6 volumes), Vol. 6, London: Henry G. Bohn, 1855.

Ponz, Antonio, *Viage de España*, (18 volumes), Vol. 1, Madrid: Viuda de Ibarra, Hijos y Compañía, 1787.

Post, Margaret, (ed.), Public Record Office, *Calendar of the Patent Rolls Elizabeth I, Volume VII, 1575–1578*, London: Her Majesty's Stationery Office, 1982.

Procopius, *History of the Wars* (7 volumes), Vol. 3, translated by H. D. Dewing, New York: Cosimo, Inc., 2007.

Ramelli, Agostino, *Le diverse et artificiose machine del Capitano Agostino Ramelli Dal Ponte Della Tresia Ingegniero del Christianissimo Re di Francia et di pollonia: nelle quali si contengono uarij et industriosi Mouimenti, degni digrandißima speculatione, per cauarne beneficio infinito in ogni sorte d' operatione*, Parigi: published by Agostino Ramelli, 1588.

Ramelli, Agostino, *The various and ingenious machines of Agostino Ramelli: a classic sixteenth-century illustrated treatise on technology / translated from the Italian and French with a biographical study of the author by Martha Teach Gnudi; technical annotations and a pictorial glossary by Eugene S. Ferguson*, New York: Dover Publications, Aldershot (England): Scholar Press, 1987.

Rymer, Thomas, *Fœdera: conventiones, litteræ, et cujuscunque generis acta publica, inter reges Angliæ et alios quosuis imperatores, reges, pontifices, principes, vel communitates, ab ingressu Gulielmi I. in Angliam, CE 1066, ad nostra usque tempora habita aut tractata: ex autographis, infra secretiores archivorum regiorum thesaurarias, asservatis, aliisque summæ vetustatis instrumentis, ad historiam anglicanam spectantibus, fideliter exscripta*, (20 volumes), Vol. 16, Londini: A. and J. Churcill, 1704–35.

Saint-Julien, *Les courriers de la fronde en vers burlesques* (2 volumes), Vol. 1, reviewed and with notes by Celestine Moreau, Paris: Chez P. Jannet, 1857.

Salazar de Mendoza, Pedro, *Origen de las Dignidades Seglares de Castilla y Leon: con relacion Sumanaria de los Reyes de estos Reynos*, Toledo: D. Rodriguez de Valdivielso, 1618.

Sauval, Henri, *Histoire et recherches des antiquités de la ville de Paris*, written c.1660 (3 volumes), Vol. 1, Paris: G. Moette, 1724.

Severim de Faria, Manuel, "Itinerario Hispánico del Chantre de Évora en 1604," *Separata, Revista de Estudios Extremeños*, Tomo XLII, No. 1, Enero-Abril, (1986), pp. 139–185.

Severim de Faria, Manuel, *Peregrinaçao de Balthazar de Faria Severim, Chantre de Évora, ao mosterio de Guadalupe, no anno de 1604*, Biblioteca Nacional de Portugal, Fundo Geral, Cód. 7642, ff. 118r–119v.

Shakespeare, William, "The Merry Wives of Windsor," in William Shakespeare, Alexander Pope, *The Works of Shakespeare* (9 volumes), Vol. 1, Birmingham: R. Martin, 1768, pp. 221–308.

Sitoni, Giovanni Francesco, *Trattato delle virtù et proprietà delléacque, del trouvarle, eleggerle, livellerale, et condurle, et di alcunéaltre sue circonstanze: manuscript*, 1599, The Huntington Library, San Marino, (CA), catalogued under HM 71895

Smart, John (of Guildhall), *A Short Account of the Several Wards, Precincts, Parishes, & c. in London*, London: n.a., 1741.

Sorbière, Samuel Joséph, *Relation d'un voyage en Angleterre, où sont touchées plusieurs choses, qui regardent l'estat des Sciences, et de la Religion, et autres matieres curieuses*, Cologne: Pierre Michel, 1666.

Speed, John, *The theatre of the empire of Great Britaine: presenting an exact geography of the kingdomes of England, Scotland, Ireland, and the iles adioyning: with the shires, hundreds, cities and shire-townes, within ye kingdome of England*, London: William Hall, 1611.

Stovv, Iohn [John Stow], *The survey of London containing the original, increase, modern estate and government of that city, methodically set down: with a memorial of those famouser acts of charity, which for publick and pious uses have been bestowed by many worshipfull citizens and benefactors: as also all the ancient and modern monuments erected in the churches, not only of those two famous cities, London and Westminster, but (now newly added) four miles compass / begun first by the pains and industry of John Stow, in the year 1598; afterwards inlarged by the care and diligence of A.M. in the year 1618; and now compleatly finished by the study & labour of A.M., H.D. and others, this present year 1633; whereunto, besides many additions (as appears by the contents) are annexed divers alphabetical tables, especially two, the first, an index of things, the second, a concordance of names*, London: Printed by Elizabeth Purslow, for Nicholas Bovrne, 1633.

Stow, John, *A survay of London: Conteyning the originall, antiquity, increase, moderne estate, and description of that city, written in the yeare 1598. by Iohn Stow citizen of London. Since by the same author increased, with diuers rare notes of antiquity, and published in the yeare, 1603. Also an apologie (or defence) against the opinion of some men, concerning that citie, the greatnesse thereof. With an appendix, contayning in Latine Libellum de situ & nobilitate Londini: written by William Fitzstephen, in the raigne of Henry the second*, London: Iohn Windet, 1603.

Stow, John, *The Annales, or a General Chronicle of England, Begun First by Maister Iohn Stow, and After Him Continued and Augmented with Matters Forreyne, and Domestique, Auncient and Modern, Unto the Ende of this Present Yeere 1624, by Edmond Howes, Gentleman*, Londini: Impensis Thomæ Adams, 1615.

Strabo (c.63 BCE–c.24 CE), *Geography*, written c.20 CE, literally translated (the first six books) by H. C. Hamilton, the remainder by W. Falconer, (3 volumes), Vol 3, London: Henry H. Bohn, 1857.

Strada, Jacobus de and Octavius de Strada, (eds), *Künstliche Abrisz allerhand Wasser-Wind- Rosz- und Handt Muhlen: beneben schönen und nützlichen Pompen auch andern Machinen damit das Wasser in Höhe zuerheben, auch lustige Brunnen und Wasserwerck, dergleichen vor diesem nie gesehen worden ... Nunmehr ... durch den Truck publicirt ... durch*, (2 volumes), Vol. 1, Frankfurt am Main: Paulum Iacobi, 1617.

Strype, John, *A Survey of the Cities of London and Westminster, Containing the Original, Antiquity, Increase, Modern Estate and Government of Those Cities* (2 volumes), Vol. 1, London: A. Churchill, 1720.

Switzer, Stephen, *An Introduction to a General System of Hydrostaticks and Hydraulicks, Philosophical and Practical*, London: T. Astley, 1729, pp. 317–320.

Trusler, John, *The London adviser and guide: containing every instruction and information useful to persons living in London*, London: printed by the author, 1786.

Tuetey, Alexandre, (ed.), *Registres des Déliberations du Bureau de la Ville de Paris*, (15 volumes), Vol. 2 (1527–1539), Paris: Imprimerie nationale, 1886.

Tuetey, Alexandre, (ed.), *Registres des Déliberations du Bureau de la Ville de Paris*, (15 volumes), Vol. 11 (1594–1598), Paris: Imprimerie Nationale, 1902.

Valturium, Robertum, *De Re Militari, Libris XII*, Parisiis: Apud Christianum Wechelum, fub infigni fcuti Bafilienfis, 1535

Vaughan, William, *Naturall and artificial directions for health: deriued from the best philosophers, as well moderne, as auncient.* London: Richard Bradocke, 1600.

Venner, Tobias, *Via recta ad vitam longam, or a plaine philosophical discourse of the nature ... and effects of all such things, as by way of nourishments and dieteticall obseruations make for the preseruation of health ... Wherein also ... the true vse of our famous Bathes of Bathe is ... demonstrated*, London: E. Griffin for R. Moore, 1620.

Villalba y Estaña, Bartolomé, *El pelegrino curioso y grandezas de España*, originally published in 1577, Pascual de Gayángos, (ed.), Madrid: Imprenta de M. Ginesta, 1886.

Villers, Philippe de, *Journal d'un voyage à Paris en 1657–1658*, Paris: A. P. Faugère, Paris, 1862.

Vinchant, François, *Voyage de François Vinchant en France & en Italie, du 16 septembre 1609 au 18 février 1610*, Bruxelles: Societe Generale D'Imprimirie, 1897.

Vitruvii Pollionis, Marcus, *De Architectura, Libri Decem*, cum commentariis, Danielis Barbari [translation and commentary], Venetiis: Apud Franciscum Franciscum Senensem, & Ioan, Crugher Germanum, 1567.

Vitruvius Pollio, Marcus, *I dieci libri dell'architettura*, tradutti e commentati da Monsignor [Daniele] Barbaro, Venegia: Francesco Marcolini, 1556.

Vitruvius Pollio, Marcus, *The Ten Books on Architecture*, translated by Morris Hicky Morgan, Cambridge (MA): Harvard University Press, 1914.

Voltaire [François-Marie Arouet], (under the pen name L'abbe Bigay), *Histoire du Parlement de Paris* (2 volumes), Vol. 2, Amsterdam: Jean Jaques du Fay, 1769.

Von Stetten, Paul (der Jüngere), *Künst-, Gewerb- und Handwerks-Geschichte der Reichs-Stadt Augsburg*, Augsburg: C. H. Stage, 1779.

Von Wecker, Ian Iaques, *Les Secrets et merveilles de Nature*, Rouen: Claude le Villain, 1620.

Walter, Caspar, *Hydraulica Augustana, Das ist: Ausführliche Beschreib- und Auslegung alles dessen was in des Heil. Röm. Reichs-Stadt Augspurg in den daselbst befindlichen Drey obern Haupt-Wasser-Thürnen so wohl, als in den Brunnen-Häusern, darinnen die Wasser-Druck-Wercke stehen, sammt den auf- und absteigenden Haupt-Wasser-Röhren, Röhr-Kästen oder Reservoirs, von welchen das Wasser durch einen weiten District der Stadt, nicht allein in die schöne publique Spring-Brunnen auf den Haupt-Plätzen, sondern auch in die privat-Häuser mittelst unter der Erden ligenden Canaelen oder Theihel geleitet wird, den Fremden, Passagiers und Liebhabern pflegt gezeiget zu werden: Woneben unterschiedliche nach dem verjüngten Maßstab verfertigte Modell von Wasser-Machinen, hölzernen Bruggen, Schnecken- oder Wändel-Treppen, gemahlte Taflen, worauf alles, was zu dem Brunnen- und Röhr-Wasser-Werck gehörig vorgestellet zu sehen*, Augsburg: Detleffsen, 1754.

Zeising, Heinrich, *Theatri machinarum erster [-letzter] Theill ... auss italienische und französische Sprach mit Fleiss transferirt durch Hieronymum Megiserum ... in Druck geordnet und vorlegt von Henning Grossen dem Jüngern*, Leipzig: [s.n.], 1613–1614.

Zonca, Vittorio, *Novo Teatro Di Machine Et Edificii / Per varie et sicure operationi / Con le loro figure tagliate in Rame / e la dichiaratione, e dimostratione / di ciascuna. / Opera necesaria ad Architetti, et a quelli, / chi di tale studio si dilettano*, Pietro Bertelli, 1607.

Iconographical and Cartographical Sources

Agas, Ralph, "Civitas Londinum: A Survey of the Cities of London and Westminster, The Borough of Southwark and Parts Adjacent in the Reign of Queen Elizabeth," paper on linen (75×185 cm), c.1560. Facsimile published by Adams and Francis, London, 1874, of an engraving from the original held in the Guildhall Library.

Al-Jazari, "Third water-raising machine (c.1200)," Picture of the pump for raising water of al-Jazari, preserved in the manuscript copy held in Topkapi Sarayi Libray, Ahmet III collection, MS 3472, extracted from Richard Covington, "The Art and Science of Water," *Saudi Aramco World*, vol. 57, number 3 (May/June 2006), pp. 14–23, p. 19. Copy of one of Al-Jazari's extant manuscripts can be seen in Schioler, *Roman and Islamic ... Wheels*, pp. 184–185.

Anonymous, "Conceptual device to raise water from the Tajo to the *Alcázar*," 1561, AGS, *Mapas, Planos y Dibujos*, XXVII-3. Ministerio de Educacion, Cultura y Deporte. Archivo General del Simancas. MPD,027,003.

Anonymous, "Plan of the northern half of the old London Bridge, incorporating a view of the water works, on the west side of the bridge opposite St Magnus the Martyr to enable Thames water to be piped to citizens' homes," watercolor/pen, paper (111× 41 cm.), c.1634, London Metropolitan Archives, Main Print Collection.

Anonymous, "Portrait of Juanelo Turriano," oil on canvas (68.5×54.3 cm), Civic Museum "A la Ponzone," Cremona (Italy).

Anonymous, "View of the Old London Bridge," c.1600. The original is in the Pepys's Collection, Magdalene College, Cambridge. Extracted from John Richard Green, *A Short History of the English People, Illustrated Edition*, edited by Mrs. J. R. Green and Mrs. Kate Norgate (4 volumes), Vol. 3, London, Macmillan and Co, 1903, frontispiece.

Anonymus (after Taccola and Francesco di Giorgio), "Two Archimedes' screws working a scoop-wheel, which in turn pulls the screws," in *Drawings of machines, Ms. Palatino 767*, Biblioteca Nazionale Centrale di Firenze, p. 147.

Anonymus, "Elévation et coupe transversale de l'ancienne pompe Samaritaine," aquarelle, (41.8×30.5 cm), 1645, Bibliothèque Nationale de France, département Estampes et photographies, FOL-VA-224(B), photo by the author.

Anonymus, "Plan des arches du Pont Neuf, plan de la Samaritaine, quay de l'Escolle," aquarelle, (36,7×53,9 cm), 1645, Bibliothèque Nationale de France, département Estampes et photographies, FOL-VA-224(B), photo by the author.

Anonymus, "View of Paris, close to the Pont Neuf, detail (c.1635)," Microfilm A 19957 in Bibliothèque nationale de France, Site Richelieu, photo by the author. The original painting is kept in the Musée Impérial de Versailles.

Anonymus, "Vista del Palacio de Toledo," engraving, c.1650, extracted from Luis Moreno Nieto y Ángel Moreno Santiago, *Juanelo y su Artificio. Antología*, Toledo: d. b. Comunicación, 2006, p. 138–139.

Arroyo Palomeque, "Panoramic view of Toledo," c.1720, extracted from Julio Porres Martin-Cleto, *Toledo a traves de sus planos*, Toledo: Diputación Provincial, Instituto Provincial de Investigaciones y Estudios Toledanos, 1989, fig. 10.

Baillieul, Gaspard, publisher, "Vue de la fameuse machine de Marly qui élève l'eau de la Seyne 535 pieds de haut servant à faire jouer les fontaines du Palais-Royal, de Versailles et de Marly," estampe, (52.8×63.1 cm), c.1700?, Bibliothèque nationale de France, département Estampes et photographie, Paris, Collection Michel Hennin, catalogued under Nr. 5257.

Boisseau, Jean, "Plan general de la ville, cité, université, isles et faubourgs de Paris, dedié a messieurs les gouverneur prevost des marchands et eschevins de la dite ville / par leurs très humble et très affectione serviteur Jean Boisseau, enlumineur du roy pour les cartes géographiques," Paris, 1654, extracted from: Adolphe Alphand, *Atlas des anciens plans de Paris*, (2 volumes), Vol. 1, Paris: Imprimerie nationale, 1880, plates XVII bis & XVII ter.

Bosse, Abraham, "Le marchand de bois," c.1640, estampe, Bibliothèque nationale de France, département Estampes et photographie, Paris, Collection Michel Hennin, Estampes relatives à l'Histoire de France, tome 33, pièces 2964, p. 379.

Brambilla, Ambrogio "Toledo," engraving, (48×72.5 cm), Romae: Joannes Orlandi formis, 1602 (original map depicted in 1585).

Bretez, Louis, and Michel-Etienne Turgot, "Plan de Paris commencé l'Année 1734" (269×325 cm), levé et dessiné par Louis Bretez, gravé par Claude Lucas, 1739.

Buck, Samuel and Nathaniel Buck, "Bird's eye view of the north end of London Bridge from the Thames, showing the waterworks" fraction of a larger print, etching, (12.6×15.7 cm), 1749, BM DPD, Crace Collection, Registration number: 1880,1113.1536.

Bullet, Pierre and Nicolás Blondel, "Plan de Paris, levé par l'ordre du Roy & par les soins de Messieurs les Prevost des Marchands & Eschevins en l'année 1676," Paris, 1676.

Caus, Isaac de, "Water-driven machine for trunks drilling," in *Nouvelle invention de lever l'eau plus haut que sa source avec quelques machines movvantes par le moyen de l'eau, et un discours de la conduit d'icelle: avec beacoup de figures en taille douce / par Isaac de Caus, ingeneur & architecte à Charles le Premier, Roy de la Grand Bretaigne*, London: Thomas Davies, 1657, Plate XII.

Champaigne, Philippe de, *Le Prévôt des marchands et les échevins de la ville de Paris*, 1648, Oil on canvas, (200×271 cm), Musée du Louvre, numéro d'inventaire MI 911.

Da Vinci, Leonardo, "Drawings of Water-lifting Devices," *Codex Atlanticus*, (12 volumes), Vol. 1, fol. 26, v., Pen and ink on paper, (61×44 cm), Biblioteca Ambrosiana, Milan.

Da Vinci, Leonardo, "Machines for lifting and pumping water, towers and buildings on top of which water is lifted by Archimedean spirals," *Codex Atlanticus* (12 volumes), Vol. 12, fol. 1069, v., Biblioteca Ambrosiana, Milan.

De Ram, Joannes, after Gabriel Perelle, "La pompe de la Samaritaine," paper, etching, c.1670–1695, (12.5×16.5 cm), British Museum, Registration number: 1981, U.3308.

Diderot, Denis, *Recueil De Planches Sur Les Sciences, Les Arts Liberaux Et Les Arts Méchaniques: Avec Leur Explication* (11 volumes—plates), Vol. 2, Paris: Briasson, David, Le Breton et Durand, 1763.

Diderot, Denis, *Recueil De Planches Sur Les Sciences, Les Arts Liberaux Et Les Arts Méchaniques: Avec Leur Explication* (11 volumes—plates), Vol. 4, Paris: Briasson, David et Le Breton, 1767.

Godefroy, Félix, d'après Pernot, "La Pompe de La Samaritaine, en 1635," Musée Carnavalet—Histoire de Paris, XIXe siècle, Lithographie, G 39161.

Hoefnagel, Joris, "Toletum," in Georg Braun and Franz Hogenberg, *Civitates Orbis Terrarum* (6 volumes), first Latin edition of volume 1, Coloniae Agrippinae: Typis Theodori Graminae, 1572, plate 3, incorrectly marked 1566.

Hoefnagel, Joris, "Toletum," in Georg Braun and Franz Hogenberg, (6 volumes), first Latin edition of volume 5, Coloniæ Agrippinæ: Apud Bertramum Bochholtz sumptibus auctorum, 1598, plate 15.

Hoiamis, Franciscus, "Map of Paris," detail, engraved by François Van Hoey, Biblioteque Nationale, Est. Va 212, 4 planches (41×212 cm) assemblées, Paris, 1619. Photo by the author in the Richelieu site of the Biblioteque Nationale.

Hollar, Wenceslaus, "Bird's-eye plan of the west central district of London, with Lincoln's Inn Fields at the top right the Covent Garden piazza left of centre, Holborn along the top, St Martin's Lane along the left, and the Thames from the Savoy to Essex Stairs at the bottom," (1660–1665), etching on paper (34.4×45.5 cm.), BM, DPD, Registration number: Q,6.136.

Hollar, Wenceslaus, "London [the long view]," bird's eye view of the City of London from Hampstead to St. Dunstan in the East, paper, etching (46.6×39 cm), Amsterdam, 1647, BM, DPD, Registration number: 1864,0611.438.

Hollar, Wenceslaus, "London before and after the Fire: two long views extending from the Temple on the west to the Tower on the right, with buildings including Baynard's castle, Stiliard, Guildhall, St Paul's Cathedral, and London Bridge; a later state with Overton's address," (25.4×70.8 cm), 1666, British Museum. Registration number: Q,6.58.

Hollar, Wenceslaus, "Ye Waterhouse" [The new River Head in Islington], print (12× 8.3 cm), London: John Overton, 1665, BM, DPD, Registration number: Q,6.51.

Iselin, Henri-Frédéric, "Statue of François Miron," Hôtel de ville, Paris, 1881.

Janssonius, Johannes, "Lutetia Parisiorum, Vulgo Paris," in *Theatrum Urbium Germaniae Superioris* (8 volumes), Vol.1 (43×52.5 cm), coloured, Amsterdam, 1657.

Laroon, Marcellus, "New River water," The Cryes of the City of London Drawne after the Life, etching, engraving, published in London, 1688–1750, plate 61 (24.7×16.3cm), BM, DPD, Registration number: 1868,0808.13614.

Lassus, J. B. A. and Alfred Darcel, (eds), *Album de Villard de Honnecourt*, facsimile of a XIII century manuscript by Villard de Honnecourt, Paris: Imprimerie Impériale, 1858, planche XIII.

Martin, Pierre-Denis, "View of the Marly Machine and the Aqueduct at Louveciennes, 1722," oil on canvas (115×165 cm), Musée Promenade, Marly-le-Roi, France.

Merian, Caspar, "View of Paris," in *Topographia Galliae* (4 volumes), Vol. 4 (33.1× 19.7 cm), Frankfurt: Verlag Caspar Merians, 1665.

Merian, Matthäus, (L'ancien), "Le plan de la ville, cité, université et Fauxbourgs de Paris avec la description de son antiquité et singularités" (1615), engraved by Merian and published in Paris by Nicolás de Mathoniére. East-southeast is at the top (50× 101 cm), Bibliothèque historique de la Ville de Paris, catalogued under A 104 f.

Merian, Matthäus, "A bird's eye view from Whitehall to beyond the Tower, with Southwark in the foreground," paper, engraving, 1638, two plates, joined (22.7× 70.4 cm), BM, DPD, Registration number: 1880,1113.1132.

Mönch, Philipp, *büch der stryt und buochßen*, manuscript (1496), Universitätsbibliothek, Heidelberg.

Norden, John, "A long view of old London bridge with houses built along it, with three columns of explanatory letterpress printed beneath;" This is an antiquarian copy, made c.1804, of the original plate by Norden of 1597, engraving (29.9×52.5 cm). BM, DPD, Registration number: 1859,1210.1004.

Norden, John, "Civitas Londini," paper, engraving, London, 1600. Bird's eye view of the City of London from Hampstead to St. Dunstan in the East; (36.5×69.3 cm), Photograph taken from the original by the author. BM, DPD, Registration number: 1880,1113.1120.

Parcerisa, Francisco Javier, "Puente de Alcántara," Litography by J. Donon, in José María Quadrado, *Recuerdos y bellezas de España—Castilla La Nueva* (2 volumes), Vol. 1, Madrid: Imprenta de D. José Repullés, 1853.

Pasquier, Robin et Bourron, "Maquette de la pompe de la Samaritaine," Wood and bronze, painted, (98×75×56 cm), 1772, Musée Carnavalet.

Pippa, Luigi (costruttore), "Reconstruction of the astrarium," Museo della Scienza e della Tecnologia, Milan (Built 1961–1963).

Pisanello, Antonio, "Piston-pump driven by a waterwheel," (15.9×23.5 cm), *Codex Vallardi*, drawing No. 2286 (c.1450), Musée Louvre, extracted from Bernhard Degenhart, *Antonio Pisanello*, Wien: Schroll, 1942, fig. 147.

Pullam, J., "A copy of the Visscher bird's eye view/long view of London of 1616," 1848. Etching printed from four plates joined as continuous strip (44.5×222.7 cm), BM, DPD, Registration number: 1880,1113.1125.

Quesnel, François, "Carte ou description nouvelle de la ville, cité, université et faubourgs de la ville de Paris" (504×384 cm), en 12 feuilles, gravure à l'eau forte, Bibliothèque nationale de France, extracted from Hilary Ballon, (ed.), *The Paris of Henri IV—Architecture and Urbanism*, Cambridge, Mass. and London: The MIT Press, 1991, pp. 233–44

Raguenet, Jean-Baptiste Nicolás, "La joute des mariniers entre le pont Notre-Dame et le pont au Change," oil on canvas laid on panel, (59×96 cm), 1756, Musée Carnavalet.

Raguenet, Jean-Baptiste Nicolás, "Le Pont Neuf et la pompe de la Samaritaine, vus du quai de la Megisserie," oil on canvas (46×83,2 cm), 1777, Musée Carnavalet.

Ryff, Walter Hermann, (whose name was Latinized to Gualtherius Hermenius Rivius,. "Vermessung der Höhe eines Gebäudes aus verschiedenen Distanzen mit einem Quadranten," *Geometrie & Vermessung & Quadrant & Gebäude*, 1547, Deutsche Fotothek 0000182.

Saint-Denis, Yves de, *Vie et martyre de saint Denis et de ses compagnons*, French manuscript from 1317, (25.5×18 cm), extracted from the website of the Bibliothèque nationale de France: http://gallica.bnf.fr/.

Sauvé, Jean, "Paris," date of publication: 1666–1690, carte, image fixe, (39,5×50.5 cm), Bibliothèque nationale de France, Collection d'Anville, catalogued as (810).

Scharf, George, "View in the late 1820s, south across the River Thames towards Southwark," watercolor (49×67 cm.), 1835, in Celina Fox, (ed.), *London—World City, 1800–1840*, New Haven (CT) and London, Yale University Press, 1992.

Schickhardt, Heinrich, "Sketch of a waterwheel driving three pumps," 1607. The original manuscript is kept in Hauptstaatsarchiv Stuttgart (Germany).

Saraceni, Carlo "Moses defending the Daughters of Jethro," 1609, Oil on copper, (28.5×35.3 cm), National Gallery, London.

Sevaistre, Eugène, "Ruins of the *Artificio*," photograph (1857), kept in the Real Biblioteca del Patrimonio Nacional de España, from Moreno Nieto, *Juanelo y su Artificio*, p. 206.

Speed, John, "View of London," in *The theatre of the empire of Great Britaine: presenting an exact geography of the kingdomes of England, Scotland, Ireland, and the iles adioyning: with the shires, hundreds, cities and shire-townes, within ye kingdome of England*, London: William Hall, 1611.

Tavernier, Melchior, "Le Plan de la Ville, Cité, Université, Faubourgs de Paris avec la Description de son Antiquité," 1628 edition, copperplate, (2 sheets, 55×73 cm), Bibliothèque historique de la Ville de Paris, catalogued as A 103.

Theotokopoulos, Domenikos (El Greco), "View of Toledo," oil on canvas, (121.3× 108.6 cm), 1597, Havemeyer Collection, The Metropolitan Museum of New York.

Theotokopoulos, Domenikos (El Greco), "Vista y Plano de Toledo," oil on canvas, (132×228 cm), c.1610, Museo El Greco, Toledo.

Treswell, Ralph, "West-cheap" (i.e. Cheapside), Bird's-eye view of London, with church and houses shown in elevation, paper, pen and ink, watercolour, (27.1×39.3 cm), 1585, Crace Collection. BM, DPD, Registration number: 1880,1113.3516.

Vassallieu, Benedit (dit Nicolay), "Portrait de la Ville. Cité et Université de Paris avec les Faubourgs dicelle," facsímile edited by Imp. Chardon aîné (Paris), 1881 (77× 62 cm), Bibliothèque nationale de France, département Cartes et plans, GE C-10778. The original consists of 4 sheets, (114×146 cm), copperplate, kept in the Bibliothèque historique de la Ville de Paris, catalogued under A 103.

Ventura Seco, "Plano de Valladolid," 1738, redrawn by Juan Agapito y Revilla in 1901, Archivo Municipal de Valladolid.

Veranzio, Fausto, "Floating mill, keeping waterwheel at a constant depth," *Machinae Novae: Cum Declaratione Latina, Italica, Hispanica, Gallica et Germanica*, Venetiis: [s.n.], 1617, Plate 18, between pp. 6–9.

Visscher, Claes Jansz, "Lutetia Parisiorum urbs, toto orbe celeberrima notissimaque, caput regni Franciae" (53×82 cm). Original version: [s.l.]: [s.n.], 1618, Bibliothèque nationale de France, département Cartes et plans, GE D-21287.

Visscher, Claes Jansz, "View of London from South bank," published by Jodocus Hondius II (43×55.5 cm each plate, full printed image 43×216.9 cm), Amsterdam, 1616, BM, DPD, Registration number: 1880,1113.1124.1–4.

Studies

Abattouy, Mohammed, "The Arabic-Latin Intercultural Transmission of Scientific Knowledge in Pre-Modern Europe: Historical Context and Case Studies," in Nayef R. F. Al-Rodhan, *The Role of the Arab-Islamic World in the Rise of the West: Implications for Contemporary Trans-Cultural Relations*, London: Palgrave Macmillan, 2012 pp. 167–219

Abbot, John William, *History of London from the Earliest Period to the Present Time*, London: A. K. Newman, 1821.

Abrams, Len, "Water for basic needs," in *First World Water Development Report*, United Nations Educational, Scientific and Cultural Organization, (2003), pp. 1–14.

Académie française, *Dictionnaire de l'Académie française* (2 volumes), Vol. 1, Paris: Firmin Didot frères, 1835.

Ackroyd, Peter, *London: The Biography*, London: Chatto and Windus, 2000.

Agapito y Revilla, Juan, *Los abastecimientos de aguas de Valladolid: apuntes históricos*, Valladolid: Imprenta La Nueva Pincia, 1907.

Aikin, John and William Enfield, Mr. Nicholson, Thomas Morgan and William Johnston, (eds), *General Biography: Or, Lives, Critical and Historical, of the Most Eminent Persons of All Ages, Countries, Conditions, and Professions, Arranged According to Alphabetical Order* (10 volumes), Vol. 2, London: G. G. and J. Robinson, 1801.

Al-Hassan, Ahmad Y, *Taqi al-Din and Arabic Mechanical Engineering* (in Arabic), Aleppo: University of Aleppo Press, 1978.

Al-Hassan, Ahmad Y. and Donald R. Hill, *Islamic Technology, an Illustrated History*, Cambridge: Cambridge University Press, 1986.

Alphand, Adolphe, *Atlas des anciens plans de Paris* (2 volumes), Vol. 1, Paris: Imprimerie nationale, 1880.

Álvarez Peláez, Raquel, "Felipe II, La Ciencia y el Nuevo Mundo," *Revista de Indias* (1999), vol. LIX, núm. 215, pp. 9–30.

Amador de los Rios, José, *Toledo Pintoresca: Ó Descripción de sus Mas Célebres Monumentos*, Madrid: Imprenta y Libreria de Don Ignacio Boix, 1845.

Anonymous, "Waterworks at Old London Bridge," *Notes and Queries*, Vol.11, 2nd S (No. 266, Feb 2, 1861), p. 90.

Antaki, George A., *Piping and Pipeline Engineering. Design, Construction, Maintenance, Integrity, and Repair*, New York and Basel: CRC Press, 2003.

Antolínez de Burgos, Juan, *Historia de Valladolid, publicada, corregida, anotada y adicionada con una advertencia por Juan Ortega y Rubio*, Valladolid: Libreria Hijos de Rodríguez, 1887.

Aranda Alonso, Fernando, *El Sistema Hidráulico Romano de Abastecimiento a Toledo*, Toledo: Diputación Provincial, Instituto Provincial de Investigaciones y Estudios Toledanos, 1997.

Aranda Pérez, Francisco José, "Los mercaderes de Toledo en el seiscientos, bases económicas y status sociopolítico," *Investigaciones históricas: Época moderna y contemporánea*, No. 12, 1992, pp. 71–96.

Aranda Pérez, Francisco José, "Nobles, discretos varones que gobernáis a Toledo. Una guía prosopográfica de los componentes del poder municipal en Toledo durante la Edad Moderna (corregidores, dignidades y regidores)," in Francisco José Aranda, (ed.), *Poderes intermedios, poderes interpuestos. Sociedad y oligarquías en la España Moderna*, Cuenca: Ediciones de la Universidad de Castilla-La Mancha, 1999, pp. 227–310.

Aranda Pérez, Francisco José, *Poder y poderes en la ciudad de Toledo: gobierno, sociedad y oligarquías en la Edad Moderna*, Cuenca: Ediciones de la Universidad de Castilla-La Mancha, 1999.

Archivo Municipal de Toledo, *Muestra Documental sobre Toledo durante el reinado de Carlos V*, 6 julio–31 diciembre de 2010.

Arenillas Parra, Miguel and Marisa Barahona Oviedo, "La presa romana de La Alcantarilla en el abastecimiento de agua a Toledo," in S. Huerta, R. Marín, R. Soler and A. Zaragozá, (eds), *Actas del Sexto Congreso Nacional de Historia de la Construcción*, Valencia, 21–24 octubre 2009, Madrid: Instituto Juan de Herrera, 2009.

Arenillas, Lidia, Miguel Arenillas, Carmen Díaz-Guerra and José María Macías, "El abastecimiento de agua a Toledo en epoca Romana," in José María Macías and Cristina Segura Graíño, (eds), *Historia del abastecimiento y usos del agua en la ciudad de Toledo*, Madrid: Confederación Hidrográfica del Tajo, 1999, pp. 35–48.

Armytage, Walter Harry Green, *A Social History of Engineering*, London: Faber and Faber, 1961.

Arnoux, Mathieu and Piere Monnet, (eds), *Le technicien dans la cité en Europe occidentale, 1250–1650*, Rome: École française de Rome, 2004.

Arul Sam, L. R., *The Love Commandment of Jesus Christ in the Gospel of Luke and Its Implication in the Indian Context*, Delhi (India): ISPCK, 2008.

Aryamontri, Deborah Chatr, "Running Water: Advances In Urban Water Supply During The Roman Empire," in *The Nature and Function of Water, Baths, Bathing, and Hygiene from Antiquity Through the Renaissance*, Volume 11 of Technology and Change in History, (ongoing series), Leiden and Boston: Brill, 2009, pp. 319–336.

Asenjo González, María, "Perfíl socioeconómico de la ciudad de Toledo en el siglo XV a traves de sus ordenanzas," *Cuadernos de historia de España*, N° 77, (2001-2002), pp. 109–144.

Ash, Eric H., *Power, Knowledge, and Expertise in Elizabetan England*, Baltimore and London: The John Hopkins University Press, 2004.

Babelon, Jean, *Germain Pilon*, Paris: Les Beaux-Arts, Edition d'etudes et de documents, 1927.

Babelon, Jean-Pierre, "L'Urbanisme d'Henri IV et Sully à Paris," in Pierre Francastel, (ed.), *L'Urbanisme de Paris et l'Europe 1600–1800*, Paris: Editions Klinchsieck, 1969, pp. 47–60.

Babelon, Jean-Pierre, *Demeures parisiennes sous Henri IV et Louis XIII*, Paris: Hazan, 1991.

Babelon, Jean-Pierre, *Henry IV*, Paris: Fayard, 1982.

Ballon, Hilary, *The Paris of Henri IV—Architecture and Urbanism*, Cambridge (MA) and London: The MIT Press, 1991.

Barbiche, Bernard, "Une révolution administrative: la charge de grand voyer de France," in André Stegmann, (ed.), *Pouvoir et institution en Europe au XVIème siècle*, Paris: Libraire Philosophique J. Vrin, 1987.

Barker, Felix, *The History of London in Maps*, London: Random House, 1990.

Barnes, Carl F., *The Portfolio of Villard de Honnecourt*, Farharm (UK): Ashgate Publishing, Ltd., 2009.

Barnes, David S., *The Great Stink of Paris and the Nineteenth-Century Struggle Against Filth and Germs*, Baltimore: John Hopkins University, 2006.

Bateman, John Frederick, "On the Present State of Our Knowledge on the Supply of Water to Towns," in *British Association for the Advancement of Science, Report of the Twenty-fifth Meeting*, London: John Murray, 1856, pp. 62–77.

Bau, Andrea María y Gabriela Fernanda Canavese, "'Agua que cura, agua que alimenta'. La dietética para sanos y el uso del agua en la sociedad española bajomedieval y moderna," *Cuadernos de historia de España*, Vol. 80, Buenos Aires (Argentina), enero/diciembre (2006), pp. 127–146.

Bautista, Emilio, José Luis Muñoz and Javier Echávarri, "Juanelo (1501–1585)," in Marco Ceccarelli, (ed.), *Distinguished Figures in Mechanism and Machine Science: Their Contributions and Legacies* (2 Parts), Part 2, pp. 95–110, Heidelberg: Springer, 2009.

Beaumont-Maillet, *L'eau a Paris*, Paris: Hazan, 1991.

Beaven, Alfred B., *The aldermen of the city of London: with notes on the parliamentary representation of the city, the aldermen and the livery companies, the aldermanic veto, aldermanic baronets and knights, etc.* (2 volumes), Vol. 2, London: Eden Fisher & Co, 1913.

Beck, Theodor, "Juanelo Turriano (1500–1585)," in *Beiträge zur Geschichte des Maschinenbaues*, Berlín: Springer, 1899, pp. 365–390.

Bedini, Silvio A. and Francis Maddison, "Mechanical Universe. The Astrarium of Giovanni de' Dondi," *Transactions of the American Philosophical Society, New Series*, vol. lvi, part 5 (October 1966), pp. 1–69.

Bell, Thomas J., *History of the Water Supply of the World*, Cincinnati: P. G. Thomson, 1882.

Belot, Victor R., *Le Pont Neuf: Histoires et Petites Histoires*, Paris: Nouvelles Editions Latines, 1970.

Beltrán Cortés, Fernando, *Apuntes para una Historia Del Frío en España*, Madrid: CSIC-Dpto. de Publicaciones, 1983.

Benedetto, Stefano A., "Macchine idrauliche e attività artigianali a Torino nel XV secolo," in Giuseppe Bracco, (ed.), *Acque, ruote e mulini a Torino* (2 volumes), Vol. 1, Torino: Archivio storico della città di Torino, 1988, pp. 174–194.

Benevolo, Leonardo, *The Arquitecture of the Renaissance*, translated by Judith Landry, (2 volumes), Vol. 2, Boulder (CO, USA): Routledge & Kegan Paul, 1978.

Benito Ruano, Eloy, "Un competidor de Juanelo Turriano (y otro proyecto de '*Artificio*' para Toledo)," *Separata of Studia Histórica et Philológica in honorem M. Batllori*, Roma: Instituto Español de Cultura, 1984, pp. 83–88.

Berry, Geoffrey. C., "Sir Hugh Myddelton and the New River," in Denis Smith, (ed.), *Water Supply and Public Health Engineering*, in Joyce Brown, general editor of the series: *Studies in the History of Civil Engineering* (12 volumes), Vol. 5, Aldershot (UK): Ashgate, 1999, pp. 46–78.

Berry, Henry, "London's Water Supply—The Coming of the Companies," a lecture given on 19 February 1943 to the Metropolitan Water Board, London, 1943, pp. 1–17.

Berthelot, Marcellin, Camille Dreyfus et al, (eds), *La grande encyclopédie: inventaire raisonné des sciences, des lettres et des arts* (31 volumes), Vol. 25, Paris: Lamiraulot, 1886.

Bertolotti, Antonino, *Artisti lombardi a Roma nei secoli XV, XVI, XVII* (2 volumes), Vol. 1, Forni (Italy): U. Hoepli, 1881.

Bertrand, Eugene, *Les Travaux souterrains de Paris* (5 volumes), Vol. 3, Paris: Dunod, 1877.

Berveglieri, Roberto, *Inventori stranieri a Venezia, 1474–1788: importazione di tecnologia e circolazione di tecnici artigiani inventori: repertorio*, Volume 58 of *Memorie, Classe di scienze morali, lettere ed arti*, Venezia; Istituto veneto di scienze, lettere ed arti, 1995.

Besant, Walter, *A History of London*, Bremen: Europäischer Hochschulverlag, 2010 (originally published in London, 1894).

Bezançon, Xavier, *2000 ans d'histoire du partenariat public-privé: Pour la réalisation des équipements et services collectifs*, Paris: Presses de l'École Nationale des Ponts et Chaussées, 2004.

Bezançon, Xavier, *Les services publics en France: du Moyen Age a la Revolution*, Paris: Presses de l'École Nationale des Ponts et Chaussées, 1995.

Biddle, Jeff E., Ross B. Emmett and Warren J. Samuels, (eds), "Further University of Wisconsin Materials: Further Documents of F. Taylor Ostrander," in *Research in the History of Economic Thought and Methodology*, Volume 23-C, pp. 59–78.

Bijker, Wiebe E., Thomas P. Hughes and Trevor Pinch, *The Social Construction of Technological Systems, New Directions in the Sociology and History of Technology*, Cambridge (MA): MIT Press, 1987.

Birembaut. Arthur, "Quelques documents sur Desargues," *Revue d'histoire des sciences et de leurs applications*, Tome 14, No. 3–4. (1961), pp. 193–204.

Blaine, Bradford B., *The Application of Water Power to Industry during the Middle Ages*, Ann Arbor (MI): Univ. Microfilms International, 1984.

Bloch, Marc, "Avènement et conquêtes du moulin à eau," *Annales d'histoire économique et sociale*, T. 7, No. 36, (1935), pp. 538–563.

Bloch, Marc, "Medieval Inventions," in *Land and Work in Medieval Europe*, translated by J. E. Anderson, 1967, (1935), pp. 169–185.

Blomfield, Reginald Theodore, *A history of French architecture from the reign of Charles VIII till the death of Mazarin*, (2 volumes), Vol. 2, London: G. Bell and sons, 1911.

Bloundelle-Burton, John, *The Fate of Henry of Navarre: A True Account of how He was Slain, with a Description of the Paris of the Time and Some of the Leading Personages*, London: Everett, 1911.

Bogucka, María, "The Largest Polish Towns' Water Supply in the XVIth—first Half of the XVIIIth Centuries," in *Le Acque Interne*, Secc. XII–XVIII, Atti della XV Settimana di Studi, 15–20 aprile 1983, Istituto Internazionale di Storia Economica F. Datini, Prato, May 2000, pp. 1–15.

Bolton, Francis, *London Water Supply*, London: W. Clowes & Sons, 1888.

Bond, James, "Canal Construction in the Early Middle Ages: An Introductory Review," in John Blair, *Waterways and Canal-Building in Medieval England*, Oxford: Oxford University Press, 2014, pp. 157–206.

Bonnaffe, Edmond, *Dictionnaire des amateurs français au XVIIe siècle*, Paris: A. Quantin, 1884.

Bonnardot, Alfred, *Études archéologiques sur les anciens plans de Paris des XVIe, XVIIe et XVIIIe siècles*, Paris: Deflorenne, 1851.

Bonnardot, Alfred, *Iconographie du vieux Paris*, Paris: s.n., 1856.

Boon, George C. and Colin Williams, "The Dolaucothi Drainage Wheel," *Journal of Roman Studies*, Volume 56, Issue 1–2 (November 1966).

Botella Llusiá, José and Antonio Fernández Molina, (eds), *Marañón en Toledo: Sobre "Elogio y Nostalgia de Toledo,"* Cuenca: Universidad de Castilla-La Mancha, 1999.

Boucher, Françoise, *Le Pont-Neuf* (2 volumes), Vol. 1, Paris: Chez Le Goupy, 1925.

Boulet, François, *Leçon d'histoire de France: Saint-Germain-en-Laye: des antiquités nationales à une ville international*, Paris: Presses franciliennes, 2006.

Boulton, Jeremy, "London 1540–1700," in Peter Clark, (ed.), *The Cambridge Urban History of Britain*, (3 volumes), Vol. 2, Cambridge, 2000, pp. 315–346.

Bousquet-Bressolier, Catherine, "Matthäus Merian's 1615 Map of Paris: Its Structure, Decoration and Message," *Imago Mundi*, 58: 1 (2006), pp. 48–69.

Bowie, Theodore, *The sketchbook of Villard de Honnecourt*, Bloomington (Indiana): Indiana University Press, 1959.

Boyer-Xambou, Marie-Thérèse, Ghislain Deleplace and Lucien Gillard, *Private Money and Public Currencies. The 16th Century Challenge*, Armond (NY) and London: M. E. Sharpre, 1994.

Brandstetter, Thomas, "The Most Wonderful Piece of Machinery the World Can Boast of: The Water-works at Marly, 1680–1830," *History and Technology*, Vol. 21, No. 2, (2005), pp. 205–220.

Brashear, Ronald, "Ramelli machines: Original drawings of 16th century machines," *Website of the Smithsonian Institution Libraries*, http://www.sil.si.edu/ondisplay/ramelli/intro.htm, accessed 19 May, 2010.

Braudel, Fernand, *Capitalism and Material Life, 1400–1800*, translated by Miriam Kochan, New York: Harper and Row, 1974.

Braudel, Fernand, *Civilization and Capitalism, 15th–18th Century, The Wheels of Commerce* (3 volumes), Vol. 2, Berkeley and Los Angeles: University of California Press, 1982.

Brett-James, Norman G., *The Growth of Stuart London*, London: G. Allen & Unwin, 1935.

Britten, F. J., *Old clocks and watches & their makers, being an historical and descriptive account of the different styles of clocks and watches of the past, in England and abroad, to which is added a list of ten thousand makers*, London: Batsford, 1904.

Broderip, William John and Robert Callis, *The Reading of upon the Statute of Sewers, 23 Hen. VIII. C. 5*, London: J. Butterworth & Son, 1824.

Bromehead, Cyril Edward Nowill, "The Influence of Its Geography on the Growth of London," *The Geographical Journal*, Vol. 60, No. 2 (1922), pp. 125–135.

Bromehead, Cyril Edward Nowill, "The Early History of Water Supply," *The Geographical Journal*, Vol. 99, No. 3, (1942), pp. 142–151.

Bromehead, Cyril Edward Nowill, "The Early History of Water-Supply (Continued)," *The Geographical Journal*, Vol. 99, No. 4 (April, 1942), pp. 183–193.

Brooks, George, "The 'Vitruvian Mill' in Roman and Medieval Europe," in Steven A. Walton, (ed.), *Wind and Water in the Middle Ages, Fluid Technologies from Antiquity to the Renaissance*, Tempe (AZ): Arizona Center for Medieval and Renaissance Studies, 2006, pp. 1–38.

Brown, Cedric C., "Sons of Beer and Sons of Ben: Drink as a Social Marker in Seventeenth-Century England," in Adam Smyth, (ed.), *A Pleasing Sinne: Drink and Conviviality in Seventeenth-Century England*, Cambridge: D. S. Brewer, 2004, pp. 3–20.

Bruun, Christer, "Roman emperors and legislation on public water use in the Roman Empire: clarifications and problems," *Water History* 4 (2012), pp. 11–33.

Bruun, Christer, "Water Supply, Drainage and Watermills," in P. Erdkamp, (ed.), *The Cambridge Companion to the City of Rome*, Cambridge: University Press, 2013, pp. 297–313.

Buckler, William Hepburn, *The Origin and History of Contract in Roman Law Down to the End of the Republican Period: Being the Yorke Prize Essay for the Year 1893*, London: C. J. Clay, 1895.

Buisseret, David, *Henry IV*, London: Taylor & Francis Group, 1989.

Buisseret, David, *Sully and the growth of centralized government in France 1598–1610*, London: Eyre & Spottiswoode, 1968.

Burian, Steven J., Findlay G. Edwards, "Historical Perspectives of Urban Drainage," in E. W. Strecker, W. C. Huber, (eds), *Global Solutions for Urban Drainage, Proceedings of Ninth International Conference on Urban Drainage*, Portland (OR), USA, September 2002, pp. 1–16.

Caballero, Fermín, *Manual geográfico-administrativo de la monarquía española: Obra útil para empleados, hombres de negocios, viajeros y curiosos, con 1.750 artículos en orden alfabético*, Madrid: Imprenta de A. Yenes, 1844.

Calvo Barco, Ángel María "Los neveros, una actividad desaparecida en nuestras montañas," *Zainak, Cuaderno de Antropología y Etnografía*, No. 14 (1997), pp. 203–213.

Cámara Muñoz, Alicia, "La profesión de ingeniero: los ingenieros del rey," in Manuel Silva Suárez, (ed.), *Técnica e Ingeniería en España I: El Renacimiento*, (6 volumes), Vol. 1, Zaragoza: Real Academia de Ingeniería, Institución Fernando El Católico y Prensas Universitarias de Zaragoza, 2004, pp. 125–164.

Cantor, Norman F., *In the Wake of the Plague*, New York: Free Press, 2001.

Carr, Cecil T., "Introduction," in Cecil T. Carr, (ed.), *Select Charters of Trading Companies CE 1530–1707*, London: Bernard Quaritch, 1913, pp. xi–cxxxvi.

Carstens, Patrick Richard and Timothy L. Sanford, *The Republic of Canada Almost*, Bloomington (Indiana): Xlibris Corporation, 2013.

Carus-Wilson, E. M., "An Industrial Revolution of the Thirteenth Century," *The Economic History Review*, Volume 11, No. 1, (1941), pp. 39–60.

Casey, James, *Early Modern Spain: A Social History*, London: Routledge, 1999.

Cast, Andrea Snowden, "Women Drinking in Early Modern England" (2001), unpublished Ph.D. dissertation, Department of History, University of Adelaide.

Castillo Martos, Manuel, "De mano e imprenta: textos científicos y tecnológicos, (siglos XVI y XVII)," in Antonio Acosta Rodríguez, Adolfo Luis González Rodríguez, Enriqueta Vila Vilar, (eds), *La Casa de la Contratación y la navegación entre España y las Indias*, Sevilla: Universidad de Sevilla, Consejo Superior de Investigaciones Científicas, 2003, pp. 567–604.

Cavanilles, Antonio José, *Memoria sobre el fuero de Madrid, del año de 1202*, Madrid: Real Academia de la Historia, 1852.

Cech, Thomas V., *Principles of Water Resources: History, Development, Management, and Policy, New York*: John Wiley & Sons, 2010.

Cerrada, Ana Isabel, José María Macías, Juan Carlos de Miguel and Cristina Segura, "Renacimiento e ilustración, el abastecimiento industrial," in José María Macías and

Cristina Segura Graíño, (eds), *Historia del abastecimiento y usos del agua en la ciudad de Toledo*, Madrid: Confederación Hidrográfica del Tajo, 1999, pp. 93–114.

Cerrada, Ana Isabel, Juan Carlos de Miguel y Cristina Segura, "De la conquista Cristiana a la modernidad," in José María Macías and Cristina Segura Graíño, (eds), *Historia del abastecimiento y usos del agua en la ciudad de Toledo*, Madrid: Confederación Hidrográfica del Tajo, 1999, pp. 70–90.

Cervera Vera, Luis, *Documentos biográficos de Juanelo Turriano*, Madrid: Fundación Juanelo Turriano, 1996.

Chalmers, Alexander, *The General biographical dictionary, containing an historical and critical account of the lives and writings of the most eminent persons in every nation; particularly the British and Irish; from the earliest accounts to the present time*, (32 volumes), Vol. 32, London, J. Nichols and Son, 1812.

Chambers, Robert, *The Book of Days: A Miscellany of Popular Antiquities in Connection with the Calendar, Including Anecdote, Biography, & History, Curiosities of Literature and Oddities of Human Life and Character*, (2 volumes), Vol. 2, Volume London & Edinburgh: W. & R. Chambers Limited, 1832.

Chant, Colin and David C. Goodman, (eds), *Pre-industrial Cities & Technology*, London: Routledge, 1999.

Chant, Colin, (ed.), *The Pre-Industrial Cities and Technology Reader*, New York: Routledge, 1999.

Chapelle, Francis, *Wellsprings: a Natural Story of Spring Bottled Waters*, New Brunswick: (NJ): Rutgers University Press, 2005.

Ciecieznski, N. J., "The Stench of Disease: Public Health and the Environment in the Late-Medieval English towns and cities," *Health Culture Society*, 4, (2013), pp. 91–104.

Cipolla, Carlo María, *Before the Industrial Revolution: European Society and Economy 1000–1700*, New York and London: W. W. Norton & Company, 1994,

Cipolla, Carlo María, "The Diffusion of Innovations in Early Modern Europe," *Comparative Studies in Society and History*, Vol. 14, No. 1 (Jan., 1972), pp. 46–52.

Cipolla, Carlo María, *Máquinas del tiempo y de la guerra*, Barcelona: Crítica, 1999.

Ciriacono, Salvatore, *Building on water: Venice, Holland, and the construction of the European Landscape in the Early Modern Times*, translated by Jeremy Scott, New York: Berghahn Books, 2006.

Ciriacono, Salvatore, "Investimenti capitalistici e cultura irrigue: La congiuntura agricola nella Terraferma veneta (secoli XVI–XVII)," in *Venezia e la Terraferma attraverso le relazioni dei rettori*, Milan: Giuffrè, 1981, pp. 123–158;

Ciriacono, Salvatore, *Land drainage and irrigation*, Aldershot (UK): Ashgate, 1998 and Clive Holmes, "Drainage Projects in Elizabethan England: The European Dimension," in *Eau et développement dans l'Europe moderne*, edited by Salvatore Ciriacono, Paris: Ed. de la Maison des sciences de l'homme, 2004, pp. 87–102.

Clark, Peter, "Water Supply and Drink Consumption Patterns in English Towns in the Early Modern Period," in *Le Acque Interne*, Secc. XII–XVIII, Atti della XV Settimana di Studi, 15–20 aprile 1983, Istituto Internazionale di Storia Economica F. Datini, Prato, May 2000, pp. 1–16.

Classen, Albrecht, "Introduction," in Albrecht Classen, (ed.), *Urban Space in the Middle Ages and the Early Modern Age*, Berlin: Walter de Gruyter, 2009, pp. 1–146.

Clay, Willian Bart, *Remarks on the water supply of London*, London: James Ridway, 1849.

Clemencin, Diego, *Memorias de la Real Academia de la Historia: leido en la Junta Pública que celebró la Real Academia de la Historia el día 31 de julio de 1807*, (14 volumes), Vol. 6, Madrid: Imprenta de I. Sancha, 1821.

Clifford, Frederick, *A History of Private Bill Legislation*, (2 volumes), Vols. 1 and 2, London: Butterworths, 1887.

Cockayne, Emily, *Hubbub: Filth, Noise & Stench in England 1600–1770*, New Haven and London: Yale University Press, 2007.

Congosto, Yolanda, *Aportación a la historia lingüística de las hablas andaluzas (siglo XVII): Descripción de una sincronía*, Sevilla: Universidad de Sevilla, 2002.

Cortés Peña, Antonio Luis and Bernard Vincent, *Historia de Granada: La Edad Moderna. Siglos XVI, XVII y XVIII*, Granada: Don Quijote, 1986.

Cosman, Madeleine Pelner and Linda Gale Jones, *Handbook to Life in the Medieval World*, (3 volumes), Vol. 1, Infobase Publishing, 2009–126.

Coudert, Allison P., "Sewers, Cesspools, and Privies: Waste as Reality and Metaphor in Pre modern European Cities," in Albrecht Classen, (ed.), *East Meets West in the Middle Ages and Early Modern Times: Transcultural Experiences in the Premodern World*, Berlin and Boston: Walter de Gruyter, 2013, pp. 713–733.

Covington, Richard, "The Art and Science of Water," *Saudi Aramco World*, vol. 57, number 3, (May/June 2006), pp. 14–23.

Coxe, Henry (pseud. [i.e. John Millard.]), *Picture of Italy; being a guide to the antiquities and curiosities of that classical and interesting country: containing sketches of manners, society and customs*, London: Sherwood, Neely & Jones, 1815.

Cozzi, Gaetano, Michael Knapton and Giovanni Scarabello, *La Repubblica di Venezia nell'età moderna* (*Storia d'Italia*, ed. Giuseppe Galasso, (12 volumes), Vol. 12–2, Turin: Unione tipografico-editrice torinese, 1992.

Crace, Frederick, *A Catalogue of Maps, Plans, and Views of London, Westminster & Southwark*, London: Spottiswoode & Co., 1878.

Craik, George Lillie, "London Bridge," in Charles Knight, (ed.), *London* (6 volumes), Vol. 1, London: H. G. Bohn, 1851, pp. 73–96.

Crespo Delgado, Daniel, "Un episodio de la historia de la conservación del patrimonio tecnológico de España: La destrucción del artificio de Juanelo en 1868," in *Fundación Juanelo Turriano 1987–2012. 25 años*, Madrid: Fundación Juanelo Turriano, 2012, pp. 57–67.

Cresy, Edward, in *An Encyclopaedia of Civil Engineering, Historical, Theoretical and Practical* (2 volumes), Vol. 1, London: Longman, Brown, Green, and Longmans, 1847.

Cuffel, Alexandra, "Polemicizing Women's Bathing Among Medieval and Early Modern Muslims and Christians," in *The Nature and Function of Water, Baths, Bathing, and Hygiene from Antiquity Through the Renaissance*, Volume 11 of Technology and Change in History, (ongoing series), Leiden and Boston: Brill, 2009, pp. 171–190.

Cunningham, Peter, *Handbook of London, past and present*, London: John Murray, 1850.

Curtis, D. R. and M. Campopiano, "Medieval land reclamation and the creation of new societies: comparing Holland and the Po Valley, c.800-c.1500," *Journal of Historical Geography*, Volume 44, (2014), pp. 93–108.

Cuthbert Hare, Augustus John, *Paris* (2 volumes), Vol. 2, London: George Allen, 1887.

Damler, Daniel, "The Modern Wonder and its Enemies: Courtly Innovations in the Spanish Renaissance," in Claus Zittel, Gisela Engel, Romano Nanni and Nicole C. Karafyllis, (eds), *Philosophies of technology: Francis Bacon and his contemporaries* (2 volumes), Vol. 1, Leiden: Brill, 2008, pp. 429–456.

De Berardino, Santino Eugenio, "The history of sanitation and wastewater management in Portugal, in Andreas N. Angelakis and Joan B. Rose, *Evolution of Sanitation and Wastewater Technologies Through the Centuries*, 1st. Edition, London: IWA, 2014, pp. 209–226.

De Boer, Wietse, "A Comparative Approach to Social Discipline," in John W. O'Malley, Kathleen M. Comerford and Hilmar M. Pabel, (eds), *Early Modern Catholicism: Essays in Honour of John W. O'Malley, S. J.*, Toronto: University of Toronto Press, 2001, pp. 84–96.

De Feo, Giovanni, George Antoniou, Hilal Franz Fardin, Fatma El-Gohary, Xiao Yun Zheng, Ieva Reklaityte, David Butler, Stavros Yannopoulos and Andreas N. Angelakis, "The Historical Development of Sewers Worldwide," *Sustainability*, 6, (2014), pp. 3936–3974.

De Feo, Giovanni, Georgios Pericles Antoniou, Larry Wesley Mays, Walter Dragoni, Hilal Franz Fardin, Fatma El-Gohary, Pietro Laureano, Eleni Ioannis Kanetaki, Xiao Yun Zheng, and Andreas N. Angelakis, "The Historical Development of Wastewater Management," *in* Saeid Eslamian, (ed.), *Handbook of Engineering Hydrology: Environmental Hydrology and Water Management*, Boca Raton (FL): CRC Press, 2014, pp. 163–218.

De Feo, Giovanni, P. Laureano, R. Drusiani and A. N. Angelakis, "Water and wastewater management technologies through the centuries," *Water Science & Technology: Water Supply—WSTWS*, 10.3 (2010), pp. 336–349.

De la Escosura y Morrogh, Luis, *El Artificio de Juanelo y el Puente de Julio César, Memoria publicada por la Real Academia de Ciencias Exactas, Físicas y Naturales de Madrid*, Madrid: Imprenta de Don Luis Aguado, 1888.

De la Pastora y Nieto, Isidoro, *Una verdad histórica relativa al uso del agua por los médicos españoles en el tratamiento de las enfermedades: Discurso que en el acto de recibir la solemne investidura de doctor en Medicina y Cirujia pronunció ante el claustro de la Universidad Central Central de Madrid*, Facultad de Medicina, Madrid: Imprenta de Diaz y Compañia, 1854.

De Miranda, Adriana, *Water Architecture in the Lands of Syria: The Water-Wheels*, Roma: L'erma di Bretschneider, 2007.

De Péréfixe. de Beaumont, Hardouin, *Histoire du roi Henri-le-Grand*, Paris: Antoine-Augustin Renouard, 1816.

De Vries, Jan, *European Urbanization 1500–1800*, Cambridge (MA): Harvard University Press, 1984.

De Waele, Michel, *Les relations entre le parlement de Paris et Henri IV*, Paris: Publisud, 2000.

Degenhart, Bernhard, *Antonio Pisanello*, Wien: Schroll, 1942.

Del Campo y Francés, Ángel, *Juanelo Turriano, semblanza iconográfica*, Madrid: Fundación Juanelo Turriano, 1997.

Del Val Valdivieso, María Isabel and Olatz Villanueva Zubizarreta, (eds), *Musulmanes y Cristianos frente al agua en las ciudades medievales*, Santander (Spain): Universidad de Castilla-La Mancha, 2008.

Del Val Valdivieso, María Isabel, "Un exponente del buen gobierno urbano: el abastecimiento de agua en la Castilla medieval," in María Isabel del Val Valdivieso y Olatz Villanueva Zubizarreta, (eds), *Musulmanes y cristianos frente al agua en las ciudades medievales*, Santander (Spain): Universidad de Castilla-La Mancha, 2008, pp. 359–380.

Descimon, Robert, "L'Echevinage Parisien sous Henri IV (1594–1609). Autonomie urbaine, conflits politiques et exclusives socials," in Neithard Bulst and J.-Ph. Genet, (eds), *La Ville la bourgeoisie et la Genese de L'Etat*, Paris: Éditions du Centre National de la Recherche Scientifique, 1988.

Dickens, Charles, (ed.), "London Water," *All the Year Around*, Vol. 6, London (1862), pp. 137–140.

Dickinson, Henry Winram, *Water supply of Greater London*, London: Leamington Spa, printed for the Newcomen Society, 1954.

Diefendorf, Barbara B., *Paris City Councillors in the Sixteenth Century*, Princeton (NJ): Princeton University Press, 1983.

Díez Borque, José María, "El año del Quijote en España: 1605," *Revista de Literatura, Universidad Complutense de Madrid* (2006), enero-junio, vol. LXVIII, No. 135, pp. 123–150.

Dillehay, Tom D., Herbert H. Eling, Jr. and Jack Rossen, "Preceramic irrigation canals in the Peruvian Andes," *Proceedings of the National Academy of Sciences of the United States of America*, (2005), vol. 102, no. 47, pp. 17241–17244.

Dobson, Mary J., *Contours of Death and Disease in Early Modern England*, Cambridge: Cambridge University Press, 2003.

Dolan, Brian, "Embodied Skills and Travelling Savants: Experimental chemistry in Eighteenth Century Sweden and England," in Ana Simões, Ana Carneiro, María Paula Diogo, (eds), *Travels of learning: a geography of science in Europe*, Dordrecht, The Netherlands: Springer (2003), pp. 115–142.

Dominguez Bordona, Jesús, "Federico Zuccaro en España," *Archivo Español de Arte y Arqueología*, no. 7, 1927, pp. 77–89.

Donners, K., M. Waelkens and J. Deckers, "Water Mills in the Area of Sagalassos: A Disappearing Ancient Technology," *Anatolian Studies*, Vol. 52 (2002), pp. 1–17.

Douglas, Ian, *Cities: An Environmental History*, London: I. B.Tauris, 2013.

Drake, Stillman. "An Agricultural Economist of the Late Renaissance," in Noel M. Swerdlow, Trevor Harvey Levere, (eds), *Essays on Galileo and the history and philosophy of science* (3 volumes), Vol. 3, Toronto: University of Toronto Press, 1999, pp. 170–189.

Droysen, Johann Gustav, *Geschichte des Hellenismus*, (2 volumes), Vol. 1, (originally written in 1836), Cambridge: Cambridge University Press, 2011.

Drusiani, Renato, Giancarlo Leoni, Danilo Demaria and Nicoletta Lemb, "Water supply of Bologna by Roman aqueduct: history, morphology and hydraulic, from ancient time to nowadays," *Water Science and Technology: Water Supply*, No. 10 (4), (2010), pp. 554–560.

Du Camp, Maxime, "Le service des eaux à Paris," *Revue des Deux Mondes*, Tome 105 (1873), pp. 275–308.

Du P., and A. Koenig, "History of water supply in pre-modern China," in Andreas N. Angelakis, Larry W. Mays, Demetris Koutsoyiannis and Nikos Mamassis, *Evolution of water supply throughout the millennia*, London and New York: IWA Publishing, 2012, pp. 169–226.

Du Plessis, Paul J., "The Protection of the Contractor in Public Works Contracts in the Roman Republic and Early Empire," *Journal of Legal History*, Vol. 25, Nr. 3 (2004), pp. 287–314.

Duby, Georges and Robert Mandrou, *A History of French Civilization*, New York: Random House, 1964.

Duff Traill, Henry and James Saumarez Mann, *Social England: a record of the progress of the people in religion, laws, learning, arts, industry, commerce, science, literature and manners, from the earliest times to the present day*, London: Casell and Company, Ltd., 1902.

Dulaure, Jacques Antoine, *Histoire de Paris, (Continuée jusqu' à nos jours par Camille Leynadier)* (5 volumes), Vol. 4, Paris: H. Boisgard, 1857.

Dumas, Alexandre, *Les trois mousquetaires*, Paris: Michel Lévy frères, 1860.

Dyer, Christopher, *Everyday Life in Medieval England*, London: A&C Black, 2001.

Eamon, William and Victor Navarro Brotons, in "Separata," in *Más allá de la Leyenda Negra: España y la Revolución Científica*, Valencia: Instituto de Historia de la Ciencia y Documentación Lopez Piñero, Universitat de València—C.S.L.C., 2007, pp. 8–14.

Elliott, J. H., *Imperial Spain, 1469–1716*, New York: Mentor books, 1966.

Elliott, J. H., *Spain and its World, 1500 - 1700*, New Haven and London: Yale University Press, 1989.

Elmer, Peter, "The Early Modern City," in Colin Chant and David C. Goodman, (eds), *Pre-industrial Cities & Technology*, London: Routledge, 1999, pp. 186–241.

Encyclopædia Britannica, 11th. Edition, Cambridge: Cambridge University Press, 1911.

Epstein, Stephan R, "Property Rights to Technical Knowledge in Premodern Europe, 1300–1800," *The American Economic Review*, Vol. 94, No. 2, Papers and Proceedings of the One Hundred Sixteenth Annual Meeting of the American Economic Association San Diego, CA, January 3–5, 2004 (May, 2004), pp. 382–387.

Epstein, Stephan R., "The Generation and Transmision of Technical Knowledge in Pre-modern Europe, c.1200-c.1800," *Global Economic History Network*, Conference 4, Leiden, The Netherlands, 16–18 September, 2004, pp. 1–39.

Epstein, Stephan R., "Labour mobility, journeyman organisations and markets in skilled labour Europe, 14th–18th centuries," in Mathieu Arnoux et Pierre Monnet, (eds), *Le technicien dans la cité en Europe occidentale, 1250–1650*, pp. 251–269, Rome: École française de Rome, 2004.

Epstein, Stephan R., "Transferring Technical Knowledge and Innovating in Europe, c.1200–1800," part of the series: *Working Papers on The Nature of Evidence: How Well Do 'Facts' Travel?* No. 01, London: Department of Economic History, London School of Economics, 2005, pp. 1–40.

Erlande-Brandenburg, Alain, *The Cathedral Builders of the Middle Ages*, London: Thames and Hudson, 1995.

Ertman, Thomas, *Birth of the Leviathan: Building States and Regimes in Medieval and Early Modern Europe*, Cambridge: Cambridge University Press, 1997.

Escobar, Sergio, "Il controllo delle acque: problemi tecnici e interessi economici," in G. Micheli, (ed.), *Storia d'Italia. Annali, III, Scienza e tecnica nella cultura e nella società dal Rinascimento ad oggi* (12 volumes), Vol. 3, Torino: G. Einaudi, 1980, pp. 83–153.

Espinosa, Aurelio, *The empire of the cities: Emperor Charles V, the comunero revolt, and the transformation of the Spanish system*, Leiden: Brill, 2009.

Euzen, Agathe and Jean-Paul Haghe, "What kind of water is good enough to drink? The evolution of perceptions about drinking water in Paris from modern to contemporary period," *Water History*, July 2012, pp. 1–14.

Evans, Harry B., *Water distribution in ancient Rome: the evidence of Frontinus*, Ann Arbor (MI): University of Michigan Press, 1997.

Evans, William David, Anthony Hammond and Thomas Colpitts Granger, *A Collection of Statutes Connected with the General Administration of the Law: Arranged According to the Order of Subjects, with Notes*, (10 volumes), Vol, 6, London: W. H. Bond, 1836.

Ewbank, Tomas A., *A Descriptive and Historical Account of Hydraulic and Other Machines for Raising Water*, New York: D. Appleton and Company, 1842.

Fagniez, Gustave, *L' économie sociale de la France sous Henri IV, 1589–1610*, Paris: Hachette et Cie., 1897.

Fantoni, Giuliana, *L'acqua a Milano*, Milano: Cappelli, 1990.

Farnsworth Gray, Harold, "Sewerage in Ancient and Medieval Times," *Sewage Works Journal*, Volume 12, No. 5 (1940), pp. 939–946.

Favier, Jean, *Paris au XVe siecle, 1380–1500*, part of the series *Nouvelle histoire de Paris* (11 volumes), Vol. 3, Paris: Association pour la publication d'une histoire de Paris, Diffusion Hachette, 1974.

Feldhaus, Franz Maria, *Die Technik der Antike und des Mittelalters*, Potsdam: Akademische verlagsgesellschaft Athenaion m.b.h, 1931.

Fernández Collado, Ángel, "Juanelo Turriano y la aportación Española a la Reforma del Calendario Gregoriano," *Toletum* 23 (1990), pp. 151–159.

Feros, Antonio, *El Duque de Lerma. Realeza y privanza en la España de Felipe III*, Madrid: Marcial Pons, 2002.

Finch, James Kip, *The Story of Engineering*, Garden City (NY): Doubleday, 1960.

Finley, M. I., "Technical Innovation and Economic Progress in the Ancient World," *The Economic History Review*, New Series, Vol. 18, No. 1, Essays in Economic History Presented to Professor M. M. Postan (1965), pp. 29–45.

Finley-Croswhite, S. Annette, *Henry IV and the Towns: The Pursuit of Legitimacy in French Urban Society, 1589–1610*, Cambridge: Cambridge University Press, 1999.

Fisher Smith, Norman Alfred, "The Origins of Water Power: A Problem of Evidence and Expectations," *Transactions of the Newcomen Society*, No. 55 (1), (1983), pp. 67–84.

Fisher, Frederick Jack, "Tawney's Century," in F. J. Fisher, (ed.), *Essays in the Economic and Social History of Tudor and Stuart England*, Cambridge: Cambridge University Press, 1961, pp. 1–14.

Fischer, Karl H.: "Die Wasserversorgung der Reichsstadt," in Stadtmagistrat Nürnberg (ed.): *Die Wasserversorgung der Stadt Nürnberg von der reichsstädtischen Zeit bis zur Gegenwart. Festschrift zur Eröffnung der Wasserleitung von Ranna*, Nürnberg: Sebald, 1912, pp. 1–130.

Fisher, Frederick Jack, "The Development of London as a Centre for Conspicuous Consumption in the Sixteenth and Seventeenth senturies," in Frederick Jack Fisher, *London and the English Economy 1500–1700*, edited by P. J. Corfield and N. B. Harte, London: Hambledon Press, 1990, pp. 105–118.

Fitchen, John, *The Construction of Cathedrals*, Chicago: University of Chicago Press, 1981.

Fletcher, Joséph, "Historical and Statistical Account of the present System of Supplying the Metropolis with Water," *Journal of the Statistical Society of London*, Vol. 8, No. 2 (June 1845), pp. 148–181.

Foord, Alfred Stanley, *Springs, Streams and Spas of London; History and Associations*, New York: Frederick A. Stokes Company, 1910.

Forbes, Robert James, *Studies in Ancient Technology* (9 volumes), Vol. 2, 3rd. edition, Leyden: Brill, 1993.

Forbes, Robert James, *Studies in Ancient Technology* (9 volumes), Vol. 1, Leiden: E. J. Brill, 1964.

Foucaud, Edward, *The book of illustrious mechanics, of Europe and America*, Hartford (CT): W. J. Hamersley, 1845.

Fournier, Édouard, *Histoire du Pont-Neuf*, Paris: E. Dentu, 1862.

Fox, Celina, (ed.), *London—World City, 1800–1840*, New Haven (CT) and London: Yale University Press, 1992.

Franceschi, Franco, "La grande manifattura tessile," in *La trasmissione dei saperi nel Medioevo (secoli XII–XV)*, Atti del XIX Convegno Internazionale di Studi (Pistoia, 16–19 maggio 2003). Pistoia, 2005, pp. 355–390.

Francis, Alan David, *The Wine Trade*, London: Adam & Charles Black, 1972.

Francis, John Michael, (ed.), *Iberia and the Americas: Culture, Politics, and History: a Multidisciplinary Encyclopedia*, (3 volumes), Vol. 1, Santa Barbara (CA): ABC-CLIO, 2006.

Frankel, Rafael, "Water Mills in Israel," in Jean-Pierre Brun and Jean-Luc Fiches, (eds), *Énergie Hydraulique et Machines Élévatrices D'eau Dans L'antiquité*, Napoli: Collection du Centre Jean Bérard, 2007, pp. 215–224.

Franklin, Alfred, *Les anciens plans de Paris, notices historiques et topographiques*, (2 volumes), Vol. 2, Paris: Willem, 1878.

Friedrichs, Christopher R., "Urbanization in Early Modern Europe: Change or Continuity?" in Colin Chant, (ed.), *The Pre-Industrial Cities and Technology Reader*, New York: Routledge, 1999, pp. 142–144.

Fulgence Girard, "La pompe du pont Notre-Dame," *Le Monde illustré*, No. 10, 20/06/1857.

Galán García, José Luis, *Sistemas de unidades físicas*, Barcelona: Reverté, 1987.

Galera Mendoza, María Esther, "La casa y el agua. El ejemplo granadino," *Cuadernos de arte de la Universidad de Granada*, No. 37, 2006, pp. 9–28.

Galignani, A. and W., *The History of Paris from the Earliest Period to the Present Day: Containing a Description of Its Antiquities, Public Buildings, Civil, Religious, Scientific, and Commercial Institutions*, (3 volumes), Vol. 3, Paris: A. and W. Galignani, 1832.

Galignani, A. and W., *The History of Paris: From the Earliest Period to the Present Day; Containing a Description of Its Antiquities, Public Buildings, Civil, Religious, Scientific and Commercial Institutions ... To which is Added, an Appendix: Containing a Notice of the Church of Saint Denis; an Account of the Violation of the Royal Tombs; Important Statistical Tables*, Paris: A. and W. Galignani, 1825.

Galluzzi, Paolo, "An interview with Paolo Galluzzi," towards the London inauguration of the Istituto e Museo di Storia della Scienza, Florence, Italy travelling exhibition *The Art of Invention: Leonardo and Renaissance Engineers*, Science Museum, London, England (October 15, 1999 to April 24, 2000). Website of the IMPA—Instituto Nacional de Matematica Pura e Aplicada (Brazil): http://w3.impa.br/~jair/einter1.html, accessed 18 June, 2012.

Galluzzi, Paolo, "Art and Artifice in the Depiction of Renaissance Machines," in Wolfgang Lefèvre, Jiirgen Renn and Urs Schoepflin, (eds), *The Power of Images in Early Modern Science*, Basel, Boston and Berlin: Birkhäuser, 2003, 1989, pp. 47–64.

Galluzzi, Paolo, *Gli ingegneri del Rinascimento da Brunelleschi a Leonardo da Vinci*, Firenze: Giunti, 1996.

Galluzzi, Paolo, "Portraits of machines in Fifteenth century Siena," in Renato G. Mazzolini, (ed.), *Non-verbal communication in science prior to 1900*, Firenze: L. S. Olschki, 1993, pp. 53–90.

Galluzzi, Paolo, (ed.), *Prima di Leonardo: cultura delle macchine a Siena nel Rinascimento*, Milano: Electa, 1991.

Galluzzi, Paolo, (ed.), *Renaissance engineers from Brunelleschi to Leonardo da Vinci*, Firenze: Giunti Editore, 1996.

Gamero, Antonio Martín, *Historia de la Ciudad de Toledo: Sus claros varones y monumentos*, Toledo: Severiano López Fando, 1862.

Gandy, M., "The Paris Sewers and the Rationalization of Urban Space," *Transactions of the Institute of British Geographers*, 24, (1999), pp. 23–44.

García Martín, Francisco, *La Comisión de Monumentos de Toledo (1836–1875)*, Toledo: Ledoria, 2008, pp. 230–237.

García Mercadal, José, *España vista por los extranjeros* (2 volumes), Vol. 2, Madrid: Biblioteca Nueva, 1920.

García Olmedo, Bernardo, *En torno a la Universidad española. Crónica no autorizada de un conflicto, Appendix "C,"* pp. 174–175. Website of Yumpu Digital Magazine, https://www.yumpu.com/es/document/view/37126013/bernardo-garcia-olmedo-en-torno-a-la-universidad-espanola, accessed 2.7.2016.

García Ruiperez, Maríano, "El Archivo Municipal de Toledo y la investigacion histórica sobre la Edad Moderna," *Cuaderno de historia Moderna*, No. 22 (1999), pp. 201–227.

García Tapia, Nicolás, "Ciencia y técnica en la España de los Austrias. Una visión desde la perspective de las investigaciones actuales," *Cuadernos Historia Moderna*, No. 15, pp. 199–209, Madrid: Editorial Complutense, 1994.

García Tapia, Nicolás, "Descubrimientos e invención técnica. La actividad de los inventores españoles," in Ana María Carabias Torres, (ed.), *Las relaciones entre Portugal y Castilla en la época de los descubrimientos y la expansión colonial* (Ponencias presentadas al Congreso Hispano-Portugués celebrado en Salamanca, 1992), Salamanca, 1994, pp. 165–180.

García Tapia, Nicolás, "El ingenio de Zubiaurre en el Pisuerga," in *Conocer Valladolid 2011/2012. V Curso de patrimonio cultural*, Valladolid: Real Academia de Bellas Artes de la Purísima Concepción, 2012, pp. 77–98, pp. 94–98.

García Tapia, Nicolás, "El ingenio de Zubiaurre para elevar el agua del rio Pisuerga a la huerta y palacio del Duque de Lerma," *Boletín del Seminario de Estudios de Arte y Arqueología*, Tomo 50, 1984, pp. 299–324.

García Tapia, Nicolás, "Nuevos datos técnicos sobre los *artificios* de Juanelo," *Anales Toledanos* XXIV, Diputación Provincial, Toledo (1987), pp. 141–159.

García Tapia, Nicolás, "Pedro Juan de Lastanosa y Pseudo Juanelo Turriano," *Llull Magazine*, Vol. 10, (1987), pp. 51–74.

García Tapia, Nicolás, *Ingeniería y arquitectura en el renacimiento español*, Valladolid: Secretariado de Publicaciones, Universidad de Valladolid, 1989.

García Tapia, Nicolás, *Patentes de Invención Españolas en El Siglo de Oro*, Madrid: Oficina Española de Patentes, 1994.

García Tapia, Nicolás, *Técnica y poder en Castilla durante los siglos XVI y XVII*, Salamanca: Junta de Castilla y Leon, Consejeria de Cultura y Bienestar Social, 1989.

García-Diego, José Antonio and Alexander G Keller, (eds), *Giovanni Francesco Sitoni, ingeniero renacentista al servicio de la Corona de España: con su códice inédito "Trattato delle virtù et proprietà delléacque," en su idioma original y traducido al castellano* (originally published in 1599 in Milan: [s.n.]), Madrid: Fundación Januelo Turriano, Ed. Castalia, 1990.

García-Diego, José Antonio, "Cinco documentos relativos a Juanelo Turriano," *Toletum, Boletín de la Real Academia de Bellas Artes y Ciencias Históricas de Toledo*, No. 8, (1977), pp. 245–284.

García-Diego, José Antonio, "El manuscrito atribuido a Juanelo Turriano de la Biblioteca Nacional de Madrid," in *Actas del I Congreso de Historia de las Ciencias* (*Separata*), Madrid: Sociedad Española de Historia de las Ciencias, 1981, pp. 449–461.

García-Diego, José Antonio, "Giovanni Francesco Sitoni, an hydraulic engineer of the Renaissance," in Salvatore Ciriacono, "Irrigation and Drainage," in the series *Studies in the history of civil engineering*, edited by Joyce Brown (12 volumes), Vol. 3, Aldershot (UK): Ashgate, 1998, pp. 59–82.

García-Diego, José Antonio, "The Chapter on Weirs in the Codex of Juanelo Turriano: A Question of Authorship," *Technology and Culture*, Vol. 17, No. 2 (Apr., 1976), pp. 217–234.

García-Diego, José Antonio, "Una muerte y un artificio," *Anales Toledanos IX*, Toledo: Diputación Provincial, Instituto Provincial de Investigaciones y Estudios Toledanos (1974), pp. 289–304.

García-Diego, José Antonio, *Juanelo Turriano, Charles V's Clockmaker. The man and his legend*, translated into English by Charles David Ley, Madrid: Editorial Castalia, 1986.

Garnett, William, *A Little Book on Water Supply*, Cambridge: Cambridge University Press, 1922.

Garrison, Ervan G., *A History of Engineering and Technology: Artful Methods*, Boca Raton (FL): CRC Press, 2nd edition, 1999.

Gavroglu, Kostas, Manolis Patiniotis, Faidra Papanelopoulou, Ana Simões, Ana Carneiro, Maria Paula Diogo, José Ramón Bertomeu Sánchez, Antonio García Belmar and Agustí Nieto-Galan, "Science and Technology in the European Periphery: Some Historiographical Reflections," *History of Science*, Vol. 46, 2008, p. 153–175.

Geddes, Jane, "Iron," in *English Medieval Industries: Craftsmen, Techniques, Products*, in John Blair, W. John Blair and Nigel Ramsay, (eds), London and Rio Grande: Hambledon Press, 1991, pp. 167–188.

Getzler, Joshua, *A History of Water Rights at Common Law*, Oxford: Oxford University Press, 2004.

Gibbs, Jack, "Federico Zuccaro y el artificio de Juanelo en 1586," *Anales Toledanos VIII*, 1973, pp. 49–51.

Gies, Frances and Joseph, *Cathedral, Forge and Waterwheel, Technology and Invention in the Middle Ages*, New York: HarperCollins, 1995.

Gille, Bertrand, "The Fifteenth and Sixteenth Centuries in the Western World" in Maurice Daumas, (ed.), *A History of Technology and Invention*, translated from the French edition by Eileen B. Hennessy (2 volumes), Vol. 2, New York: Crown, 1969.

Gille, Bertrand, *Histoire des Techniques*, translated by P. Southgate and T. Williamson, (2 volumes), Vol. 1, New York: Gordon and Breach, 1986.

Gille, Bertrand, *The Renaissance engineers*, London: Lund, Humphries and Company, 1966.

Gille, Paul, "Construction and Building," in Maurice Daumas, (ed.), *A History of Technology & Invention* (3 volumes), Vol. 2, translated by Eileen B. Hennessy, New York: Crown, 1969, pp. 493–585.

Gimpel, Jean, *The Medieval Machine*, New York: Holt, Rinehart and Winston, 1976.

Girard, Pierre-Simon, *Mémoire pour servir d'introduction au devis général ... pour la distribution des eaux du canal de l'ourcq dans l'intérieur de Paris*, Paris: Imprimerie Impériale, 1812.

Girard, Pierre-Simon, *Recherches sur les eaux publiques de Paris, les distributions successives qui ont été proposés pour en augmenter le volume*, Paris: Imprimerie Impériale, 1812.

Gladstone, Adam, *History of the London Water Supply, from the creation of Man to CE 1884*, London: [s.n.], 1884.

Gleick, Peter H., "Basic Water Requirements for Human Activities: Meeting Basic Needs," *Water International*, Vol. 21 (1996), No. 2, pp. 83–92.

Glick, Thomas F. and Helena Kirchner, "Hydraulic Systems and Technologies in Islamic Spain: History and Archaeology," in Paolo Squatriti, (ed.), *Working with Water in*

Medieval Europe, Technology and resource-use, Leiden, Boston and Koln: Brill, 2000, pp. 267–330.

Glick, Thomas F., *Islamic and Christian Spain in the early Middle Ages*, Princeton: Princeton University Press, 1979.

Glick, Thomas, "Technology in the Middle Ages," in Albrecht Classen, (ed.), *Handbook of Medieval Studies: Terms—Methods—Trends*, 3 volumes, vol. 2, Berlin and New York: De Gruyter, 2010, pp. 1305–1310.

Goicochea Zala, Javier, "Azudes, molinos y otros aspectos de los Veintiún Libros de los Ingenios y Máquinas," *Revista de Obras Públicas*, No. 3396 (March 2000), pp. 59–68.

González Conde, Lucas, Miguel Bermejo Herrero, María Gloria del Rio Cidoncha and Juan Martínez Palacios: "Reconstrucción Virtual del Artificio de Juanelo Turriano para Elevar Agua del Río Tajo a Toledo. De la Tradición al Futuro," *Actas del Congreso Internacional Conjunto XVII INGEGRAF—Congreso Internacional de Ingeniería Gráfica*, Sevilla, España, 2005, pp. 1–9.

González Conde, Lucas, *Reconstrucción Virtual del Artificio de Juanelo Turriano para elevar agua del Rio Tajo a Toledo*, proyecto fin de carrera, Escuela Superior de Ingenieros, Universidad de Sevilla, 2005.

González Tascón, Ignacio, *Fábricas Hidráulicas Españolas*, Madrid: Ministerio de Obras Públicas y Urbanismo, Centro de Publicaciones, 1987.

González Tascón, Ignacio, "La Difusión Medieval del Molino Hidráulico," in Pedro Navascués Palacio e Ignacio González Tascón, (eds.), *Ars mechanicae: ingeniería medieval en España*, Madrid: Fundación Juanelo Turriano, Ministerio de Foment, (2008), pp. 99–130.

Gonzalvez, Ramón, "El clima Toledano en los siglos XVI y XVII," *Boletín de la Real Academia de la Historia*, Tomo CLXXIV, No. II (1977), pp. 305–333.

Goode, William O., "Moving West: Three French Queens and the Urban History of Paris," *The French Review*, Vol. 73, No. 6 (May 2000), pp. 1116–1129.

Goodman, David C., *Power and Penury: Government, Technology and Science in Philip II's Spain*, Cambridge: Cambridge University Press, 2002.

Gorton, John, *A general biographical dictionary* (4 volumes), Vol. 3, London: Henry G. Bohn, 1851.

Gossett, Suzanne, (ed.), *Thomas Middleton in Context*, Cambridge: Cambridge University Press, 2011.

Goubert, Jean-Pierre, *The Conquest of Water*, translated by Andrew Wilson, Cambridge: Polity Press, 1989.

Gould, W. T. S., *Population and Development*, New York: Routledge, 2013.

Graber, Frédéric, "Inventing needs: expertise and water supply in late eighteenth and early nineteenth-century Paris," *The British Journal for the History of Science*, Vol. 40, Issue 3, Cambridge (2007), pp 315–332.

Gracia Rivas, Manuel, "En el IV Centenario del fallecimiento de Pedro Zubiaur, un marino vasco del siglo XVI," *Itsas Memoria. Revista de Estudios Marítimos del País Vasco*, 5, Untzi Museoa-Museo Naval, Donostia-San Sebastián, (2006), pp. 157–171.

Green, John Richard, *A Short History of the English People, Illustrated Edition*, edited by Mrs. J. R. Green and Mrs. Kate Norgate (4 volumes), Vol. 3, London: Macmillan and Co., 1903.

Green, Peter, *Hellenistic History and Culture*, Los Angeles and London: University of California Press, 1993.

Greene, J. Patrick, *Medieval Monasteries*, London and New York: Continuum, 2005.

Greene, Kevin, "Perspectives in Roman Archeology," *Oxford Journal of Archaeology*, Volume 9, Issue 2 (1990), pp. 209–219.

Greene, Kevin, "Technological Innovation and Economic Progress in the Ancient World: M. I. Finley Re-Considered," *The Economic History Review*, New Series, Vol. 53, No. 1 (2000), pp. 29–59.

Greene, Kevin, *The Archaeology of the Roman Economy*, Berkeley and Los Angeles: University of California Press, 1990.

Greengrass, Mark, *France in the Age of Henri IV, the struggle for stability*, London: Longman, 1984.

Grewe, Klaus, "La máquina romana de serrar piedras. La representación en bajorrelieve de una sierra de piedras de la antigüedad, en Hierápolis de Frigia y su relevancia para la historia técnica," translated by Miguel Ordóñez, in Isaac Moreno Gallo, (ed.), *Las técnicas y las construcciones de la Ingeniería Romana*, Madrid: V Congreso de las Obras Públicas Romanas, 2010, pp. 381–401.

Grenet-Delisle, Claude, *Louis de Foix: horloger, ingénieur, architecte de quatre rois*, Bordeaux: Fédération Historique du Sud-Ouest, 1998.

Guess, Alice C., "The machines of Francesco Di Giorgio: demonstrations of the world," a Master's thesis submitted to the McGill university (Montreal), 1998.

Guiffrey, V. J., "Liste alphabétique des artistes et artisans employés à l'embellissement et à l'entretien des châteaux royaux du Louvre, des Tuileries, de Fontainebleau, de Saint-Germain etc., de 1605 à 1656, avec la mention de leurs gages," in *Nouvelles Archives de l'art français*, Paris: F. de Nobele, 1872, pp. 1–54.

Hale, J. R., "Girolamo Maggi: a Renaissance scholar and military buff," *Italian Studies*, 40 (1985), pp. 31–50.

Haliwell, James, (ed.), *Books of Characters, illustrating the Habits and Manners of Englishmen from the Reign of James I. to the Restoration*, London: J. E. Adlard, 1857.

Hanotaux, Gabriel, *La France en 1614; La France et la Royauté avant Richelieu*, Londres, Edimbourg et New-York: Nelson, 1913.

Hanson, David J., *Preventing Alcohol Abuse: Alcohol, Culture, and Control*, Westport (CT): Greenwood Publishing Group, 1995.

Harben, Henry Andrade, "Water Supply of London," in *A Dictionary of London*, London: H. Jenkins, 1918.

Harding, Vanessa, *The Dead and the Living in Paris and London, 1500–1670*, Cambridge: Cambridge University Press, 2002.

Hardy, Anne, "Water and the search for Public Health in London in the Eighteen and Nineteenth Centuries," *Medical History*, Vol. 28 (1984), 250–282.

Harkness, Deborah F., *The Jewel House, Elizabethan London and the Scientific Revolution*, New Haven and London: Yale University Press, 2007.

Harley, John Brian, "Deconstructing the Map," *Cartographica: The International Journal for Geographic Information and Geovisualization*, Vol. 26, No. 2, Summer 1989, pp. 1–20.

Harris, A. Katie, From *Muslim to Christian Granada: Inventing a City's Past in Early Modern Spain*, Baltimore: Johns Hopkins University Press, 2007.

Harris, Lawrence Ernest, *The Two Netherlanders: Humphrey Bradley and Cornelius Drebbel*, Leiden: E. J. Brill, 1961.

Harris, John Raymond, "Law, Espionage and Transfer of Technology from Eighteen Century Britain," in Robert Fox, (ed.), *Technological Change: Methods and Themes in the History of Technology*, Amsterdam: Routledge, 1996, pp. 123–136.

Hartmeier, Winfried., and Monika Reiss, "Production of beer and wine," in Karl Esser, Joan W. Bennett, Paul A. Lemke and H. D. Osiewacz, (eds), *Industrial Applications*, Berlin, Heidelberg, New York: Springer, 2002, pp. 49–66.

Hatcher, John, *The history of the British coal industry* (5 volumes), Vol. 2, Oxford: Clarendon Press, 1993.

Hayes, John Richard, *The Genius of Arab Civilization: Source of Renaissance*, Cambridge (MA): Taylor & Francis, 1983.

Heller, Henry, *Labour, Science and Technology in France, 1500–1620*, Cambridge: Cambridge University Press, 2002.

Higham, Charles, *The Civilization of Angkor*, Berkeley and Los Angeles: University of California Press, 2004.

Hill, Donald, *A History of Engineering in Classical and Medieval Times*, London: Routledge, 1984.

Hind, Arthur Mayger, *Wenceslaus Hollar and his views of London and Windsor in the seventeenth century*, London: John Lane, 1922.

Hirt, Alfred Michael, *Imperial Mines and Quarries in the Roman World: Organizational Aspects 27 BCE–CE 235*, Oxford: Oxford University Press, 2010.

Hoffmann, Albrecht, "Zum Stand der Städtischen Wasserversorgung in Mitteleuropa vor dem Dreissigjährigen Krieg", in Frontinus-Gessellschaft e.V. (eds), *Die Wasserversorgung in der Renaissancezeit, Vol 5 of Geschichte der Wasserversorgung*, (5 Volumes), Mainz am Rhein, 2000, pp. 101–144.

Holmes, Clive, "Drainage Projects in Elizabethan England: The European Dimension," in Salvatore Ciriacono, (ed.), *Eau et développement dans l'Europe moderne*, Paris: Ed. de la Maison des sciences de l'homme, 2004, pp. 87–102.

Holt, Mack P., "Wine, Community and Reformation in Sixteenth Century Burgundy," *Past and Present*, (1993), No. 38, pp. 58–93.

Holt, Richard and Gervase Rosser, "Introduction: The English Town in the Middle Ages," in Richard Holt and Greased Rosser, (eds), in *The Medieval Town in England 1200–1540*, London and New York: Routledge, 2014, pp. 1–18.

Holt, Richard, "Medieval England's Water-Related Technologies," in Paolo Squatriti, (ed.), *Working with Water in Medieval Europe: Technology and Resource-use*, Leiden: Brill, 2000, pp. 51–100.

Home, Gordon, *Old London Bridge*, London: John Lane, 1931.

Hone, William, *The Table Book* (2 volumes), Vol. 1, London: Hunt and Clarke, 1827.

Hornsey, Ian S., *A History of Beer and Brewing*, Cambridge: Royal Society of Chemistry, 2003.

Horrox, Rosemay, *The Black Death*, Manchester: Manchester University Press, 1994.

Hounsell, Peter, *London's Rubbish*, Electronic edition, chapter I, Stroud, Gloucestershire: Amberley Publishing Limited, 2014.

Hughson, David, *London; Being an Accurate History and Description of the British Metropolis and Its Neighborhood: To Thirty Miles Extent, from an Actual Perambulation*, (6 volumes), Vol. 6, London: W. Stratford, 1809.

Hulme, E. Wyndham, "History of the Patent System Under the Prerogative and at Common Law" *Law quarterly review*; volume XII (1896), pp. 141–154.

Hulme, E. Wyndham, "History of the Patent System Under the Prerogative and at Common Law—a Sequel" *Law quarterly review*, Vol. XVI (1900), pp. 44–56.

Humphrey, John W., Ancient Technology, Westport (CN) and London: Greenwood Press, 2006.

Hunt Yungblut, Laura, *Strangers settled here amongst us: policies, perceptions, and the presence of aliens in Elizabethan England*, London: Routledge, 1996.

Hussey, Andrew, *Paris, the Secret History*, New York: Bloomsbury, 2007.

Ibbetson, David J., *A Historical Introduction to the Law of Obligations*, Oxford: Oxford University Press, 1999.

Jackson, Catherine Hannah Charlotte Elliot, *Old Paris: its Court and Literary Salons*, (2 volumes), Vol. 1, Boston: Joséph Knight Company, 1895.

Jacob, Margaret C., *The Cultural Meaning of the Scientific Revolution*, Philadelphia: Temple University Press, 1988.

Jenkins, Rhys (in collaboration with E. Wyndham Hulme), "Notes on the London Bridge Waterworks," in *Links in the History of Engineering and Technology from Tudor Time*, London: Ayer Publishing, 1971, pp. 131–140.

Jenkins, Rhys, "A Chapter in the History of the Water Supply of London: A Thames-side Pumping Installation and Sir Edward Ford's Patent from Cromwell," *Transactions of the Newcomen Society*, Vol. 9, pp. 43–51.

Jenkins, Rhys, "Bevis Bulmer," in *Links in the History of Engineering and Technology from Tudor Time*, Freeport (NY): Books for Libraries Press, 1971, pp. 24–27.

Jenkins, Rhys, "George Sorocold: A Chapter in the History of Public Water Supply," *The Engineer*, Vol. 126 (1918), pp. 333–4.

Jenkins, Rhys, "Hollar's Map II," in T. Fairman Ordish, (ed.), *London Topographical Record*, illustrated (1901) (5 volumes), Vol. 2, London, 1903, pp. 110–111.

Jenner, Mark S. R., "'Another epocha'?: Hartlib, John Lanyon and the improvement of London in the 1650s," in Mark Greengrass, Michael Leslie and Timothy Raylor, (eds), *Samuel Hartlib and Universal Reformation: Studies in Intellectual Communication*, Cambridge: Cambridge University Press, 1994, pp. 343–356.

Jenner, Mark S. R., "From Conduit Community to Commercial Network?" in Mark S. R. Jenner and Paul Griffiths, (eds), *Londinopolis: Essays in the Cultural and Social History of Early Modern London*, Manchester: Manchester University Press, 2000, pp. 250–265.

Jenner, Mark S. R., "L'eau changée en argent? La vente de l'eau dans les villes anglaises au temps de l'eau rare," *XVVe siècle* (2003/4), No. 221, pp. 637–651.

John Calvert, *The gold rocks of Great Britain and Ireland*, London: Chapman and Hall, 1853.

Jones, Colin, *Paris, Biography of a City*, New York: Penguin Books, 2006.

Jones, William M., "Two Learned Italians in Elizabethan England," *Italica*, Vol. 32, No. 4 (Dec. 1955), pp. 242–247.

Jorgensen, D., "Cooperative sanitation: Managing streets and gutters in the late medieval England and Scandinavia," *Technology and Culture*, 49, (2008), pp. 547–567.

Jorgensen, D., "Local government response to urban river pollution in late medieval England," *Water History*, 2, (2010), pp. 35–52.

Juan García, Natalia, "Prácticas alimentarias en los siglos XVII y XVIII en el clero regular de Aragón—los manjares de la comunidad de monjes de San Juan de la Peña," *Studium: Revista de Humanidades*, No. 15 (2009), pp. 165–198.

Jubert, Gérard, *Ordonnances Enregistrees au Parlement de Paris Sous le Règne de Henri IV. Inventaire analitique des registres X1a 8640 a 8646*, Paris: Archives Nationales, 1993.

Jufre García, Françesc Xavier, "Nuevo Modelo del Artificio de Juanelo," presentation to The II Congreso Internacional de Patrimonio e Historia de la *Ingeniería*, June 2007, Las Palmas de Gran Canaria.

Jufre García, Françesc Xavier, *El artificio de Juanelo Turriano para elevar agua al Alcázar de Toledo (s. XVI). Modelo con escaleras de Valturio*, Lleida (Spain): Editorial Milenio, 2008.

Kagan, Richard L., "Contando vecinos. El censo toledano de 1569," *Studia Histórica. Historia moderna*, No. 12, 1994, pp. 115–136.

Karmon, David, "Restoring the Ancient Water Supply System in Renaissance Rome: The Popes, the Civic Administration, and the Acqua Vergine," *The Waters of Rome*, No. 3, (2005), pp. 1–13.

Kaylor, Noel Harold and Philip Edward Phillips, *A Companion to Boethius in the Middle Ages*, Leiden: Brill, 2012.

Keene, Derek, "Issues of water in medieval London to c.1300," *Urban History*, 28 (2001), pp. 161–179.

Kellenbenz, Hermann, "Technology in the Age of the Scientific Revolution, 1500–1700," in Carlo M. Cipolla, (ed.), *The Fontana Economic History of Europe*, London, 1974, pp. 177–272.

Keller, Alexander G., "Renaissance Waterworks and Hydromechanics," *Endeavour*, Vol. XXV, (1966), pp. 141–145.

Keller, Alexander G., "The Missing Years of Jacques Besson, Inventor of Machines, Teacher of Mathematics, Distiller of Oils, and Huguenot Pastor," *Technology and Culture*, Vol. 14, No. 1 (Jan. 1973), pp. 28–39.

Keller, Alexander G., "Spain and Italy: interchange reflected in two late Renaissance treatises on hydrotechnics," in Alessandra Fiocca, Daniela Lamberini, Cesare Maffioli, *Arte e Scienza delle Acque nel Rinascimento*, Venezia: Marsilio Editori, 2003, pp. 95–103.

Kip Finch, James, *The Story of Engineering*, Garden City (NY): Doubleday, 1960.

Kirby, Richard Shelton et al, *Engineering in History*, New York, Toronto, London: McGraw-Hill Book Company, 1956.

Klemm, Friedrich, *A History of Western Technology*, translated by Dorothea Waley Singer, Cambridge (MA): The MIT Press, 1991.

Knittl, Margaret Albright, "The design for the initial drainage of the Great Level of the Fens: an historical whodunit in three parts," *Agricultural History Review* 55 (2007), pp. 23–50.

Knoespel, Kenneth J., "Gazing on Technology: Theatrum Mechanorum and the Assimilation of Renaissance Machinery," in Mark L. Greenberg and Lance Schachterle, (eds), *Literature and Technology*, London and Toronto: Lehigh University Press, 1992, pp. 99–124.

Kollmann, Franz Joséph and F. A. Oldenburg, *Die Wasserwerke von Augsburg*, Augsburg: Rieger, 1850.

Kucher, Michael P., *The Water Supply System of Siena, Italy: The Medieval Roots of the Modern Networked City*, New York: Routledge, 2005.

Laborde, Leon de, *Le Parlement de Paris, sa compétence et les ressources que l'érudition trouvera dans l'inventaire de ses archives. Préface de l'Inventaire des actes du Parlement de Paris*, Paris: H. Plon, 1863.

Lacour, François, "La pompe du pont Notre-Dame," *Le Monde illustré*, No. 48, 13/3/1858.

Lacunza, J. M., "Pedro Zubiaur y su relación con Rentería," *Revista Oarso*, Ayuntamiento de Errenteria, No. 49, (2004), pp. 89–91.

Laffont, Robert, *Histoire de Paris et des Parisiens*, Paris: Editions du Pont Royal, 1957.

Lafont Mateo, German, *Pampliega, Torrepadierne y Santiuste. Mil años de Historia. Siglos VII al XVII*, Salamanca: German Lafont Mateo, 2010.

Lambert, Bernard, *The History and Survey of London and its Environs* (4 volumes), Vol. 2, London: T. Hughes, 1806.

Lanaro, Paola e Gian Maria Varanini, "I ponti 'costruiti': riflessioni sull'esempio veronese," in D. Calabi e Cl. Conforti, (eds), *I ponti delle capitali d'Europa dal Corno d'oro alla Senna*, Milano: Electa, 2002, pp. 58–74.

Landels, John Gray, *Engineering in the Ancient World*, Berkeley and Los Angeles: University of California Press, 1978.

Langdon, John, *Mills in the Medieval Economy: England 1300–1540*, Oxford: Oxford University Press, 2004, extracted from *eBook Collection* (*EBSCOhost*), accessed 7 June 2016.

Lavedan, Pierre, *Histoire de l'urbanisme a Paris*, Paris: Hachette, 1975.

Law, John and Vicky Singleton, "Performing Technology's Stories: On Social Constructivism, Performance, and Performativity," *Technology and Culture*, Vol. 41, No. 4 (2000), pp. 765–775.

Lawton, Bryan, *Various and Ingenious Machines, Power Generation and Transport* (2 volumes), Vol. 1, Leiden and Boston: Brill, 2004.

Lazardzig, Jan, "The Machine as Spectacle: Function and Admiration in Seventeenth Century Perspectives on Machines," in Helmar Schramm, Ludger Schwarte and Jan Lazardzig, (eds), *Instruments in Art and Science. On the Architectonics of Cultural Boundaries*, Berlin and New York: Walter de Gruyter, 2008, pp. 153–175.

Lazare, Félix and Louis Lazare, *Dictionnaire historique des rues et monuments de Paris* [*en*] *1855: avec les plans des 48 quartiers*, Paris: Servedit, 1855.

Le Coff, Jaques, *Time, Work, and Culture in the Middle-Ages*, translated by Arthur Coldhammer, Chicago and London: University of Chicago Press, 1980.

Le Roi, Joséph-Adrien, *Curiosités historiques sur Louis XIII, Louis XIV, Louis XV, Mme de Maintenon, Mme de Pompadour, Mme du Barry, etc.*, Paris: Henry Plon, 1864.

Le Roy Ladurie, Emmanuel, *The Peasants of Languedoc*, translated by John Day, Chicago: University of Illinois Press, 1976.

Lefèvre, Wolfgang, "Introduction," in Wolfgang Lefèvre, (ed.), *Picturing Machines 1400–1700* (*Transformations: Studies in the History of Science and Technology*), Cambridge (MA): The MIT Press, pp. 1–12.

Legrand, J. G. and Charles Paul Landon, *Description de Paris et de ses édifices*, (2 volumes), Vol. 1, Paris: Treuttel, 1818.

Lemeunier, Guy, "Drenaje y crecimiento agrícola en la España mediterránea (1500–1800)," *Areas: revista de Ciencias Sociales*, Vol. 17 (1998), pp. 31–41.

Lethaby, W. R., "Hollar's Map I," in T. Fairman Ordish, (ed.), *London Topographical Record*, illustrated, (5 volumes), Vol. 2, London, 1903, pp. 110–111.

Lewis, Michael Jonathan Taunton, *Millstone and Hammer: The Origins of Water Power*, Hull (UK): Hull University Press, 1997.

Lindberg, David C., "The Transmission of Greek and Arabic Learning to the West," in David C. Lindberg, (ed.), *Science in the Middle Ages*, Chicago: Chicago University Press, pp. 52–90.

Llaguno y Amirola, Eugenio and Juan Agustín Ceán Bermúdez, *Noticias de los arquitectos y arquitectura de España desde su restauración*, (4 volumes), Vol. 2, Madrid: Imprenta Real, 1829.

Lloyd, Herbert Alan, *Giovanni de' Dondi's horological masterpiece, 1364*, Oxted (Surrey): [s.n.], 1956.

Lo Cascio, Elio and Paolo Malanim, "Mechanical Energy and Water Power in Europe, A Long Stability?" in Ella Hermon, (ed.), *Vers une gestion intégrée de l'eau dans l'Empire romain: actes du Colloque international*, Université Laval, October 2006, Issue 180, Roma: L'erma di Bretschneider, 2008, pp. 201–208.

Lombardi, Leonardo, "Camillo Agrippa's Hydraulic Inventions on the Pincian Hill (1574–1578)," translated by Katherine W. Rinne, *The Waters Of Rome*, No. 5, (2008), pp. 1–10.

Long, Pamela O., "Hydraulic Engineering and the Study of Antiquity: Rome, 1557–70," *Renaissance Quarterly*, Volume 61, Number 4, Winter 2008, pp. 1098–1138.

Long, Pamela O., "Invention, Authorship, 'Intellectual Property,' and the Origin of Patents: Notes toward a Conceptual History," *Technology and Culture*, Vol. 32, No. 4, Special Issue: Patents and Invention (Oct., 1991), pp. 846–884.

Long, Pamela O., *Openness, secrecy, authorship: technical arts and the culture of knowledge from antiquity to the Renaissance*, Baltimore, 2001.

López Castellón, Enrique, *Historia de Castilla y León* (10 volumes), Vol. 5, Valladolid: Reno, 1984.

López de Ayala Álvarez de Toledo y del Hierro, Jerónimo, Conde de Cedillo, "Discursos leídos ante la Real Academia de la Historia en la recepción pública al Ilmo. Señor D. Jerónimo López Ayala y Álvarez de Toledo el día 23 de Junio de 1901," in *Toledo en el siglo XVI, después del vencimiento de las comunidades*, Madrid: Imprenta de los hijos de M. G. Hernández, 1901.

Lorente Toledo, Enrique, "Carlos I y su relación con Toledo: de la rebelión a la Corte," *Archivo secreto: revista cultural de Toledo*, No. 1 (2002), pp. 210–227.

Lota Brown, Meg and Kari Boyd McBride, *Women's Roles in the Renaissance*, Westport (CT): Greenwood Publishing Group, 2005.

Lucas, Adam Robert, "Industrial Milling in the Ancient and Medieval Worlds: A Survey of the Evidence for an Industrial Revolution in Medieval Europe," *Technology and Culture*, 46 (1), (2005), pp. 1–30.

Lucas, Adam Robert, "Narratives of Technological Revolution in the Middle Ages," in A. Classen, (ed.), *Handbook of Medieval Studies: Terms—Methods –Trends*, Berlin: De Gruyter, pp. 967–990.

Lucas, Adam Robert, "The Role of the Monasteries in the Development of Medieval Milling," in Steven A. Walton, (ed.), *Wind & Water in the Middle Ages: Fluid Technologies from Antiquity to the Renaissance*, Tempe (AZ): Arizona Center for Medieval and Renaissance Studies, 2006, pp. 89–128.

Lucas, Adam Robert, *Ecclesiastical lordship, seigneurial power and the commercialization of milling in medieval England*, Farnham Surrey: Ashgate, 2014.

Lucas, Adam Robert, *Wind, water, work: ancient and medieval milling technology*, Leiden: Brill, 2006, Technology and change in history, (ongoing series), v. 8.

Macfarlane, Alan, *The savage wars of peace: England, Japan and the Malthusian trap*, Oxford: Wiley-Blackwell, 1997.

Macías, José María and Cristina Segura Graíño, (eds), *Historia del abastecimiento y usos del agua en la ciudad de Toledo*, Madrid: Confederación Hidrográfica del Tajo, 1999.

Magnusson, Roberta and Paolo Squatriti, "The Technologies of Water In Medieval Italy," in, Paolo Squatriti, (ed.), *Working with Water in Medieval Europe: Technology and Resource-use*, Leiden: Brill, 2000, pp. 217–266.

Magnusson, Roberta J., "Public and Private Urban Hydrology: Water Management in Medieval London," in Steven A. Walton, (ed.), *Wind and Water in the Middle Ages: Fluid Technologies from Antiquity to the Renaissance*, Tempe (AZ): Arizona Center for Medieval and Renaissance Studies (2006), pp. 171–187.

Magnusson, Roberta J., "Water and wastes in medieval London," in Terje Tvedt, (ed.), *A history of water* (3 volumes), Vol. 3, London: I. B. Tauris, 2006, pp. 299–313.

Magnusson, Roberta J., "Water Supplies and Sewerage," in Thomas F. Glick, Steven J. Livesey and Faith Wallis, (eds), *Medieval Science, Technology, and Medicine: An Encyclopedia*, New York and London: Routledge, 2005, pp. 505–508

Magnusson, Roberta J., *Water Technology in the Middle Ages: Cities, Monasteries, and Waterworks after the Roman Empire*, Baltimore and London: JHU Press, 2003.

Mahoney, Paul G., "Contract or Concession? An Essay on the History of Corporate Law," *Georgia Law Review*, Nr. 34 (2000), pp. 873–893.

Maier, Jorge, "Catálogo e Índices," *Comisión de Antigüedades de la Real Academia de la Historia Castilla-La Mancha*, Madrid: Publicaciones del Gabinete de Antigüedades de la Real Academia de la Historia, 1999, pp. 97–100.

Major, Kenneth J., "Waterworks of London Bridge," *International Molinology*, No. 65, (2002), pp. 23–25.

Mallett, M. E. and J. R. Hale, *The Military Organisation of a Renaissance State: Venice C. 1400 To 1617*, Cambridge: Cambridge University Press, 2006.

Manderscheid, Hubertus, "The Water Management of Greek and Roman Baths," in Örjan Wikander, (ed.), *Handbook of Ancient Water Technology*, Technology and Change in History, (ongoing series), Vol. 2, Leiden: Brill (2000), pp. 467–538.

Marchetti, Raffaele, *Sulle acque di Roma antiche e moderne*, (2 volumes), Vol. 1, Roma: Sinimberghi, 1886.

Marinoni, Augusto and Carlo Pedretti, (eds), *Il Codice atlantico della Biblioteca Ambrosiana di Milano*, (3 volumes), Vol. 1, Firenze: Giunti, 2000.

Marion, Marcel, (ed.), *Dictionnaire Des Institutions De La France Aux Dix-Septieme Et Dix-Huitieme*, Paris: Ayer Publishing, 1923.

Markham, Adam, *A Brief. History of Pollution*; London: Earthscan Publications, 1994.

Marthorez, Jules Michel Henri, *Les étrangers en France sous l'ancien régime, histoire de la formation de la population française* (2 volumes), Vol. 2, Paris: E. Champion, 1919–1921.

Martín, Alberto Marcos, *España en los siglos XVI, XVII y XVIII: economía y sociedad*, Barcelona: Editorial Critica, 2000.

Martin, Henri, *Martin's history of France: the age of Louis XIV*, translated by Mary L. Booth (2 volumes), Vol. 2, Boston: Walker, Wise, 1865.

Martinez, Montse, Pere Malgrat, Cristina Vila, Roman Llagostera and Miguel Salgot, "History of the Sewerage System in Barcelona, Spain: From its Origenes to Garcia Faria Plan," in Andreas N. Angelakis and Joan B. Rose, *Evolution of Sanitation and Wastewater Technologies Through the Centuries*, 1st. Edition, London, IWA, 2014, pp. 383–400.

Martínez de Pisón, Eduardo, Carlos Caballero Casado, Sonia Fernández Esteban, Nuria de Andrés de Pablos, David Palacios Estremera, Carlos Caballero Casado, Sonia Fernández Esteban, Amparo Martín Espinosa y Ana Mas Hernández, *Encerrar el frío—El Pozo de Nieve de La Granja y el aprovechamiento tradicional de la nieve en el Guadarrama*, Segovia: Diputación de Segovia, Ayuntamiento del Real Sitio de San Ildefonso, 2014.

Martz, Linda, "Implementation of pure-blood statutes in sixteenth century Toledo," in Bernard Dov Cooperman, *In Iberia and beyond: Hispanic Jews between cultures: proceedings of a symposium to mark the 500th anniversary of the expulsion of Spanish Jewry*, Cranbury (NJ) and London: Associated University Presses, 1998, pp. 245–272.

Martz, Linda, "Toledo y los Toledanos en 1561," *Vol. 5 of Monografías*, Toledo: Instituto Provincial de Investigaciones y Estudios Toledanos, 1974.

Martz, Linda, *Poverty and Welfare in Habsburg Spain*, Cambridge: Cambridge University Press, 1983.

Mathew, Arnold Harris and Annette Calthrop, *The Life of Sir Tobie Matthew, Knight*, London: Elkin Mathews, 1907.

Matthews, William, *Hydraulia, An Historical and Descriptive Account of the Water Works of London*, London: Simpkin, Marshal, and Co., 1835.

Mays, L. W., M. Skivaniotis and A. N. Angelakis, "Water for human consumption through history," in Andreas N. Angelakis, Larry W. Mays, Demetris Koutsoyiannis and Nikos Mamassis, *Evolution of water supply throughout the millennia*, London; New York: IWA Publishing, 2012, pp. 19–42.

Mays, Larry W., (ed.), *Ancient Water Technologies*, Dordrecht, Heidelberg, London and New York: Springer, 2010.

Mays, Larry W. "A Brief History of Water Technology During Antiquity: Before the Romans," in Larry W. Mays, (ed.), *Ancient Water Technologies*, Dordrecht, Heidelberg, New York, London: Springer Science & Business Media, 2010, pp. 1–28.

Mays, Larry W., "A Brief History of Roman Water Technology," in Larry W. Mays, (ed.), *Ancient Water Technologies*, Dordrecht, Heidelberg, New York, London: Springer Science & Business Media, 2010, pp. 115–134.

Mays, Larry W., "Water Technology in Ancient Egypt," in Larry W. Mays, (ed.), *Ancient Water Technologies*, Dordrecht, Heidelberg, New York, London: Springer Science & Business Media, 2010, pp. 57–66.

Mays, Larry W., M. Sklivaniotis and Andreas Nikolaos Angelakis, "Water for human consumption through history," in Andreas Nikolaos Angelakis, Larry W. Mays and Demetris Koutsoyiannis, *Evolution of Water Supply Through the Millennia*, London: IWA Publishing, 2012, pp. 19–43.

McClellan III, James E. and Harold Dorn, *Science and Technology in World History: An Introduction*, Baltimore: JHU Press, 2015.

McNeil, Ian, (ed.), *An Encyclopaedia of the History of Technology*, London and New York: Routledge, 1990.

McNeil, William H., *Plagues and Peoples*, London: Penguin Books, 1985.

Melosi, Martin V., *The Sanitary City: Environmental Services in Urban America from Colonial Times to the Present*, Pittsburgh: University of Pittsburgh Press, 2008.

Merton, Robert King, "Science, Technology and Society in Seventeenth Century England," *Osiris*, Vol. 4 (1938), pp. 360–632.

Miller, Helen, "London and Parliament in the Reign of Henry VIII," *Historical Researh*, Vol. 35, Issue 92 (November 1962), pages 128–149.

Mills, Allan A., "Leonardo da Vinci and Perpetual Motion," *Leonardo*, Vol. 41, No. 1, (February 2008), pp. 39–42.

Miron de L'Espinay, Albert, *François Miron et l'administration municipale de Paris sous Henri IV de 1604 à 1606*, Paris: E. Plon, Nourrit et Cie, 1885.

Mitchell, Rosamond Joscelyne and Mary Dorothy Rose Leys, *A history of London life*, London: Longmans, Green and Co, 1958.

Mitford, John, (ed.), *The poetical works of Jonathan Swift*, (3 volumes), Vol. 1, London: William Pickering, 1833.

Mithen, Steven, *Thirst: For Water and Power in the Ancient World*, Cambridge (MA): Harvard University Press, 2012.

Mitterauer, Michael, *Why Europe?: The Medieval Origins of Its Special Path*, translated by Gerald Chapple, Chicago and London: The University of Chicago Press, 2003.

Moiset, Charles, (ed.), "La poésie auxerroise au XVIe siècle," *Annuaire historique du département de l'Yonne*, xxi (1857), pp. 73–83.

Mokyr, Joel, *The Gifts of Athena: Historical Origins of the Knowledge Economy*, Princeton: Princeton University Press, 2001.

Molà, Luca, *The Silk Industry of Renaissance Venice*, Baltimore: JHU Press, 2000.

Mommsen, Theodor, *The history of Rome*, translated by William Purdie Dickson, (4 volumes), Vol. 4, New York: Scribner, 1871.

Monckton, Herbert Anthony, *A History of English Ale and Beer*, London: Bodley Head, 1966.

Montemayor, Julián, *Tolède entre fortune et déclin (1530–1640)*, Limoges: Presses Univ. Limoges, 1996.

Moon, Francis C., *The Machines of Leonardo Da Vinci and Franz Reuleaux: Kinematics of Machines from the Renaissance to the 20th Century*, Dordrecht (The Netherlands): Springer, 2007.

Moreno Nieto, Luis y Ángel, *Juanelo y su Artificio. Antología*, Toledo: d. b. Comunicación, 2006.

Moreno Santiago, Angel, "La técnica en el siglo XVI según los manuscritos de Juanelo Turriano," *Provincia—Revista de la Diputación Provincial de Toledo*, No. 74, Toledo (June 1971), folleto separado.

Morera, Raphaël, "Environmental change and globalization in seventeenth century France: Dutch traders and draining of French wetlands (Arles, Petit Poitou)," *International Review of Social History*, No. 55 (2010), Supplement, pp. 79–101.

Moreri, Louis, *Le grand dictionaire historique, ou Le mêlange curieux de l'histoire sacrée et profane*, (10 volumes), Vol. 4, Amsterdam: Chez Brunel, 1740.

Morris, Rupert Hugh, *Chester in the Plantagenet and Tudor Reigns*, New York: Read Books, 2008.

Mousnier, Roland, *The Institutions of France under the Absolute Monarchy, 1598–1789: The Organs of State and Society*, translated by Arthur Goldhamnmer, (2 volumes), Vol. 2, Chicago: University of Chicago Press, 1984.

Mousnier, Roland, *The Institutions of France under the Absolute Monarchy, 1598–1789: Society and the State*, translated by Brian Pearce (2 volumes), Vol. 1, Chicago: University of Chicago Press, 1979.

Mukerji, Chandra, *Territorial Ambitions and the Gardens of Versailles*, Cambridge University Press, 1997.

Mullins, Claud, *London's Story*, London: G. Bell and Sons, 1920.

Multhauf, Robert P., "Mine Pumping in Agricola's Time and Later," *U.S. National Museum Bulletin* 218 (1959), pp. 113–120.

Mumford, Lewis, *Technics and Civilization*, London: Routledge & Kegan Paul Ltd, 1955.

Munro, John H., "Money, Wages, and Real Incomes in the Age of Erasmus," in *The Correspondence of Erasmus* (12 volumes), Vol. 1 (1484–1500), Toronto: University of Toronto Press, 1974, pp. 311–347.

Munro, John H., "Industrial Energy from Water-Mills in the European Economy, 5th to 18th Centuries: the Limitations of Power," in Simonetta Cavaciocchi, (ed.), *Economia ed energia, seccoli XIII–XVIII*, Atti delle 'Settimane di Studi' e altrie Convegni, Istituto Internazionale di Storia Economica, 'Francesco Datini da Prato', vol. 34, Florence: Le Monnier (2003), pp. 215–261.

Musée Carnavalet, *Catalogue—Pont-Neuf, 1578–1978: exposition organisée par le Musée Carnavalet et la Délégation à l'action artistique de la ville de Paris: Mairie annexe du 1er arrondissement, 10 juin-9 juillet 1978, Musée Carnavalet 19 juillet-27 août 1978*, Paris: Le Musée, 1978.

Musson, Albert Edward and Eric Robinson, *Science and Technology in the Industrial Revolution*, Manchester: Manchester University Press, 1969.

Nagler, G. K., *Neues allgemeines Künstler-Lexicon*, München: E. A. Fleischmann, 1836.

Needham, Joséph, *Clerks and Craftsmen in China and the West: Lectures and Addresses on the History of Science and Technology*, Cambridge: Cambridge University Press, 1970.

Needham, Joséph, *Science and Civilization in China: Volume 4, Physics and Physical Technology, Part 2, Mechanical Engineering*, Cambridge: Cambridge University Press, 1965.

Nef, John Ulric, "The Industrial Revolution Reconsidered," *The Journal of Economic History*, Vol. 3, No. 1 (May, 1943), pp. 1–31.

Nef, John Ulric, "The Progress of Technology and the Growth of Large-Scale Industry in Great Britain, 1540–1640," *The Economic History Review*, Vol. 5, No. 1 October 1934, pp. 3–24.

Nef, John Ulric, *The Rise of the British Coal Industry*, (2 volumes), Vol. 1, Oxford: Routledge, 1966.

Newman, Karen, "Towards a Topographic Imaginary—Early Modern Paris," in Carla Mazzio and Douglas Trevor, (eds), *Historicism, Psychoanalysis and Early Modern Culture*, New York and London: Routledge, 2000, pp. 59–81.

Nicholas, David, *Urban Europe 1100–1700*, Basingstoke (UK) and New York: Palgrave Macmillan, 2003, pp. 155–167.

Nickler, Pierre, "La machine hydraulique de Palfour sous la Terrasse de Saint Germain-en-Laye. Des ingénieurs Liégeois à Sir Samuel Morland," in "André Le Nôtre (1613–1700) Art et Science des Jardins et de la Terrasse de Saint-Germain-en-Laye" *Bulletin des Amis du Vieux Saint-Germain*, No. 38, (2001), p. 25–39.

Noël, François L. and J. M. Carpentier, *Dictionnaire des inventions*, Bruxelles: J. P. Meline, Cans et Compagnie, 1837.

Oakeshott, Michael Joséph, *Political education: an inaugural lecture delivered at the London School of Economics and political Science on March 6, 1951*, Cambridge: Bowes and Bowes, 1951.

Oleson, John Peter, "Water-Lifting," in Örjan Wikander, (ed.), *Handbook of Ancient Water Technology*, Technology and Change in History, (ongoing series), Vol. 2, Leiden: Brill, 2000, pp. 217–302.

Oleson, John Peter, *Greek and Roman Mechanical Water-Lifting Devices: The History of a Technology*, Toronto: University of Toronto Press, 1984.

Ortega y Gasset, José, "Man the technician," in *History as a system*, New York: Norton, 1961.

Ostolaza Elizondo, María Isabel, "Los veintiún libros de los ingenios y de las máquinas—composición, reconstrucción arqueológica, proceso de copia, copistas, datación," *Archivo de filología aragonesa*, Vol. 48–49, 1992–1993, pp. 225–262.

Palas, Jan, Jaroslav Jasek and Ondrej Benes, "A brief history of the sewerage system in Prague and the role of William Heerlein Lindley," in Andreas N. Angelakis and Joan B. Rose, *Evolution of Sanitation and Wastewater Technologies Through the Centuries*, 1st. Edition, London, IWA, 2014, pp. 401–418.

Paley, William James Paxton and John Ware, *Natural Theology: Or, Evidences of the Existence and Attributes of the Deity, Collected from the Appearances of Nature*, London: Lincoln, Edwards and Co., 1833.

Palma-Cayet, Pierre-Victor, "Chronologie septenaire ou histoire de la paix entre les roys de France et d'Espagne," in *Suite de la Collection Universelle des Mémoires Particuliers Relatifs à l'Histoire de France* (72 volumes), Vol. 70, Paris: De l'imprimerie et fonderie d'Orizet, 1807.

Parker, Henry, *Ancient Ceylon*, New Delhi and Madras: Asian Educational Services, 1999, specially Chapter X: "The Earliest Irrigation Works," pp. 347–412.

Parro, Sisto Ramón, *Toledo En La Mano, Ó: Descripción Histórico-Artística De La Magnífica Catedral Y De Los Demas Célebres Monumentos Y Cosas Notables Que Encierra Esta Famosa Ciudad* (2 volumes), Vol. 2, Toledo: Imprenta y Librería de Severiano López Fando, 1857.

Parsons, William Barclay, *Engineers and Engineering in the Renaissance*, Cambridge (MA) and London: M.I.T. Press, 1976.

Partington, Charles F., *British cyclopedia of the arts and sciences* (2 volumes), Vol. 1, London: Orr & Smith, 1835.

Paula Mellado, Francisco de, *Diccionario universal de Historia y de Geografía: Contiene: historia propiamente dicha* (7 volumes + 1 suplement), Vol. 2, Madrid: Francisco de Paula Mellado, 1846.

Pavon Maldonado, Basilio, "En torno al acueducto y la rueda hidráulica árabe de Toledo según Idrisi," *Al-Andalus Magreb: Estudios árabes e islámicos*, No. 5, (1997), pp. 273–294.

Pavon Maldonado, Basilio, *Tratado de arquitectura hispanomusulmana, I. Agua (aljibes, puentes, qanats, acueductos, jardines, ruedas hidráulicas, baños, corachas)* (4 volumes), Vol. 1, Madrid: Consejo Superior de Investigaciones Cientificas, 1990.

Peachey, Stuart, *The tipler's guide to drink and drinking in the early 17th century*, Bristol (UK): Stuart Press, 1992.

Peces Ventas, Juan Luis, "El Artificio de Juanelo Turriano," *Revista de Informacion de la Junta de Comunidades de Castilla-La Mancha*, No. 23, 1997.

Phillips, Jeremy, "The English patent as a reward for invention: The importation of an idea," *The Journal of Legal History* (1982), Vol. 3, Issue 1, pp. 71–79.

Pierce, Patricia, *Old London Bridge: The Story of the Longest Inhabited Bridge in Europe*, London: Headline Book Publishing, 2002.

Pinto Crespo, Virgilio, Rafael Gili Ruiz and Fernando Velasco Medina, *Historia del Proyecto de investigación Saneamiento de Madrid*, Madrid: Centro de Documentación para la Historia de Madrid. Colaboración entre la Fundación Canal de Isabel II y la Universidad Autónoma de Madrid, marzo 2012 - diciembre 2014.

Pitts, Vincent J., *Henri IV of France, His Reign and Age*, Baltimore: John Hopkins University Press, 2009.

Platt, J. C. and J. Saunders, "Underground," in Charles Knight, (ed.), *London* (6 volumes), Vol. 1, London: Henry G. Bohn, 1851, pp. 225–240.

Popplow, Marcus, "Hydraulic Engines in Renaissance Privileges for Inventions and 'Theatres of Machines'," in Alessandra Fiocca, Daniela Lamberini, Cesare Maffioli, *Arte e Scienza delle Acque nel Rinascimento*, Venezia: Marsilio Editori, 2003, pp. 73–83.

Popplow, Marcus, *Models of machines: a 'missing link' between early modern engineering and mechanics?*, Preprint, Berlin: Max-Planck-Institut für Wissenschaftsgeschichte, 2002.

Porres Martin-Cleto, Julio, "Consideraciones y nuevos datos sobre el *Artificio* de Juanelo," *Boletín de la Real Academia de Bellas Artes y Ciencias Históricas de Toledo*, 40 (1999), pp. 119–131.

Porres Martin-Cleto, Julio, "El artificio de Juanelo en 1639," Separata de *Anales Toledanos*, XIV, Toledo, 1982, pp. 175–186.

Porres Martin-Cleto, Julio, "El Artificio de Juanelo," Separata de *Temas Toledanos*, No. 47, Toledo: Diputación Provincial, Instituto Provincial de Investigaciones y Estudios Toledanos, 1987.

Porres Martin-Cleto, Julio, "El final del artificio de Juanelo," *Toletum*, No. 10 (1980), pp. 171–176.

Porres Martin-Cleto, Julio, Rafael Juan del Cerro Malagón y José Luis Isabel Sánchez, *Panorámica de Toledo de Arroyo Palomeque*, Toledo: Instituto Provincial de Investigaciones y Estudios Toledanos, 1992.

Porres Martin-Cleto, Julio, *Toledo a traves de sus planos*, Toledo: Diputación Provincial, Instituto Provincial de Investigaciones y Estudios Toledanos, 1989.

Porter, Roy, *London, a Social History*, Cambridge (MA): Harvard University Press, 1995.

Portuondo, Marià, *Secret science: Spanish cosmography and the New World*, Chicago and London: The University of Chicago Press, 2009.

Pounds, Norman John Greville, *An Economic History of Medieval Europe*, New York: Longman Publishing, 1994.

Prager, Frank D., "A Manuscript of Taccola, Quoting Brunelleschi, on Problems of Inventors and Builders," *Proceedings of the American Philosophical Society*, Vol. 112, No. 3 (Jun. 21, 1968), pp. 131–149.

Pratt, Julius W., "Machinery in Sixteenth-Century English Industry," *The Journal of Political Economy*, Vol. 22, No. 8 (Oct., 1914), pp. 775–790.

Pringle Nichol, John, *Cyclopædia of the physical sciences: comprising acoustics, astronomy, dynamics, electricity, heat, hydrodynamics, magnetism, philosophy of mathematics, meteorology, optics, pneumatics, statics, &c. &c*, London: Charles Griffin and Company, 1868.

Quadrado, José María, *Recuerdos y bellezas de España—Castilla La Nueva* (2 volumes), Vol. 1, Madrid: Imprenta de D. José Repullés, 1853.

Rabin, Dan and Carl Forget, *The dictionary of beer and brewing*, Chicago: National Book Network, 1998.

Rahn Phillips, Carla, "Time and Duration: A Model for the Economy of Early Modern Spain," *The American Historical Review*, Vol. 92, No. 3 (Jun., 1987), pp. 531–562.

Raithby, John, (ed.), *The statutes at large, of England and of Great Britain: from Magna Carta to the union of the kingdoms of Great Britain and Ireland* (20 volumes), Vol. 3, London: G. Eyre and A. Strahan, 1811.

Raithby, John, (ed.), *The Statutes at Large, from Magna Carta to the Union of the Kingdoms of Great Britain and Ireland* (20 volumes), Vol. 4, London: G. Eyre and A. Strahan, 1811.

Raithby, John, (ed.), *The Statutes of the United Kingdom and Ireland, 3 George IV*, London: George Eyre and Andrew Strahan, 1822.

Raithby, John, (ed.), *The Statutes of the United Kingdom of Great Britain and Ireland, From CE 1819; 60 George III—to CE 1822; 3 George IV*, London: John Eyre and Andrew Strahan, 1822.

Ranum, Orest, *Paris in the Age of Absolutism: an Essay*, University Park (PA): Pennsylvania State University Press, 2004.

Rapeg, F. Jamil, "The Translation of Greek Natural Philosophy into Arabic: Background and Motivations," in David C. Lindberg and Michael H. Shank, (eds), *The Cambridge*

History of Science, (8 volumes), vol. 2, New York: Cambridge University Press, 2013, pp. 27–61.

Rappaport, Steve, *Worlds within Worlds: Structures of Life in sixteenth Century London*, Cambridge: Cambridge University Press, 1989.

Read, Geoffrey F., "The Development of Public Health Engineering," in Geoffrey F. Read and Ian Vickridge, (eds), *Sewers. Rehabilitation and New Construction. Repair and Renovation*, Burlington (MA): Butterworth-Heinemann, 1996.

Reddaway, Thomas Fiddian, *The Rebuilding of London After the Great Fire*, London: Jonathan Cape, 1940.

Redel, Enrique, *Ambrosio de Morales. Estudio biográfico*, Córdoba: Imprenta del Diario, 1909.

Rees, Abraham, *The Cyclopædia: Or, Universal Dictionary of Arts, Sciences, and Literature* (39 volumes), Vol. 23, London: Longman, Hurst, Rees, Orme & Brown, 1819.

Reid, Donald, *Paris Sewers and Sewermen: Realities and Representations*, Cambridge (MA): Harvard University Press, 1993.

Reinoso Delgado, Ricardo, *Estudio y Simulación del Artificio de Juanelo Turriano*, proyecto fin de carrera, Escuela Técnica Superior de Ingenieros, Universidad de Sevilla, 2003.

Reti, Ladislao, (ed.), *The Unknown Leonardo*, London: Hutchinson, 1974.

Reti, Ladislao, "El Artificio de Juanelo en Toledo: su Historia y su Técnica," Separata de *Provincia, Revista de la Diputación Provincial de Toledo*, No. 60 (1967), pp. 1–46.

Reti, Ladislao, "Francesco di Giorgio Martini's Treatise on Engineering and Its Plagiarists," *Technology and Culture*, Vol. 4, No. 3 (Summer 1963), pp. 287–298.

Reti, Ladislao, "The Codex of Juanelo Turriano (1500–1585)," *Technology and Culture*, Vol. 8, No. 1 (Jan., 1967), pp. 53–66.

Revuelta Pol, Bernardo y Dolores Romero Muñoz, "Juanelo Turriano, relojero e ingeniero cremonés" in "Realismo y espiritualidad: Campi, Anguissola, Caravaggio y otros artistas cremoneses y españoles en los siglos XVI–XVIII," catalogue of the exhibition that took place in the Castillo de Alaquàs (March–July, 2007), Valencia: Ajuntament d'Alaquàs, 2007, pp. 73–81.

Reynolds, Terry S., *Stronger than a Hundred Men*, Baltimore and London: John Hopkins University Press, 1983.

Rich, E. E. and C. H. Wilson, (eds), The Cambridge Economic History of Europe: Volume V, The Economic Organization of Early Modern Europe, Cambridge: CUP Archive, 1977

Richards, Charles Henry, William Henry Christopher Payne and John Philpott Henry Soper, *London Water Supply: Being a Compendium of the History, Law, & Transactions Relating to the Metropolitan Water Companies from Earliest Times to the Present Day*, London: P. S. King & Son, 1899.

Richardson, John, *The Annals of London*, Berkeley (CA): University of California Press, 2000.

Riley, Henry Thomas, (ed.), "The French Chronicle of London: Edward I," in *Chronicles of the Mayors and Sheriffs of London: 1188–1274*, London: [s.n.], 1863, pp. 237–248.

Riley, Henry Thomas, (editor and translator), *Memorials of London and London life, in the XIIIth, XIVth and XVth Centuries*, London: Longmans, Green and Co., 1868, pp. 295–300.

Ringrose, David, "The Impact of a New Capital City: Madrid, Toledo, and New Castile, 1560- 1660," *The Journal of Economic History* Vol. 33, No. 4 (1973), pp. 761–791.

Rinne, Katherine Wentworth, "Between precedent and experiment: restoring the Acqua Vergine in Rome (1560–70)," in L. Roberts, S. Shaffer and P. Dear, (eds), *The mindful hand: inquiry and invention from the late Renaissance to early industrialization*, Amsterdam: Edita, 2007, pp. 94–115.

Rinne, Katherine Wentworth, "Fluid Precision: Giacomo della Porta and the Acqua Vergine fountains of Rome", in Jan Birksted, (ed.), *Landscapes of Memory and Experience*, London: Taylor and Francis, 2000, pp. 183–201.

Rinne, Katherine Wentworth, *The Waters of Rome: Aqueducts, Fountains, and the Birth of the Baroque City*, New Haven and London: Yale University Press, 2010.

Rinne, Katherine Wentworth, "Water: the Currency of Cardinals in Late Renaissance Rome," in A. Calzona and D. Lamberini, (eds), *La Civilta` delle Acque tra Medioevo e Rinascimento*, Florence: Olshki, 2010, pp. 367–387.

Rittiez, François, *L'Hôtel de ville et la bourgeoisie de Paris: origines, moeurs, coutumes, institutions municipales depuis les temps les plus reculés jusqu'à 1789*, Paris: Schlesinger Frères, 1863.

Robertson, Joseph Clinton, (ed.), *The Mechanics' Magazine, Museum, Register, Journal and Gazette*, Vol. LI, London, 1849.

Robertson, Joseph Clinton, (ed.), *The Mechanics' Magazine, Museum, Register, Journal and Gazette*, Vol. IX, London, 1828.

Robertson, Hector Menteith, "Sir Bevis Bulmer: A Large-Scale Speculator of Elizabetan and Jacobean Times," *Journal of Economic and Business History*, Vol. 4 (1931), p. 99–120.

Roche, Daniel, "Le temps de l'eau rare du Moyen Age a l'époque moderne," *Annales. Histoire, Sciences* Sociales, 39ᵉ Année, No. 2 (Mar.–Apr., 1984), pp. 383–399.

Roche, Daniel, *A History of Everyday Things: The Birth of Consumption in France, 1600–1800*, translated by Brian Pearce, Cambridge: Cambridge University Press, 2000.

Roche, Daniel, *The people of Paris: an essay in popular culture in the 18th century*, Berkeley and Los Angeles: University of California Press, 1987.

Romero Muñoz, Dolores (coordinación, Daniel Crespo Delgado), *La navegación del Manzanares: el proyecto Grunenbergh*, Madrid: Fundación Juanelo Turriano, 2015.

Rousseau, Pierre, *Histoire des techniques et des inventions*, Paris: Fayard, 1958.

Rubio Mañé, Ignacio José, *El Virreinato, Orígenes y jurisdicciones y dinámica social de los virreyes* (4 volumes), Vol. 1, Mexico: Universidad Nacional Autónoma de Mexico, 1983.

Rudden, Bernard, *The New River: A Legal History*, Oxford: Oxford University Press, 1985.

Ruhland, Florian, "Power, pleasure, and pollution: Water use in pre-industrial Nuremberg and Prague," *Klaudyán: Internet Journal of Historical Geography and Environmental History*, Vol. 4, (2007), No. 2, pp. 5–18.

Rumble, Victoria R., *Soup through the Ages: A Culinary History with Period Recipes*, Jefferson (NC): McFarland, 2009.

Russo, Lucio, *The Forgotten Revolution: How Science Was Born in 300 BCE and Why It Had to Be Reborn*, translated by Silvio Levy, Berlin and Heidelberg: Springer, 2004.

Rynne, Colin, "Technological Continuity, Technological 'Survival': the Use of Horizontal Mills in Western Ireland, c.1632–1940," *Industrial Archaeology Review*, 33: 2, (2011), pp. 96–105.

Sabine, Ernest L., "City Cleaning in Mediaeval London," *Speculum*, Volume 12, (1937), pp 19–43.

Sabine, Ernest L., "Latrines and cesspools of Mediaeval London," *Speculum*, 9, (1934), pp. 303–321.

Sabra, A. I., "Situating Arabic Science: Locality versus Essence," *Isis*, 87 (1996), pp. 654–70.

Saint-Julien, *Les courriers de la fronde en vers burlesques*, reviewed and with notes by Celestine Moreau (2 volumes), Vol. 1, Paris: Chez P. Jannet, 1857.

Sáiz, J. Patricio, "El peluquero de la Reina," *Serie teoría económica e historia económica*, Departamento de análisis económico, Universidad Autónoma de Madrid, (January 2005), p. 15.

Saliba, George, *Islamic science and the making of the European Renaissance*, Cambridge (MA): MIT Press, 2007.

Salzman, James, "Is it safe to drink the water? *Duke Journal of Environmental Law & Policy*, Vol. 19–1 (Fall 2008), pp. 1–42.

Salzman, James, "Thirst: A Short History of Drinking Water," *Social Science Network Electronic Paper Collection*, Research Paper Series, paper No. 92, Duke Law School Legal Studies, December 2005, pp. 1–33.

Salzmann, Louis Francis, "Scaffoldings, Cranes and Other Machines," in *Building in England down to 1540*, Oxford: Oxford University Press, 1992.

Salzmann, Louis Francis, *English industries of the Middle Ages being an introduction to the Industrial history of mediaeval England*, London: Constable and Company, 1913.

Samman, Mahmod, "Rediscovering the Waterwheel—'Noria Al-Muhammadiyya,' an ASME International Historic Landmark," a presentation to the American Society of Mechanical Engineers, Northwest Houston Subsection meeting, August 2005.

Sánchez Gómez, Julio, "La minería," in Silva Suárez, (ed.), *Técnica e Ingeniería en España*, pp. 439–476.

Sánchez López, Elena and Javier Martínez Jiménez, *Los acueductos de Hispania: construcción y abandono*, Madrid: Fundación Juanelo Turriano, 2016.

Sánchez Mayendia, José Cristóbal, "El Artificio de Juanelo en la Literatura Española," *Cuadernos Hispanoamericanos* 103 (1948), pp. 73–92.

Sanchez Palencia, Almudena, "En torno a Juanelo Turriano," *Anales Toledanos XIX*, Toledo (1984), pp. 70–91.

Sangrador y Vitores, Matias, *Historia de la muy noble y leal ciudad de Valladolid desde su más remota antigüedad hasta la muerte de Fernando VII*, (2 volumes), Vol. 1, Valladolid: Imprenta de D. M. Aparicio, 1854.

Santos Vaquero, Ángel, "Alcalde mayor de alzadas en el ayuntamiento de Toledo," *Espacio, Tiempo y Forma, Serie IV, Historia Moderna*, tomo 16 (2003), págs. 11–29.

Saunders, E. Stewart, "Louis XIV: Patron of Science and Technology," *Purdue University Libraries Research Publications*, Paper 46 (1984).

Sawday, Jonathan, *Engines of the Imagination: Renaissance Culture and the Rise of the Machine*, New York: Taylor & Francis, 2007.

Say, Horace, *Etudes sur l'administration de la ville de Paris, et du department de la Seine*, Paris: Guillaumin, 1846.

Scaglia, Gustina, *Francesco di Giorgio: checklist and history of manuscripts and drawings in autographs and copies from ca. 1470 to 1687 and renewed copies (1764–1839)*, Bethlehem (PA): Lehigh University Press and London: Associated University Presses, 1992.

Schioler, Thorkild, *Roman and Islamic Water-Lifting Wheels*, Odense: Odense University Press, 1973.

Schomberg, Anette, "Ancient Water Technology: between Hellenistic Innovation and Arabic Tradition," *Syria*, 85, (2008), pp. 119–128.

Scott Dixon, C., "Introduction," in C. Scott Dixon, Dagmar Freist, Mark Greengrass, (eds), *Living with Religious Diversity in Early-modern Europe*, Farnham (UK): Ashgate Publishing, Ltd., 2009, pp. 1–20.

Scott, Virginia, *Molière: A Theatrical Life*, Cambridge: Cambridge University Press, 2002.

Scott, William Robert, *The constitution and finance of English, Scottish and Irish joint-stock companies to 1720* (3 volumes), Vol. 3, Cambridge: Cambridge University Press, 1911.

Seaborne Davies. David Richard, "Further Light on the Case of Monopolies," *Law Quarterly Review*, Vol. 48, Issue 3 (July 1932), pp. 394–414.

Secord, James A., "Knowledge in transit," *Isis*, 95, (2004), pp. 654–672.

Shadwell, Arthur, *The London Water Supply*, London: Longmans, Green, and Co., 1899.

Shapin, Steven, "Understanding the Merton Thesis," *Isis*, Vol. 79, No. 4 (Dec., 1988), pp. 594–605.

Shapiro, Sheldon, "The Origin of the Suction Pump," *Technology and Culture*, Vol. 5, No. 4 (Autumn 1964), pp. 566–578.

Sharpe, Reginald Robinson, (ed.), *Calendar of letter-books preserved among the archives of the Corporation of the City of London at the Guildhall*, London: John Edward Francis, 1912.

Sharpe, Reginald Robinson, (ed.), "Folios 110b–135b," in *Calendar of Letter-Books of the City of London: 1275–1298*, London: Her Majesty's Stationery Office, 1899, pp. 207–230.

Sharpe, Reginald Robinson, *London and the Kingdom* (3 volumes), Vol. 2, London and New York: Longmans, Green, and Co., 1894.

Sheppard, Francis, *London, a History*, Oxford: Oxford University Press, 1998.

Shrewsbury, John Findlay Drew, *The plague of the Philistines: and other medical-historical essays*, London: V. Gollancz, 1964.

Silva Suárez, Manuel, "Introduccion," in Silva Suárez, (ed.), *Técnica e Ingeniería*, pp. 9–22.

Silva Suárez, Manuel, (ed.), *Técnica e Ingeniería en España I: El Renacimiento*, (6 volumes), Vol. 1, Zaragoza: Real Academia de Ingeniería, Institución Fernando El Católico y Prensas Universitarias de Zaragoza, 2004.

Simms, D. L., "Water-driven saws, Ausonius, and the Authenticity of the *Mosella*," *Technology and Culture*, Vol. 24, (1983), pp. 635–634.

Simms, D. L., "Water-driven Saws in Late Antiquity," *Technology and Culture*, Vol. 26, (1985), pp. 275–276.

Simons, Eric N., *Henry VII, the First Tudor King*, London: Muller, 1968.

Singer, Charles and E. J. Holmyard, A. R. Hall and Trevor I. Williams, *A History of Technology* (3 volumes), Vol. 2, Oxford: Clarenton Press, 1956.

Skempton, Alec W., *A Biographical Dictionary of Civil Engineers in Great Britain and Ireland: 1500–1830*, London: Thomas Telford, 2002.

Smiles, Samuel, "Life of Sir Hugh Myddelton," in *Lives of the Engineers, with an Account of their Principal Works: Comprising also a History of Inland Communication in Britain*, (3 volumes), Vol. 1, London: John Murray, 1862, pp. 85–152.

Smiles, Samuel, *Lives of the Engineers, with an Account of Their Principal Works*, (3 volumes), Vol. 1, London: John Murray, 1861.

Smith, John Thomas, *The Cries of London: Exhibiting Several of the Itinerant Traders of Antient and Modern Times*, London: John Bowyer Nichols And Son, 1839.

Smith Denis, (ed.), *Water Supply and Public Health Engineering*, in Joyce Brown, general editor of the series: *Studies in the History of Civil Engineering*, (12 volumes), Vol. 5, Aldershot (UK), 1999.

Smith, Bruce R., *The Acoustic World of Early Modern England: Attending to the O-Factor*, Chicago and London: University Of Chicago Press, 1999.

Smith, Norman Alfred Fisher, "The History of Spanish Dam-Building," *Transactions of the Newcomen Society*, Volume 41 (1), (1968), pp. 45–47.

Smith, Norman Alfred Fisher, "The Origins of Water Power: A Problem of Evidence and Expectations," *Transactions of the Newcomen Society*, 55 (1), (1983), pp. 67–84.

Soria Mesa, Enrique, "Los estudios sobre las oligarquías municipales en la Castilla moderna. Un balance en claroscuro," *Manuscrits* No.18, (2000), pp. 185–197.

Soulié, Eudoxe, *Notice du Musée Impérial de Versailles* (3 volumes), Vol. 1, Paris: Charles de Mourgues frères, 1859.

Spar, Debora and Krzysztof Bebenek, "To the Tap: Public versus Private Water Provision at the Turn of the Twentieth Century," *Business History Review* 83 (Winter 2009), Cambridge (MA), pp. 675–702.

Sprague DeCamp, Lyon, *The Ancient Engineers*, New York: Ballantine Books, 1963.

Squatriti, Paolo, (ed.), *Working with Water in Medieval Europe, Technology and resource-use*, Leiden, Boston, Koln: Brill, 2000.

Staley, Cady and George Spencer Pierson, *The separate system of sewerage: its theory and construction*, New York: D. Van Nostrand, 1899.

Stegmann, André, (ed.), *Pouvoir et institutions en Europe au XVIème siècle*, Paris: Librairie Philosophique J. Vrin, 1987.

Stevenson, Angus, (ed.), *Oxford Dictionary of English Oxford*: Oxford University Press, 2010.

Stoyle, Mark, *Water in the City:The Aqueducts and Underground Passages of Exeter*, Exeter: University of Exeter Press, 2014.

Stuart, Kathy, *Defiled Trades and Social Outcasts: Honor and Ritual Pollution in Early Modern Germany*, Cambridge and New York: Cambridge University Press, 2006.

Suárez, Edna, "El organismo como máquina: Descartes y las explicaciones biológicas," in Carlos Álvarez and Rafael Martínez Enríque, (eds), *Descartes y la ciencia del siglo XVII*, México D. F.: Siglo XXI, 2000.

Sullivan, Ceri, "Thomas Middleton's View of Public Utility," *Review of English Studies*, Vol. 58, Issue 234 (April 2007), pp. 162–174.

Sunderland, Septimus, *Old London's spas, baths, and wells*, London: J. Bale, Sons & Danielsson, 1915.

Sutcliffe, Anthony, *Paris, An architectural History*, New Haven and London: Yale University Press, 1993.

Swift, Jonathan, "A Description of a City Shower" (1710) in: *The poetical works of Jonathan Swift*, compiled and commented by John Mitford (3 volumes), Vol. 1, London: William Pickering, 1833, pp. 93–96.

Tamburrino, Aldo, "Water Technology in Ancient Mesopotamia," in Larry W. Mays, (ed.), *Ancient Water Technologies*, Dordrecht, Heidelberg, New York, London: Springer Science & Business Media, 2010, pp. 29–52.

Tamizey de Larroque, Philippe, *Louis de Foix et la Tour de Cordouan*, Bordeaux (France): Chaumas, 1864.

Taylor, C., "The disposal of Human Waste: A comparison between Ancient Rome and Medieval London," *Past Imperfect*, 11, (2005), pp. 53–72.

Tempelhoff, Johann W. N., "Historical Perspectives on Pre-Colonial Irrigation in Southern Africa," *African Historical Review*, Vol 40, No. 1, pp. 121–159.

Terrell, Thomas, *The law and practice relating to letters patent for inventions*. 6th. ed., revised and re-written by Courtney Terrell and Arthur Jaffé, London: Sweet and Maxwell, 1921.

Terrell, Thomas, *The law and practice relating to letters patent for inventions*. 3rd. ed., edited by William Peter Rylands, London: Sweet and Maxwell, 1896.

The Encyclopaedia Britannica, a dictionary of arts, sciences, literature and general information, 11th ed. (32 volumes), Vol. 10, New York: Encyclopaedia Britannica, 1910.

Thirsk, Joan, *Alternative Agriculture: A History: from the Black Death to the Present Day*, Oxford: Oxford University Press, 1997.

Thirsk, Joan, *Economic policy and projects: the development of a consumer society in Early Modern England*, Oxford: Clarendon Press, 1978.

Thomas, Charles R., *Practical Wine Talk: A Physician-Winemaker Examines Wine*, Bloomington (IN): AuthorHouse, 2013.

Thomas, Henry, *The wards of London: comprising a historical and topographical description of every object of importance within the boundaries of the city. With an account of all the companies, institutions, buildings, ancient remains ... and biographical sketches of all eminent persons connected therewith*, (2 volumes), Vol 1, London: J. Gifford, 1828.

Thompson, Edward P., "The Moral Economy of the English Crowd in the Eighteenth Century," *Past and Present*, no. 50 (1971), pp. 76–136.

Thompson, I. A. A., "The Purchase of Nobility in Castile, 1552–1700," in *War and Society in Habsburg Spain*, Aldershot (UK): Variorum, pp. 313–378.

Thompson, I. A. A., *War and Government in Habsburg Spain, 1560 - 1620*, London: Athlone Press, 1976.

Thompson, Ian, *The Sun King's Garden: Louis XIV, André Le Nôtre and the Creation of the Gardens of Versailles*, New York: Bloomsbury Publishing USA, 2006.

Thomson, Richard, *Chronicles of London Bridge*, London: Thomas Tegg, 1839.

Thornbury, Walter, *Old and New London* (6 volumes), Vol. 1, p. 280, London: Cassell, Petter, Galpin & Co, 1878.

Timbs, John, *Curiosities of London: exhibiting the most rare and remarkable objects of interest in the metropolis*, London: D. Bogue, 1855.

Tomory, Leslie, "London's water supply before 1800 and the roots of the networked city," *Technology and culture*. 56 (2015), 3, pp. 704–737.

Torres Balbas, Leopoldo, "La Albolafia de Córdoba y la gran *noria* toledana," *Al-Andalus*, 7, (1942), pp. 464–465.

Touchard-Lafosse, George, *Histoire de Paris, Composée sur un Plan Nouveau* (4 volumes), Vol. 4, Paris: P.-H. Krabbe, 1834.

Trevor Hodge, A., "Engineering Works," in Örjan Wikander, (ed.), *Handbook of Ancient Water Technology*, Technology and Change in History, (ongoing series), Vol. 2, Leiden: Brill 2000, pp. 67–94.

Truman, Ronald W., *Spanish Treatises on Government, Society and Religion in the Times of Philip II*, Leiden, Boston and Koln: Brill, 1999.

Turriano, Juanelo, *Breve discurso a S.M. el Rey Católico en torno a la reducción del año y reforma del calendario*, edited by José Antonio García-Diego y J. M. González Aboin, Madrid: Fundación Juanelo Turriano, 1990.

Tynan, Nicola, "London's Private Water Supply, 1582–1902," in Paul Seidenstat, David Haarmeyer and Simon Hakim, (eds), *Reinventing Water and Wastewater Systems: Global Lessons for Improving Water Management*, Hoboken (NJ): J. Wiley, 2002, pp. 341–360.

Unger, Richard W., *Beer in the Middle Ages and the Renaissance*, Philadelphia: University of Pennsylvania Press, 2007.

Usher, Abbott Payson, *A History of Mechanical Inventions*, revised edition, Cambridge, (MA): Harvard University Press, 1962.

Varanini, Gian Maria, "Energia idraulica e sviluppo urbano nella Verona comunale" in R. Comba, (ed.), *Paesaggi urbani dell'Italia padana nei sec. VIII–XIV*, Bologna: Cappelli, 1988, pp. 331–372.

Vicente Maroto, Isabel and Maríano Esteban Piñeiro, *Aspectos de la ciencia aplicada en la España del Siglo de Oro*, Salamanca: Junta de Castilla y León, 1990.

Vidal Lescuyer, Virginia, "Historia del Uso del Agua en Una Gran Ciudad: Caso de París," *Ambient, Curso 2002–2003*, Barcelona (July 2003), pp. 1–5.

Viganò, Marino, "'Parente et alieuo del già messer Janello'—Primeras notas sobre Bernardo y Turriano," in Alicia Cámara, Rafael Moreira y Marino Viganò, *Turriano ingeniero del rey*, Madrid: Fundación Juanelo Turriano, Madrid: 2010, pp. 203–227.

Vilà Valentí, Juan, *Professor Joan Vilà Valentí: el seu mestratge en la geografia universitària*, Barcelona: Edicions Universitat Barcelona, 1999.

Villalba y Guitarte, Joaquín de, *Epidemiología española o Historia cronológica de las pestes, contagios, epidemias y epizootias que han acaecido en España desde la venida de las cartagineses hasta el año 1801: Con noticia de algunas otras enfermedades de esta especie que han sufrido los españoles en otros reynos, y de los autores nacionales que han escrito sobre esta materia*, Madrid: Imprenta de Fermín Villalpando, 1803.

Viñas, Carmelo y Ramón Paz, *Relaciones de los Pueblos de España Ordenadas por Felipe II, Reino de Toledo*, (3 partes), parte 3, Madrid: Instituto Balmes de Sociología, Instituto Juan Sebastián Elcano de Geografía, Consejo Superior de Investigaciones Científicas, 1963.

Vincent, Bernard, "Economia y sociedad en el reino de Granada en el siglo XVI," in *Historia de Andalucía*, Vol. 4, *La Andalucía del Renacimiento*, edited by Antonio Dominguez Ortiz, Madrid: Editorial Cupsa, 1982, pp. 160–223.

Vuorinen, Heikki S. and others, "History of water and health from ancient civilizations to modern times," *Water Science & Technology*, Vol. 7, No. 1 (2007), pp. 49–57.

Wagner, John A. and Susan Walters Schmid, *Encyclopedia of Tudor England,* Santa Barbara (CA): ABC-CLIO, 2012.

Ward, Daniel, and K. McKague, "Water Requirements for Livestock Factsheet," Ontario (Canada): Ministry of Agriculture and Food and Rural Affairs, May 2007.

Wear, Andrew, *Knowledge and Practice in English Medicine, 1550–1680*, Cambridge and New York: Cambridge University Press, 2000.

Weinstein, Rosemary, "New Urban Demands in Early Modern London," *Medical History*, Supplement No. 11, (1991), pp. 29–40.

Weiss, David A., *The Great Fire of London*, London: Trafford Publishing, 2012.

Welch, Charles, *History of the Tower Bridge and of other bridges over the Thames built by the Corporation of London. Including an account of the Bridge House Trust from the Twelfth century, based on the records of the Bridge House Estates Committee*, London: Smith, Elder and Co., 1894.

Wells, Samuel, *The History of the Drainage of the Great Level of the Fens called Bedford Level*, London: R. Pheney, 1830, (2 volumes), Vol. 2.

Wheatley, Henry B., *London past and present, its history, associations, and traditions*, (3 volumes), Vol. 3, London: John Murray, 1891.

Wheeler, William Henry and Leonard Charles Batty, *A History of the Fens of South Lincolnshire*, Boston: J. M. Newcomb, 1897.

White, K. D., "'The Base Mechanic Arts'? Some Thoughts on the Contribution of Science (Pure and Applied) to the Culture of the Hellenistic Age," in Peter Green, (ed.), *Hellenistic History and Culture*, Los Angeles and London: University of California Press, 1993, pp. 211–237.

White, Lynn Townsend Jr., "Jacopo Aconcio as an Engineer," in Lynn Townsend White Jr., *Medieval religion and technology: collected essays*, Berkeley: University of California Press, 1978, pp. 149–174.

White, Lynn Townsend Jr., *Medieval Technology and Social Change*, London and Oxford: Oxford University Press, 1962.

White, Lynn Towsend Jr., "Kyeser's *Bellifortis*: The First Technological Treatise of the Fifteenth Century," *Technology and Culture*, Vol. 10, No. 3 (Jul., 1969), pp. 436–441.

Whitney, William Dwight,*The Century dictionary and cyclopedia; a work of universal reference in all departments of knowledge, with a new atlas of the world*, 10 Volumes, Vol. 1, New York: The Century Co., 1904, p. 297.

Whittaker, G. B., (publisher), *The History of Paris from the Earliest Period to the Present Day: Containing a Description of Its Antiquities, Public Buildings, Civil, Religious, Scientific, and Commercial Institutions* (3 volumes), Vol. 3, Paris, 1825.

Wikander, Örjan, "Industrial Applications," in Örjan Wikander, (ed.), *Handbook of Ancient Water Technology*, Technology and Change in History, (ongoing series), Vol. 2, Leiden: Brill (2000), pp. 401–410.

Wikander, Örjan, "The Water-Mill," in Örjan Wikander, (ed.), *Handbook of Ancient Water Technology*, Technology and Change in History, (ongoing series) Vol. 2, Leiden: Brill, 2000, pp. 371–400.

Williams, Gwyn, *Medieval London: from Commune to Capital*, London: University of London, Athlone Press, 1970.

Williams, John, *An historical account of sub-ways in the British metropolis, for the flow of pure water and gas into the houses of the inhabitants*, London: Carpenter and Son, 1828.

Williams, Patrick, *The Great Favourite: the Duke of Lerma and the court and government of Philip III of Spain, 1598–1621*, Manchester: Manchester University Press, 2006.

Williamson, Frederick, "George Sorocold of Derby: a pioneer of water supply," *Journal of the Derbyshire Archaeological & Natural History Society*, Vol.57 (1936), pp. 43–93.

Wilson, Andrew I., "Classical water technology in the early Islamic world," in C. Bruun and A. Saastamoinen, (eds), *Technology, ideology, water: from Frontinus to the Renaissance and beyond*, Roma: Instituti Romani Finlandiae, 2004, pp. 115–141.

Wilson, Andrew I., "Drainage and Sanitation," in Örjan Wikander, (ed.), *Handbook of Ancient Water Technology*, Technology and Change in History, (ongoing series), Vol. 2, Leiden: Brill (2000), pp. 151–179

Wingate, Charles Frederick, "The Water-supply of Cities," *North American Review*, Vol. 136, No. 317 (Apr., 1883), pp. 364–374.

Wolf, Abraham, *A History of Science, Technology and Philosophy in the 16th and 17th Centuries*, New York: Harper Torchbooks, Harper & Brothers, 1959.

Wright, Rita P., *The Ancient Indus—Urbanism, Economy, and Society*, New York: New York University, 2010.

Wrigley, E. Edward Anthony and Roger S. Schofield, *The Population History of England: 1541–1871: a Reconstruction*, Cambridge: Cambridge University Press, 1989.

Yu, Sa, "Political Privilege, Legal Right, or Public Policy Tool? A History of the Patent System," *International Association for the Advancement of Teaching and Research in Intellectual Property (ATRIP) Essay Competition 2009*, 31 August 2009.

Zanetti, Cristiano, *Juanelo Turriano. De Cremona a la Corte: formación y red social de un ingeniero del Renacimiento*, Madrid: Fundación Juanelo Turriano, 2015.

Zanoni, Felice, "Un brevetto pontificio d'invenzioni del '500. Janello Torriano e un documento dell'Archivo segreto vaticano," *Bolletino storico cremonese* (1940), pp. 145–154.

Zeller, Gaston, *Les institutions de la France au XVIe siècle*, Paris: Presses Universitaires de France, 1949.

Zeller. Gaston, "Royal Administration before the Intendants: Parlements and Governors," in H. J. Cohn, (ed.), *Government in Reformation Europe*, London: Macmillan, 1972, pp. 225–256.

Zilsel, Edgar, *The Social Origins of Modern Science*, Dordrecht: The Netherlands, Springer, 2003.

Zupko, Ronald Edward, *Revolution in Measurement: Western European Weights and Measures Since the Age of Science, Memoirs of the American Philosophical Society*, Volume 186, Philadelphia: American Philosophical Society, 1990.

Websites

Chateau de Modave:
http://www.modave-castle.be/
City of London Corporation: http://www.cityoflondon.gov.uk/
City of London Library & Art Gallery Electronic (COLLAGE)—a computerized information system providing access to large areas of Guildhall's Gallery visual collections.
http://collage.cityoflondon.gov.uk/collage/app
El *Artificio* de Juanelo Turriano en Toledo, created and maintained by the late Juan Luis Peces Ventas:
http://home.worldonline.es/agallar/
El *Artificio* de Juanelo Turriano, "Modelo con escaleras de Valturio del para elevar agua al Alcázar de Toledo (s. XVI)," created and maintained by Françesc Xavier Jufre García:
http://www.artificiodejuanelo.org
Encyclopædia Britannica Online:
http://www.britannica.com/
Global Economic History Network:
http://www2.lse.ac.uk/economicHistory/Research/GEHN/Home.aspx
History of Science and Technology in Islam, authored by Ahmad Yousef al-Hassan Gabarin:
http://www.history-science-technology.com/default.htm
Il website cartografico e grafico della Biblioteca Nazionale Marciana di Venezia:
http://geoweb.venezia.sbn.it/
Instituto Nacional de Matematica Pura e Aplicada (IMPA—Brazil):
http://w3.impa.br/~jair/einter1.html.
Joconde, Portail des collections des musées de France.
http://www.culture.gouv.fr/
La Machine de Marly:
http://www.marlymachine.org/
Lawrence H. Officer and Samuel H. Williamson's purchasing-power calculator at their website:
http://www.measuringworth.com
LMA shelves.
http://www.aim25.ac.uk/

Museo Galileo—Institute and Museum of the History of Science, Florence:
http://www.museogalileo.it/en/index.html.

Paulus Swaen old Map auction and galleries
http://www.swaen.com/

Roman aqueducts and related items, created and maintained by Cees Passchier (Mainz, Germany) and Wilke Schram (Utrecht, the Netherlands):
http://www.cs.uu.nl/people/wilke/introduction/

The Archimedes Project, Database Machine Drawings, Max Planck Institute for the History of Science (MPIWG), Berlin:
http://dmd.mpiwg-berlin.mpg.de/

The Dibner Library of the History of Science and Technology—The Smithsonian Institution Libraries:
http://www.sil.si.edu/libraries/Dibner/index.htm.

The Foundation for Science, Technology and Civilization (FSTC).
http://www.muslimheritage.com/Default.aspx

The Hispanic Seminary of Medieval Studies (HSMS):
http://www.hispanicseminary.org/

The International Association for the Advancement of Teaching and Research in Intellectual Property (ATRIP):
http://www.atrip.org/.

The International Water History Association:
http://www.iwha.net.

The Library of the University of Castilla-La Mancha:
http://www.biblioteca.uclm.es/

The Newberry Chicago's Independent Research Library: http://www.biblioserver.com/newberry/index.php

The Smithsonian Institution Libraries:
http://www.sil.si.edu/

The Taylor & Francis Group, an international academic publisher:
http://www.tandfonline.com/.

The University Library System of the Chinese University of Hong Kong:
http://library.cuhk.edu.hk/.

The University of London & History of Parliament Trust—British History online:
http://www.british-history.ac.uk/

Université de Liège, Belgium, 1996.
http://www.hec.be/

Venipedia, an English website about the city of Venice.
http://www.venipedia.org/wiki/index.php?title=Sewage_disposal.

Website of the Bibliothèque nationale de France,
http://gallica.bnf.fr/

Website of Eduardo Sánchez Butragueño, Collection of photographs of 'Forgotten Toledo',
http://www.flickr.com/photos/65595512@N00/6850700051/sizes/o/in/photostream/.
Website of the Consorcio de Toledo,
https://consorciotoledo.wordpress.com/
Website of the La Academia de Ciencias Matemáticas, Físico-Químicas y Naturales de Granada, Facultad de Ciencias de la Universidad de Granada.
http://wdb.ugr.es/~academia/otros/entornoalauniversidad.pdf, accessed 6.2.2014

Online Sources for Retrieved Images

Fig. 2.1.
Website of the Bibliothèque nationale de France:
http://gallica.bnf.fr/ark:/12148/bpt6k6212475p/f286.item.r=Villard%20de%20Honnecourt, accessed 30.10.2016.
Fig. 2.5.
StudyBlue, Website for Greek And Roman Painting Final—Art History And Archaeology, showing material from Washington University in St. Louis:
https://www.studyblue.com/notes/note/n/greek-and-roman-painting-final/deck/14472983, accessed 2 November, 2016.
Fig. 2.6.
Website of the ETH-Bibliothek, the central university library for ETH Zurich:
http://www.e-rara.ch/zut/content/titleinfo/3153756, accesed 23.11.2015.
Fig. 2.7.
Website of the ETH-Bibliothek, the central university library for ETH Zurich:
http://www.e-rara.ch/oec/content/titleinfo/6105281, accessed 24.12.2014.
Fig. 2.8.
Archimedes Screw website of the New York University:
https://www.math.nyu.edu/~crorres/Archimedes/Screw/applications/ScrewPompeii.jpg (Webmaster Prof. Chris Rorres), accessed 14.6.2016.
Fig. 2.9.
Archimedes Screw website of the New York University:
https://www.math.nyu.edu/~crorres/Archimedes/Screw/applications/ScrewFigurine.jpg., accessed 14.6.2016.
Fig. 2.10.
Max Planck Institute for the History of Science, Library website:
http://echo.mpiwg-berlin.mpg.de/ECHOdocuView?url=/permanent/library/H54G8BZR/index.meta, accessed 4.8.2015.

Fig. 2.12.
Website of the ETH-Bibliothek, the central university library for ETH Zurich: http://www.e-rara.ch/zut/content/titleinfo/3153756, accesed 23.11.2015.

Fig. 2.13.
Website of the ETH-Bibliothek, the central university library for ETH Zurich: http://www.e-rara.ch/oec/content/titleinfo/6105281, accesed 11.10.2014.

Fig. 2.15.
Website of the ETH-Bibliothek, the central university library for ETH Zurich: http://www.e-rara.ch/oec/content/pageview/6105657, accessed 24.12.2014.

Fig. 2.16.
Website of the ETH-Bibliothek, the central university library for ETH Zurich: http://www.e-rara.ch/zut/content/titleinfo/24009, accessed 24.12.2014.

Fig. 2.17.
Internet Archive: https://archive.org/details/guidetoexhibitgroobrituoft, accessed 9.9.2014.

Fig. 2.18.
https://commons.wikimedia.org/wiki/File:Al-Jazari_Automata_1205.jpg, accessed 2.12.2016.

Fig. 2.19.
Yale University Beinecke Rare Book & Manuscript Library: http://brbl-dl.library.yale.edu/vufind/Record/3443389?image_id=1214755, accessed 24.11.2015.

Fig. 2.20.
Website of the Biblioteca Nazionale Centrale, Firenze: http://www.bncf.firenze.sbn.it/Bib_digitale/Manoscritti/Palat766/main.htm p. 57, accessed 27.11.2014.

Fig. 2.21.
Website of the Universitätsbibliothek Johann Christian Senckenberg, Goethe-Universität, Frankfurt am Main, accessed 12.10.2013.
http://sammlungen.ub.uni-frankfurt.de/msma/content/pageview/3657414, accessed 14.9.2011.

Fig. 2.22.
https://fr.wikipedia.org/wiki/Codex_Vallardi#/media/File:Pisanello_-_Codex_Vallardi_2286_r.jpg, accessed 9.4.2015.

Fig. 2.23.
Yale University Beinecke Rare Book & Manuscript Library: http://brbl-dl.library.yale.edu/vufind/Record/3443389?image_id=1214755, accessed 24.11.2015.

Fig. 2.24.
E-Leo, website of the Biblioteca Leonardiana, Vinci (Italy): http://www.leonardodigitale.com/index.php?lang=ENG, accessed 28.12.2016.

Fig. 2.25.

Website of the Biblioteca Nazionale Centrale, Firenze, *Manuscrito Palatino 766*, fol. 3, p. 5, http://www.bncf.firenze.sbn.it/Bib_digitale/Manoscritti/Palat766/main .htm, accessed 3.4.2013.

Fig. 2.26.

Heidelberg Universitätsbibliothek website:

http://digi.ub.uni-heidelberg.de/diglit/cpg126/0001, fol. 11 r, accessed 14.1.2015.

Fig. 2.27.

Cornell University Library website:

http://ebooks.library.cornell.edu/cgi/t/text/pageviewer-idx?c=kmoddl;cc=kmodd l;idno=kmod049;node=kmod049%3A2;view=image;seq=111;size=50;page=root, accessed 8.8.2012.

Fig. 2.28.

The Max Planck Institute for the History of Science, Library website:

http://echo.mpiwg-berlin.mpg.de/ECHOdocuView?url=/permanent/archimedes_ repository/large/cered_disco_021_it_1567/index.meta, accessed 25.5.2014.

Fig. 2.29.

The Max Planck Institute for the History of Science, Library website:

http://echo.mpiwg-berlin.mpg.de/ECHOdocuView?url=/permanent/library/ UR271U6Y/index.meta&start=71&viewMode=image&pn=73, accessed 25.5.2014.

Fig. 2.30.

Max Planck Institute for the History of Science, Library website: http://echo.mpiwg-berlin.mpg.de/ECHOdocuView?url=/permanent/library/AoQMGXBK/pageimg &start=161&pn=163&mode=imagepath, accessed 7.11.2014.

Fig. 2.31.

Max Planck Institute for the History of Science, Library website:

http://echo.mpiwg-berlin.mpg.de/ECHOdocuView?url=%2Fpermanent%2Flibrar y%2FH54G8BZR%2Findex.meta&start=41&viewMode=auto&pn=48, accessed 1.2.2017.

Fig. 2.32.

Max Planck Institute for the History of Science, Library website:

http://echo.mpiwg-berlin.mpg.de/ECHOdocuView?url=/mpiwg/online/ permanent/library/5CTEBAHQ&st art=161&pn=163, accessed 5.6.2014.

Fig. 2.33.

The Max Planck Institute for the History of Science, Library website:

http://echo.mpiwg-berlin.mpg.de/ECHOdocuView?mode=imagepath&url=/ permanent/library/AoQMGXBK/pageimg, accessed 7.11.2014.

Fig. 2.34.

The Max Planck Institute for the History of Science, Library website:

http://echo.mpiwg-berlin.mpg.de/ECHOdocuView?mode=imagepath&url=/ permanent/library/AoQMGXBK/pageimg, accessed 7.11.2014.

Fig. 2.35.
The Max Planck Institute for the History of Science, Library website:
http://echo.mpiwg-berlin.mpg.de/ECHOdocuView?mode=imagepath&url=/permanent/library/AoQMGXBK/pageimg, accessed 7.11.2014.

Fig. 2.36.
Cornell University Library website:
http://ebooks.library.cornell.edu/k/kmoddl/pdf/059_v4thru6_Volume_V.pdf, accessed 3.5.2015.

Fig. 2.37.
Website of the Bibliothèque nationale de France:
http://gallica.bnf.fr/ark:/12148/bpt6k65250757/f102.item, accesed 9.2.2015.

Fig. 2.38.
Website of the Bibliothèque-médiathèque de Nancy:
http://bmn-renaissance.nancy.fr/items/show/1856, accessed 13.4.2015.

Fig. 2.39.
Cornell University Library website:
http://ebooks.library.cornell.edu/cgi/t/text/pageviewer-idx?c=kmoddl;cc=kmoddl;idno=kmod021b;node=kmod021b%3A24;view=image;seq=113;size=100;page=root, accessed3.12.2024.

Fig. 2.40.
Bayerische Staats Bibliothek Digital:
http://reader.digitale-sammlungen.de/de/fs1/object/display/bsb10942275_00087.html, accessed 17.7.2014.

Fig. 2.41.
Website of the Bibliothèque nationale de France:
http://gallica.bnf.fr/ark:/12148/btv1b9066031c/f93.item.r=denis%201317, accessed 14.6.2014

Fig. 2.42.
The Max Planck Institute for the History of Science, Library website:
http://echo.mpiwg-berlin.mpg.de/ECHOdocuView?mode=imagepath&url=/permanent/library/AoQMGXBK/pageimg, accessed 7.11.2014.

Fig. 2.43.
The Max Planck Institute for the History of Science, Library website:
http://echo.mpiwg-berlin.mpg.de/ECHOdocuView?url=/mpiwg/online/permanent/library/5CTEBAHQ&start=161&pn=135, accessed 5.6.2014

Fig. 2.44.
The Max Planck Institute for the History of Science, Library website:
http://echo.mpiwg-berlin.mpg.de/ECHOdocuView?thumbCols=5&start=61&viewMode=images&url=/permanent/library/ECTHTCNQ/index.meta&pn=64&thumbSize=300&thumbRows=10, accessed 12.3.2013.

Fig. 2.45.
https://commons.wikimedia.org/wiki/File:Fotothek_df_tg_0000182_Geometrie_%5E_Vermessung_%5E_Quadrant_%5E_Geb%C3%A4ude.jpg, accessed 7.8.2013.

Fig. 2.46.
Website of the Biblioteca Nazionale Centrale, Firenze:
http://www.bncf.firenze.sbn.it/Bib_digitale/Manoscritti/Palat767/main.htm, accessed 12.12.2016.

Fig. 2.47.
E-Leo, website of the Biblioteca Leonardiana, Vinci (Italy):
http://www.leonardodigitale.com/index.php?lang=ENG fol. 1069 v, accessed 9.6.2015.

Fig. 2.48.
The Max Planck Institute for the History of Science, Library website:
http://echo.mpiwg-berlin.mpg.de/ECHOdocuView?url=/permanent/library/UR271U6Y/index.meta&start=121&viewMode=image&pn=130, accessed 14.4.2016.

Fig. 2.49.
Yale University Beinecke Rare Book & Manuscript Library:
http://brbl-dl.library.yale.edu/vufind/Record/3443389, accessed 24.11.2015.

Opposite the beginning of Chapter 3 (Toledo)
Website of the Toronto Public library:
http://www.torontopubliclibrary.ca/detail.jsp?Entt=RDMDC-912-4643H57&R=DC-912-4643H57, accessed 21.7.2012.

Fig. 3.6.
https://en.wikipedia.org/wiki/Astrarium_of_Giovanni_Dondi_dell'Orologio#/media/File:Astrario_Dondi_05869_01_dia_-_Museo_scienza_e_tecnologia_Milano.jpg, accessed 8.1.2017.

Fig. 3.7.
Website of Universitätsbibliothek Heidelberg:
http://diglit.ub.uni-heidelberg.de/diglit/braun1582bd1, accessed, 17.8.2016.

Fig. 3.8.
Toronto Public library:
http://www.torontopubliclibrary.ca/detail.jsp?Entt=RDMDC-912-4643H57&R=DC-912-4643H57, accessed 21.7.2012.

Fig. 3.9.
Website of the Bibliothèque nationale de France,
http://gallica.bnf.fr/ark:/12148/btv1b550052185/f1.item.zoom, accessed 12.12.2014.

Fig. 3.10.
https://upload.wikimedia.org/wikipedia/commons/c/ca/El_Greco_-_View_of_Toledo_-_Google_Art_Project.jpg, accessed 24.12.2010.

Fig. 3.11.
https://en.wikipedia.org/wiki/View_and_Plan_of_Toledo#/media/File:El_Greco_-_View_and_Plan_of_Toledo_-_Google_Art_Project.jpg, accessed 22.10.2015.

Fig. 3.12.
Internet Archive:
https://archive.org/details/lesdelicesdelesp01alva, accessed 9.5.2013.

Fig. 3.13.
Biblioteca Digital de la Real Academia de Historia, Madrid:
http://bibliotecadigital.rah.es/dgbrah/es/consulta/registro.cmd?id=6156, accessed 30.9.2015.

Fig. 3.14.
Website of the Ayuntamiento de Toledo:
http://www.ayto-toledo.org/archivo/imagenes/pym/ciudad/planos.asp, accessed 27.2.2014.

Fig. 3.15.
A Fundacion Juanelo Turriano image:
https://www.flickr.com/photos/98983335@N04/9301628642, accessed 2.3.2013.

Fig. 3.17.
Collection—Kinematic Models for Design:
http://ebooks.library.cornell.edu/cgi/t/text/pageviewer-idx?c=kmoddl&cc=kmoddl&idno=kmod057&node=kmod057%3A35&view=image&seq=273&size=50, accessed 18.12.2015.

Fig. 3.21.
Peces Ventas' website:
http://home.worldonline.es/agallar/discussion, accessed July 12, 2012. Not available anymore, presently shown in:
https://juanellotoriano.wikispaces.com/Juan+Luis+Peces+Ventas, accessed 12.12.2016

Fig. 3.27.
Opposite the beginning of Chapter 4 (London)
Website of Universitätsbibliothek Heidelberg [VD16-B7188) Universitätsbibliothek Heidelberg:
http://digi.ub.uni-heidelberg.de/diglit/braun1582bd1/0011, accessed, 17.8.2016.

Fig. 4.1.
Website of the Institut Cartogràfic i Geològic de Catalunya, catalogued under RM.6784:
http://cartotecadigital.icc.cat/cdm/singleitem/collection/europa/id/2281/rec/1, accessed 20.4.2016.

Fig. 4.2.
http://www.wikigallery.org/wiki/painting_160194/Ralph-Treswell/Map-of-Cheapside,-London,-1585#licensing, accessed 29.9.2015.
Fig. 4.4.
Project Gutenberg EBook:
http://www.gutenberg.org/ebooks/37817, accessed 2.10.2016.
Fig. 4.6.
Website of the British Library:
http://www.bl.uk/collection-items/great-fire-of-london-map, accessed 28.6.2014.
Fig. 4.7.
Website of Glasgow University:
http://special.lib.gla.ac.uk/exhibns/month/nov2003.html, accessed 28.11.2011
Fig. 4.8.
City of London—COLLAGE ('City of London Library & Art Gallery Electronic'):
http://collage.cityoflondon.gov.uk/quick-search?q=Plan%20of%20the%20northern%20half%20of%20the%20old%20London%20Bridge&WINID=1479650595157, accessed 3.3.2016.
Fig. 4.9.
Image extracted from:
http://www.chinaoilpainting.com/htmlimg/image-26131.htm#, accessed 4.3.2016.
Fig. 4.11.
Archive.org:
http://archive.org/stream/shorthistoryofen03gree#page/n7/mode/2up, accessed 26.11.2015
Fig. 4.12.
Website of the Folger Shakespeare Library, Folger Digital Image Collection:
http://luna.folger.edu/luna/servlet/detail/FOLGERCM1~6~6~27938~102015:The-view-of-London-Bridge-from-east, accessed 1.1.2017.
Fig. 4.13.
Website of the National Library of Sweden—World Digital Library:
https://www.wdl.org/en/item/14397/view/1/1/, accessed 17.5.2016.
Fig. 4.14.
https://commons.wikimedia.org/wiki/File:London_panorama,_1616b.jpg, accessed 3.9.2016.
Fig. 4.15.
Archive.org:
http://archive.org/stream/shorthistoryofen03gree#page/n7/mode/2up, accessed 26.11.2015.
Fig. 4.16.
https://commons.wikimedia.org/wiki/File:Depiction_of_the_London_Bridge_Waterworks_from_1731.jpg, 22.1.2017.

Fig. 4.17.
Website of the National Oceanic and Atmospheric Administration:
https://docs.lib.noaa.gov/rescue/rarebooks_1600-1800/TC160B371654.pdf, accessed 30.4.2015.

Fig. 4.18.
Website of the National Library of Sweden—World Digital Library:
https://www.wdl.org/en/item/14397/view/1/1/, accessed 3.7.2016.

Fig. 4.19.
https://commons.wikimedia.org/wiki/File:London_panorama,_1616b.jpg, accessed 30.8.2014.

Fig. 4.20.
https://commons.wikimedia.org/wiki/File:Wenceslas_Hollar_-_London_before_and_after_the_fire_(State_2).jpg, accessed 14.3.2015.

Fig. 4.21.
Archive.org:
https://archive.org/details/bub_gb_CbUWAAAAQAAJ, accessed 2.2.2017.

Fig. 4.22.
Website of the Folger Shakespeare Library, Folger Digital Image Collection:
http://luna.folger.edu/luna/servlet/detail/FOLGERCM1~6~6~237390~115712:-Bird-s-eye-plan-of-the-west-central, accessed 21.9.2014

Fig. 4.23.
Website of the Folger Shakespeare Library, Folger Digital Image Collection:
http://luna.folger.edu/luna/servlet/detail/FOLGERCM1~6~6~356580~130047:Ye-Waterhouse--graphic----W-Hollar, accessed 4.11.2015.

Fig. 4.24.
https://commons.wikimedia.org/wiki/File:New_River_Water.jpg, accessed 18.7.2012.

Fig. 4.25.
Opposite the beginning of Chapter 5 (Paris)
https://commons.wikimedia.org/wiki/File:Merian_map_of_Paris_1615_-_Gallica.jpg, colorized version, accessed 2.2.2017.

Fig. 5.1.
https://commons.wikimedia.org/wiki/File:Philippe_de_Champaigne_-_Le_Pr%C3%A9v%C3%B4t_des_marchands_et_les_%C3%A9chevins_de_la_ville_de_Paris_(1648).JPG, accessed 9.12.2013.

Fig. 5.2.
Website of Historic Paris:
http://www.paristoric.com/index.php/paris-d-hier/statues/statues-de-l-hotel-de-ville, accessed 30.11.2012.

Fig. 5.4.

Cultural Heritage Website, ECHO, Max Planck Institute:

http://echo.mpiwg-berlin.mpg.de/ECHOdocuView?thumbCols=5&url=/mpiwg/online/permanent/library/XR4AU189/pageimg&viewMode=image&thumbRows=5&start=136&thumbSize=300&mode=imagepath&pn=157, accessed 3.3.2012.

Fig. 5.5.

Website of the Bibliothèque de l'Institut National d'Histoire de l'Art, collections Jacques Doucet:

http://bibliotheque-numerique.inha.fr/viewer/5750/?offset=1#page=58&viewer=picture, accessed 4.6.2014.

Fig. 5.6.

Website of the Bibliothèque nationale de France:

http://gallica.bnf.fr/ark:/12148/btv1b53102569k/f1.item.zoom, accessed 24.5.2017.

Fig. 5.7.

https://commons.wikimedia.org/wiki/File:Merian_map_of_Paris_1615_-_Gallica.jpg, accessed 7.7.2013.

Fig. 5.9.

Website of Ebay:

http://www.ebay.fr/itm/cpa-Gravure-dessin-75-Vieux-PARIS-La-POMPE-de-la-SAMARITAINE-en-1635-/232343637864?hash=item3618c28f68:g:UygAAOSwRLZT-wUB, accessed 17.1.2017.

Fig. 5.10.

Website of the Bibliothèque nationale de France:

http://gallica.bnf.fr/ark:/12148/btv1b8403618r, accessed 30.10.2016.

Fig. 5.11.

Max Planck Institute for the History of Science, Library website:

http://echo.mpiwg-berlin.mpg.de/ECHOdocuView?thumbCols=5&url=%2Fmpiwg%2Fonline%2Fpermanent%2Flibrary%2FXR4AU189%2Fpageimg&viewMode=image&thumbRows=5&start=136&thumbSize=300&mode=imagepath&pn=263, accessed 7.7.2012.

Fig. 5.12.

https://commons.wikimedia.org/wiki/File:La_pompe_de_la_Samaritaine.jpg, accessed 1.12.2016.

Fig. 5.13.

Website of the Bibliothèque nationale de France:

http://gallica.bnf.fr/ark:/12148/btv1b530580657, accessed 8.5.2016.

Fig. 5.14.

Website of the ETH-Bibliothek, the central university library for ETH Zurich,

http://www.e-rara.ch/zut/content/pageview/1611885, accesed 23.11.2015

Fig. 5.15.
https://en.wikipedia.org/wiki/Turgot_map_of_Paris#/media/File:Plan_de_Turgot.jpg, accessed 2.6.2014

Fig. 5.16.
https://commons.wikimedia.org/wiki/File:Encyclopedie_volume_2–311.png, accessed 14.2.2013

Fig. 5.17.
https://commons.wikimedia.org/wiki/File:Raguenet,_La_joute_des_mariniers.jpg, accessed 1.1.2015.

Fig. 5.18.
Chateau de Modave:
http://www.modave-castle.be/, accessed 28.9.2014.

Fig. 5.19.
Website of the Bibliothèque nationale de France:
http://gallica.bnf.fr/ark:/12148/btv1b6945480x/f1.item.zoom, accessed 30.10.2016.

Fig. 5.20.
https://en.wikipedia.org/wiki/Machine_de_Marly#/media/File:Machine_de_Marly.jpg, accessed 27.5.2013.

Index for Personal Names, Towns and Cities[1]

1 Authors indexed are the ones mentioned in the main text. All authors included in the bibliography.

Index of Subjects

www.ingramcontent.com/pod-product-compliance
Lightning Source LLC
LaVergne TN
LVHW010541300926
849162LV00002B/204